T0138224

The Mantle of the Earth

The Mantle of the Earth

Genealogies of a Geographical Metaphor

VERONICA DELLA DORA

The University of Chicago Press
Chicago and London

PUBLICATION OF THIS BOOK HAS BEEN AIDED BY A GRANT FROM THE
BEVINGTON FUND.

The University of Chicago Press, Chicago 60637
The University of Chicago Press, Ltd., London
© 2021 by The University of Chicago

29 28 27 26 25 24 23 22 21 20   1 2 3 4 5

ISBN-13: 978-0-226-74129-1 (cloth)
ISBN-13: 978-0-226-74132-1 (e-book)
DOI: https://doi.org/10.7208/chicago/9780226741321.001.0001

Library of Congress Cataloging-in-Publication Data

Names: Della Dora, Veronica, author.
Title: The mantle of the earth : genealogies of a geographical metaphor /
    Veronica della Dora.
Description: Chicago ; London : The University of Chicago Press, 2020. |
    Includes bibliographical references and index.
Identifiers: LCCN 2020022866 | ISBN 9780226741291 (cloth) |
    ISBN 9780226741321 (ebook)
Subjects: LCSH: Geography—History. | Metaphor. | Textile fabrics. | Earth (Planet)—
    Mantle. | Earth (Planet)—Surface.
Classification: LCC G81 .D45 2020 | DDC 910.9—dc23
LC record available at https://lccn.loc.gov/2020022866

♾ This paper meets the requirements of ANSI/NISO Z39.48-1992 (Permanence of Paper).

To the beloved memory of my spiritual father, Monk Apolló of Docheiariou
Αἰωνία ἡ μνήμη

Contents

INTRODUCTION

On Mantles, Maps, and Metaphors

This is a book about the earth's mantle, at once an object and a metaphor bestowing coverage and revelation, enclosure and disclosure. This book is a tale of beauty and precariousness, of discovery and doubt, of religious sensibility and shameless hubris. It is a story of planetary fascination and of global uncertainty.

*

In any contemporary English dictionary or encyclopedia, the earth's mantle is that "part of our planet's interior lying beneath the crust and above the central core."[1] This region is estimated to occupy 84 percent of the earth's volume, as opposed to the mere 1 percent of the terrestrial crust.[2] The word "mantle" (*Mantel* in German) was first introduced by the German geophysicist Emil Wiechert in 1896, when seismic wave sounding suggested that the earth's interior is bipartite, as opposed to the uniform rigid mass generally envisaged by nineteenth-century scientists.[3] The term came into general use in English in the 1940s.[4]

Outside of geology and geophysics, however, the earth's mantle is also understood to designate what lies *above* the terrestrial surface. For example, it is often taken as a synonym for "vegetal mantle." As a poetic figure, it has long been employed to describe the diversity and splendor of the natural world covering the planet and its seasonal changes (for instance, spring embellishing the earth's mantle with fresh flowers). In this sense, the mantle is taken as a synonym for "garment," that is, both a piece of clothing and an adornment. This book is concerned mostly with this second meaning—that is, with the geographical, rather than geological, mantle. More specifically, it is concerned with its origins, with its history, and with its manifold ramifications.

Unlike its nineteenth-century geological cognate, the geographical mantle has its roots in the ancient world. Greek mythology ascribes the transmutation of Chthonia (the formless earth) into Gaia (Mother Earth) to the creative power of a mantle Zeus cast on her shoulders. In Scripture and in early Christian literature, the earth and the sky repeatedly feature as cloaks and garments: biblical heavens are shaped "like a tent" (Psalm 104:2), "worn out" and rolled up "like a mantle" at the end of times (Psalm 102:26), whereas, according to the Church Fathers, the terrestrial surface is a beautiful, yet mutable and perishable, multicolored embroidery. The visibility of the mantle clothes and thereby reveals the invisibility of the Creator; it allows humans to experience God's presence palpably through his own works.

Thus understood, the mantle metaphor conveys a sense of fragile and transient beauty. Its polychrome texture drifts between the hidden profundities of the earth and the dark infinity of the universe. It naturally directs the gaze to the surface, but it also implies the existence of a hidden depth. The mantle conceals and reveals. It gives visual shape to the tension between the visible and the invisible, between the hidden and the manifest. It is a point of contact between the physical and the metaphysical, the known and the unknown, the finite and the infinite. As such, over the centuries, it has inspired the pen of philosophers, geographers, and theologians alike, and has shaped artists' and map makers' visual vocabularies. Yet, the mantle has also challenged generations of scientists and explorers to tear it apart in order to penetrate the earth's secrets. Its opening has been taken as a metaphor for discovery. More disturbingly, the wearing out of its fine texture continues to be a powerful image for environmental degradation and human harm to the planet.

This book is an inquiry into the intellectual genealogy of the earth's mantle and its metaphorical representation in art and literature, but principally in the several works and traditions of geography that come together in the map. Exploring the emergence and enduring metaphorical potency of the earth's mantle in Western intellectual history, images, and imagination reveals changing, and sometimes competing, perceptions of nature and global space. It also reveals much about the nature of geography and of human thought.

*

Across Indo-European languages, mantles are defined by their double function: to conceal and to protect. The English word "mantle," the Italian *mantello*, the French *manteau*, and the German *Mantel* all designate a "loose, sleeveless garment."[5] They are cognates with the Latin word *mantelum*, meaning "a cloth, hand towel, or napkin" (which, in turn, is deemed to derive from *manus*, or "hand"). The Greek *mandyas*, by contrast, is believed to be a Persian borrowing, and it originally designated a "woolen cloak," while its ancient analogue *himation* simply referred to an outer garment worn above the *chitōn* (or tunic). Metaphorically, *himation* also referred to the defensive walls of a city. The same meaning is found in the Old French *mantel*, from which come the military metaphor *démanteler* ("to dismantle"), originally meaning "to tear down the fortifications of a city," and its antonym *emmanteler*, or "to cover with a mantle."[6] In English, "mantle" is both noun and verb. Figuratively, "to mantle" can indicate both opening and enclosure. It can be used in the sense of "to spread" or "to extend," but also, more poetically, in the sense of "to embrace kindly." Both meanings chime with the all-encompassing vastness of the earth's mantle gently enfolding the globe.[7]

Much more frequently, however, "to mantle" means "to obscure, or protect by covering up."[8] No matter what their size, shape, material, color, or texture, across different languages mantles protect bodies and identities. Mantles remove things from sight. Mantles hide. Fictional mantles have long concealed illicit sons and guns, enabled identity exchanges, covered up intrigues.[9] Mantles veil "naked truth." Mantles wrap planetary cores and parts of biological organisms alike (think, for example, of the "cerebral mantle," or of the "mantle" covering the visceral mass of a mollusk).[10] Mantles can mask death itself. In Nikolaij Gogol's and Dino Buzzati's stories, the cloaks of deceased protagonists sinisterly move through the streets of Saint Petersburg, or pay their last visit home, concealing underneath nothing but the terrifying ghost of death.[11] Today, Cloak (the cognate of "mantle") is the name of an app that uses data from social networks to "help you avoid the people you do not want to meet." Its slogan is: "Incognito mode for real life."[12]

Mantles hide. And yet, by a strange paradox, for almost three thousand years the mantle has featured in Western culture as a precondition for *seeing* the world. The reason is that our experience and knowledge of the planet is by definition superficial, that is, bound to its surface. In its most literal sense, the Greek verb *geo-graphein* means "to write," or rather, to scratch the earth's surface—the face of the earth. We can thus imagine the earth's mantle as a planetary blanket constantly inscribed by humans with their plethora of stories, as well as with roads and canals, cities and hamlets, power lines and fiber-optic cables. Following American historian Joyce Chaplin, we can likewise imagine the mantle as a giant fabric spun out of humans' innumerable globe-girdling journeys, from Magellan to the International Space Station.[13]

Exploring the genealogy of the earth's mantle metaphor means unveiling a story of endless fascination as much as of global precariousness. It means unveiling the story of how the earth's surface and biosphere have been imagined and drastically transformed over the past three thousand years (and increasingly so today, in the era of the Anthropocene). In this sense, this book follows the humanistic tradition of *longue durée* histories of perceptions of nature set by Clarence Glacken and Keith Thomas, among others, and extends it to the present.[14] More specifically, it puts the history of Western environmental consciousness in dialogue with visual genealogies of global space.

Earthbound humans imagined and depicted the earth in its entirety long before it was photographed from space. Plato, for example, gave us the evocative description of a polychrome terrestrial globe floating in the darkness of the universe, as if seen from the moon.[15] The Greek philosopher and his contemporaries associated the spherical shape of the earth and of the universe (which was also imagined to be spherical) with perfection and eternity. Over the centuries, according to the British geographer Denis Cosgrove, the image of a unified globe has underpinned and prompted conceptions of human unity and what he called "a poetics of global space."[16]

As we shall see in the following chapters, the earth's mantle has also been associated with ideas of beauty and of human (as well as natural) unity in diversity (by way of weaving or web metaphors, for example). There is nonetheless a fundamental difference between the two images. As opposed to the sphere of the globe, frozen in time and eternally suspended in space, the mantle emphasizes mutability and unpredictability. On the one hand, its opening and closing movements reveal the expansions and contractions of global space in the geographical imagination of its inhabitants; on the other hand, its changing surface, embroidered and worn out by human action, reveals the unfolding drama of humanity.

*

As a surface inscribed by human visions, the earth's mantle fulfills the basic function of a map, that is, to make the world legible. Translating the macro-scale of the earth onto the micro-scale of the human body, the map enables me to grasp what my eyes cannot encompass in its totality—and it enables me to do so from the comfort of my room. Renaissance cartographers imagined their maps as prostheses of the eye, allowing the observer to visualize distant lands—indeed, the entire surface of the earth—from above. Abraham Ortelius, the author of the first printed atlas, for example, referred to his maps as "charts being placed as if they were certaine glasses before our eyes."[17] In this sense, maps were artificial skins, or membranes, separating and yet at the same time bridging interior and exterior worlds.[18] It is not by chance that the word "map" comes from the Latin *mappa* ("cloth"), that is, from the material of which maps were made, rather than from what they portrayed.[19]

Like written or spoken language, maps hardly draw attention to themselves as media. In Christian Jacob's words, "an effective map is transparent because it is a signified without a signifier. It vanishes in the visual and intellectual operation that unfolds its content."[20] An effective map makes no noise. And in its unquestioned transparency lies its metaphorical potency as a "mirror of the world." Cartographic transparency, however, conceals the fact that, rather than being more or less accurate representations of reality, maps are in truth pervasive, powerful, and duplicitous cultural artifacts. On them "ocular vision and the mind's eye coincide."[21] Far from being transparent windows on the world, they are indeed windows on historically and culturally specific *ways of seeing* the world. In its journey through time, this book examines the manifold ways in which maps and other cartographic work have been producing and representing forms of human inscription upon the surface of the earth. In this sense, this is ultimately a book about "ways of seeing" more than anything else.

Maps are more than representations and more than a form of communication. They are constants in humans' cognitive organization. Indeed, Jacob goes as far as regarding maps as "one of the founding mechanisms of Western thinking," while the Swedish geographer Gunnar Olsson deems them "*the intermediary without which we would be madly lost in a world unknown*."[22] Like Chthonia's mantle, each map actualizes and reifies "an attempt to impose the discipline of reason onto what is indistinct, indeterminate, and formless."[23] As such, maps have supreme agency over territory. Tracing a simple line on a map can divide neighborhoods and entire nations alike. At the same time, maps translate planners' and engineers' visions into actual cities, highways, and canals.

Maps at once trigger and enable the physical transformation of the land—to the point that today the map no longer copies territory; territory has become the map.[24] As architects around the world turn their attention to rooftops' design and artificial islands are crafted in the most fanciful shapes for the sole purpose of being admired through Google Earth, it is hard to disentangle the terrain from cartographic representation.[25] Ultimately, the mantle of the earth is nothing but a huge 1:1 map. Or, if you prefer, maps are nothing but miniature mantles of the earth.

*

The mantle is a powerful metaphor for cartographic representation, as it is for geographical knowledge. It best expresses geography's dual nature as a scientific as much as an aesthetic practice. Thirty years ago, in a pioneering attempt at mapping the history of geographical thought, Horacio Capel identified two alternating paradigms. On the one hand, he argued, there was a historicist (or idiographic) paradigm resting on description, understanding the world as a text, delighting itself in the fine detail of the surface. On the other hand, there was a positivist (or nomothetic) paradigm envisaging the world as a machine, preoccupied with the search for causes and effects, attempting to pierce the surface to unveil the hidden workings of nature and society. Capel limited his account to the history of geography as a formal academic discipline, thus covering little more than a century. Yet, if we take geography in its broader and more ancient sense, *geo-graphein*, we find that these two approaches have deep roots that ramify into other times and into other domains of knowledge—natural philosophy, to start with.[26]

Heraclitus's aphorism "Nature loves to hide" and its subsequent representation as a veiled statue of Isis (the monstrous many-breasted goddess of nature) haunted the French philosopher Pierre Hadot for more than four decades, resulting in his influential monograph *The Veil of Isis*. The book charted the history of the idea of nature from antiquity to the twentieth century as a journey underpinned by two fundamental attitudes that are somewhat reminiscent of Capel's historicist and positivist paradigms (and that will run through the following chapters of this book)—namely, what he termed the Orphic and the Promethean.

Orpheus, the legendary musician and poet, penetrates the secrets of nature "through melody, rhythm and harmony";[27] he respects nature's modesty; he contents himself with the poetic contemplation of her veil, that is, the tapestry of her visible forms. Nature shall lift her beautiful veil when and to the extent to which she shall wish. By contrast, Prometheus, the founder of experimental science, does not envisage nature's veil as an object for quiet contemplation. Nor does he see it as a medium for knowledge. On the contrary, he deems it a barrier to knowledge. For Prometheus, Isis's veil is a thick skin to be pierced, a dark mantle to be stripped away. Knowing nature is a brutal act. As Hadot observes, in sketching the program of modern experimental science, Francis Bacon uses "the vocabulary of violence, constraint, even torture."[28] The secrets of nature, argues Bacon, "are better revealed under the torture of experiments than when they follow the natural course." And this torture chamber has its own instruments, including telescopes and microscopes. But the human eye, too, can be an instrument of violence. Prometheus's gaze is that aggressive, penetrative gaze that "seeks the underground aspects of phenomena, violating the secret of things."[29]

Bridging seen and unseen, screening off and revealing, offering itself to aesthetic contemplation and merciless perforation, the mantle metaphor gives visual shape to a whole epistemic apparatus. Metaphorology, the

German philosopher Hans Blumenberg wrote, "seeks to burrow down to the substructure of thought, the underground, the nutrient solution of systematic crystallizations."[30] Blumenberg was interested in what he called "absolute metaphors"—that is, metaphors that could not be dissolved conceptually; metaphors that established an autonomous identity, rather than figures of comparison (for example, "the light of truth"). Shifts in absolute metaphors, Blumenberg believed, marked paradigmatic shifts. In other words, the replacement of one metaphor by another, or a metaphor's accretion of meaning, signaled changing ways of seeing in a society. For the German philosopher, metaphors (like maps) were thus mirrors of epochs and intellectual cultures; their "vicissitudes" brought to light "the cultural subconscious" to which they gave involuntary expression.[31]

Blumenberg's metaphorological approach can be easily applied to the history of geographical thought. In his groundbreaking monograph *The Geographical Tradition*, David Livingstone showed how, in order to legitimize its disciplinary status, throughout the nineteenth and twentieth centuries geography adopted the language of the dominating paradigm in the sciences, along with its armament of metaphors.[32] For example, when at the turn of the nineteenth century enthusiasm for the natural sciences was running high, geographical metaphors were grounded in biology and evolutionary theories. States and empires were conceived of as "biological organisms," as "predators" and "competitors" engaged in a perpetual struggle for *Lebensraum*.[33] William Morris Davis explained the "life" of a geographical area in terms of "a quick growing plant" and landscape change as a "cycle."[34] By contrast, when physics and the exact sciences took over evolutionary biology, cities became "centres of gravity" and "attraction poles," and geographical regions transmuted into "functional" regions governed by "gravity models." The spatial language of surfaces, diffusion, channels, fluxes, nodes, and networks became geography's vernacular.[35] Yet, to what extent are geographical metaphors distinctive? Which sort of metaphors are they?

The American philosopher of science Richard Boyd makes a basic distinction between literary and scientific metaphors. A literary metaphor, he argues, originates from a specific work of a specific author. "When the same metaphor is employed by other authors, a reference to the original employment is often implicit. When the same metaphor is employed often, by a variety of authors, and in a variety of minor variations, it becomes either trite or hackneyed, or it becomes 'frozen' into a figure of speech or a new literal expression."[36] Through overuse, literary metaphors lose their insightfulness; they lose their momentum. By contrast, scientific metaphors draw their power precisely from their appropriation by the entire scientific community.

"Variations on them are explored by hundreds of scientific authors without their interactive quality being lost."[37]

Good scientific metaphors persuade; good literary metaphors strike by wonder. When effective, both of them set the human mind in a productive state of instability. Theory-constitutive scientific metaphors are conceptually open-ended. They encourage the discovery of new features; they introduce "terminology for features of the world whose existence seems probable, but many of whose fundamental properties have yet to be discovered."[38] Likewise, fresh literary metaphors are not limited to expressing ideas and sensations, but, in Gaston Bachelard's words, they "try to have a future." Through the subtlety of their innovation they "revive the source"; they "renew and redouble the joy of wonder."[39]

Geographical metaphors curiously linger between literary and scientific metaphors. They at once embrace and elude both categories. They reflect the hybrid status of geography as a scientific and humanistic mode of knowledge. According to the Italian geographer Giuseppe Dematteis, "weak" geographical metaphors are usually pedagogical metaphors (the fluvial basin, the mountain chain, natural boundaries); they are commonplaces that have crystallized to the extent that they have become invisible to our eyes, and as such they conceal power "as if behind a smoky curtain."[40] "Strong" geographical metaphors, by contrast, are analogous to theory-constitutive scientific metaphors. They are those that make us see phenomena and processes with new eyes (for example, "the city as an organism" not only allows us to describe known relationships, but it invites further reflection, it unveils the existence of something akin to homeostasis, the genetic code, or self-organization). In other words, effective geographical metaphors are open-ended metaphors that allow multiplicity of meaning and practice; they are those metaphors that push the gaze beyond the tight boundaries of its comfort zone and induce thoughts to sail toward ever-shifting horizons.[41]

*

The geographical metaphor par excellence, the earth's mantle is at once a scientific and literary metaphor. In the earth sciences it holds the status of a theory-constitutive metaphor (tectonic plate theory could only be explained, or indeed even conceived, through the mantle). In other contexts, however, the earth's mantle is a poetic figure that has crystallized over the centuries. And yet, even so, the mantle is in no way a static or monolithic metaphor. By contrast, it can be defined as what literary critic Christy Wampole called a "super-metaphor," that is, a metaphor that actively incorporates other metaphors—and compels us to see the world from different perspectives.[42] In its capacious embrace, the earth's mantle bodies forth both the global and the planetary. In other words, it represents both the social and the physical, and their deeply intertwined textures. As the following chapters of this book will show, the mantle opens up and enlarges the world. It unveils a plurality of worlds.

Mantles come in different shapes and sizes, from the short *chlamys* of ancient Greek soldiers to the loose, cape-like cloaks worn in western Europe from the twelfth to the sixteenth century. Mantles come in solid colors, but they can be equally mended with hasty patches, worn out by the action of time, embroidered with rich decorative patterns. Mantles also come in different materials. Their varied textures offer themselves differently to the touch: the thick softness of velvet, the slippery ripples of precious silk, the warm roughness of raw wool. If you look closely at their texture, mantles will reveal the processes by which they were made. Rare surviving pieces of ancient fabric, for example, show that weaving was akin to modern rug making. The weft thread was not simply passed between the warp threads, as in modern cloth, but it was knotted about each of them.[43] Knotting and weaving were thus expressed by the same words in Sanskrit, Greek, and Latin, and provided the texture for metaphors of creation. Spun by the gods, humans' threads of life were woven together and knotted in the great fabric of the world—as they continue to be.

As a super-metaphor, the mantle thus encompasses surface, thread, texture, and craft metaphors. Not only do these metaphors operate at different spatial scales (the organic "tissues" of our body, the "texture" of a city, the "network" of commerce, the "ripples" of the terrestrial crust, the "fabric" of creation), but they also signpost geographical practice: geographers "weave" texts, follow "threads," pierce "veils." More broadly, Orphic and Promethean approaches are respectively expressed through "unveiling" and "weaving" metaphors. The "mantle of the earth" metaphor bodies forth materiality and transformation. It brings geographical features under human scrutiny and makes them palpable to the eye. The mantle is not a rigid plane. Unlike a flat screen, the mantle is a dynamic surface. Its velvety texture bends, molds, and ruffles like the waves of the sea (indeed, kymatology, the study of waves, was conceived as a branch of geography).[44] More remarkably, the mantle of the earth opens and closes with the slow turns of history. Its opening and closing movements unveil different perceptions of space. These movements and transformations are the focus of this book and shape its structure.

*

The book falls into four thematic parts: "Clothing Creation," "Unveiling Space," "The Surfaces of Modernity," and "Weaving Worlds." Each of these sections groups chapters arranged in a roughly chronological order. Part I, "Clothing Creation," follows the weft threads of ancient and medieval history as they weave through mythical and biblical mantles. Chapter 1 traces the origins and ramifications of the earth's mantle metaphor in the Greco-Roman world, from ancient mythology to Strabo's geographical writings, and its translation from textual metaphor to material craft: from Queen Kypros of Judea's wool cloth embroidered with Earth and Ocean to the layout of the city of Alexandria, which was described as a cloak and as a microcosm.

Chapter 2 explores garment metaphors in Scripture and their legacy in early Christian theological literature and in Byzantine hymnography and art. In the Old Testament such metaphors bear witness to the transient beauty of creation as opposed to eternal divine splendor, a common motif in theological orations and encomia, from Eusebius of Caesarea to Basil the Great and Gregory of Nazianzus. In Hebrew cosmography the firmament (*stereoma*) was the "solid" (*stereon*) perceptible boundary of the visible creation, behind which was concealed the uncreated God. In patristic writing, the veil of the firmament was symbolized by the curtain of the Temple of Jerusalem, which was also interpreted as a prefiguration of the veil of Christ's flesh. In Mary's weaving of the veil of the Temple, Byzantine commentators saw a foreshadowing of the Incarnation, for Christ clothed himself in the "robe of the flesh" woven from the body of the Virgin. As with the mantle of the heavens and the garment of the earth, Christ's mantle of flesh both enfolded and concealed his divinity. At the same time, however, it also made it accessible to humankind.

This motif is further discussed in chapter 3, the last chapter of the section, which traces the reception and visual translation of the biblical garment metaphor in the Latin West. In particular, this chapter focuses on regal cosmographic mantles and on *mappae mundi* (literally, "world cloths"), a cartographic genre recorded from the eighth to the fifteenth century. These maps wove biblical vignettes with contemporary cities and other geographical features within the closed space of a circular world island. On some of them (for instance, the Psalter *mappa mundi*) the world is enshrined in the figure of Christ, a distant echo of the biblical cosmographic mantle and the patristic conception of God encompassing the totality of creation. On the Ebstorf *mappa mundi*, a fourteenth-century exemplar produced for a Benedictine nunnery, the body of Christ is woven in the fabric of the *orbis terrarum*, with his head, hands, and feet piercing the landmass at the cardinal points. The map is nothing but a replica of Veronica's veil (vernicle): the latter displays the impression of Christ's face on a piece of linen; the former displays the Creator's handiwork—the imprint of his face on the face of the earth.

In Part I, the mantle of creation features as a divine craft. It is the product of a poetic act, the reflection and embodiment of a supreme order. The mantle envelops the reassuring closed space of the cosmos, of the earth, and of the human body. Part II, "Unveiling Space," traces the shift from "wrapping" to "unveiling." More specifically, it explores the disruption of this ancient order and the transformations the mantle underwent through the following centuries.

As chapter 4 shows, the great geographical discoveries, along with the pioneering of linear perspective and of human anatomy in the Renaissance, marked the opening of space at both a macro- and micro-scale. As explorers crossed the Atlantic and circumnavigated Africa, the closed world order of the medieval *mappa mundi* gave way to an expanding order, which was reflected in Ptolemaic maps shaped as opening mantles (for example, the Waldseemüller map of 1507). Likewise, as the fleshy mantle of the human body was for the first time pierced by the anatomist, new and unexplored territories were uncovered and charted in dedicated atlases (Andreas Vesalius's *De humanis corporis "fabrica"* was printed in 1543). At the same time, with architects and artists theorizing and implementing linear perspective, space acquired a third dimension to be penetrated by the eye—depth.

Stimulated by Bacon's experimental science, in the seventeenth century the piercing eye was empowered by optical devices, such as the telescope and the microscope. An infinitude of new worlds thus opened up—from the micro-scale of mysterious bacteria to the dark depths of the boundless universe. Ubiquitous presences on the frontispieces of scientific books, unfolding veils and heavy drapes lifted by winged putti and mythical figures became metaphors for the progress of human knowledge. In cosmographic works, veils and drapes came to articulate a powerful dialectic of lights and shadows, visibilities and invisibilities, known and unknown. Chapter 5 explores the evolution of these representations and the opening of space and nature during the scientific revolution that took place between the seventeenth and early eighteenth centuries.

Chapter 6 turns to Romantic perceptions of creation and the transformation of the mantle from a heavy drape or velvety curtain into an ethereal diaphanous veil. According to Goethe and his admirer Alexander von Humboldt, who is traditionally credited as the forefather of modern geography, poetic contemplation helped the scientist achieve an intimate, spiritual contact with the cosmos and unveil its mysteries. Unlike Renaissance explorers and anatomists, Goethe, Humboldt, and their contemporaries did not view the mantle as an opaque curtain to be pierced; rather, they saw it as a transparent veil. It did not hide but rather revealed, diffusing a transcendent glow. Atmospheric veils of haze, mist, and light shroud the landscapes described by German Romantic poets and their Anglophone counterparts (such as Coleridge and Wordsworth) and painted by artists such as Frederic Edwin Church, who followed Humboldt's footsteps to South America. Chapter 6 explores the veil as a poetic and aesthetic trope, and as a metaphor for a distinctively Romantic mode of geographical knowledge.

Part III, "The Surfaces of Modernity," considers new incarnations of the mantle metaphor from the late nineteenth century through the Cold War of the mid-twentieth century in the context of geography, scientific exploration, and environmental discourse. Chapter 7 focuses on the institutionalization and establishment of geography as an academic discipline in Europe and North America, and on shifting perceptions of space in an age increasingly dominated by global transports and communications. From a seeker of inner cosmic forces, the geographer of the twentieth century became a systematic scrutinizer of the earth's surface. This was now an almost totally "unveiled" surface. On it there were no further vast *terrae incognitae* to be explored, nor curtains to be lifted up, nor diaphanous veils to be looked through. The geographer's task became that of weaving together different areas of knowledge in a coherent disciplinary tapestry, interlacing the threads of hard science with those of culture, building patterns of physical landscapes and human activities. For the modern geographer, "mantle" was just a synonym for "surface," and "surface" a synonym for "landscape." At the same time, technological developments, such as powered flight and deep mining, added thickness to the earth's surface. By the end of World War II, human space had expanded to the stratosphere and to still untapped underground realms.

During the Cold War, new vertical domains were opened up to human exploration, including outer space and the earth's interior, along with the scientific colonization of Antarctica. A highly symbolic geopolitical feat fueled by ideological rivalry, the "assault" on these extreme environments triggered a return of the Promethean trope traditionally associated with discovery. Veils parted once again. While seventeenth-century veils, however, were pierced visually, by the technologically empowered eye, those of the modern era were trespassed in the most physical sense. Since the late 1950s, the veil of the atmosphere has been repeatedly perforated by powerful rockets carrying satellites, animals, and even human beings. In 1961, the US government sponsored a project that attempted to drill deep through the ocean crust to the earth's mantle—an earth sciences response to the space race at the height of the Cold War. Chapter 8 considers the political implications of the earth's mantle as a

scientific metaphor at this time of intense technological innovation and geopolitical confrontation.

Chapter 9 turns to a countermetaphor that became popular in those same
years, as the threats posed by nuclear fallout and chemicals shifted public
attention to the "green mantle of the earth." Previously used by Church Fathers and Romantic poets alike, this ancient metaphor was appropriated by
ecological writer Rachel Carson as a poetic figure for terrestrial vegetation. In
her bestselling (yet highly contentious) book *Silent Spring* (1962), which has
been traditionally credited as the cornerstone of the environmental movement, Carson described the harmful effects of pesticides on the environment.
In alerting the American public to the irreversible damage inflicted upon
the green mantle, Carson stressed its "textured" nature. The ravaging of the
"green mantle" continued to be discussed by later ecological writers, including E. O. Wilson, Michael Soulé, and Bruce Wilcox. In its new, "textured"
incarnation, the mantle metaphor played a crucial role in mobilizing Western
public opinion, thanks to its ability to make invisible relationships visible.

Part IV, "Weaving Worlds," turns from nature and the physical earth's
mantle to its social texture. Looking at the fabric of modern and contemporary textile artworks and metaphors, chapter 10 moves from the mantle as
a surface to the process of weaving. In particular, it shows how this creative
act described in ancient Greek and Byzantine literature has been reappropri-
ated by modern artists to produce their own artworks and political countermaps. From the late eighteenth until the early twentieth century, schoolgirls in Britain and the United States created embroidered map samplers and
even silk globes. In 1961 (the year before Carson's book was published) the
Spanish-Mexican artist and anarchist Remedios Varo Uranga produced her
own parody of that world in a surrealist painting showing a group of girls
trapped in a tower as they weave a copy of the world that ends up covering the
world itself. A range of actual cartographic textiles have since been crafted by
various artists, from Alighiero Boetti's tapestries to Mona Hatoum's carpets
and Katy Beinart's interfaith tablecloth. Considered together, all these fabrics
unveil shifting geopolitical orders and territorial imaginations, including the
transition to a post–Cold War world and an increasingly globalized present.

Today, in a world saturated with flickering images and plasma screens,
there seems to be an increasing anxiety to recover the materiality of surfaces.
A new mantle, what human geographer Nigel Thrift called "an ecology of
screens," has wrapped the earth—a vast geographical web of perception that
crowds our daily lives, informing, entertaining, affecting life, or "simply providing ground." Not without a certain irony, we have become increasingly

dependent on, if not intertwined in, this texture of screens. Today, geographers talk about mixed realities, digital envelopes, and augmented places. TV screens, computer screens, smartphone screens, tablet screens, NAVSAT screens, mega-screens permeate our daily routines, operating like a second skin. The digital is no longer a kind of covering, and the "skin" is of course an organ. As our lives become increasingly imbricated in social networks, navigated through Google Maps, or affected by global news, we are left wondering whether plasma is the next incarnation of the ancient mantle of the earth, or simply part of its fabrics. This question is the focus of the last chapter of the book.

*

In its multiple incarnations, the mantle features as one of the most resilient and yet mutable geographical metaphors in Western spatial history. Tracing the genealogies and evolution of the metaphor means interrogating the relationship "between surface and that which lies beyond or behind it, between appearance and underlying reality, between sensory appreciation and intellectual understanding."[45] Each variant of the mantle metaphor explored in the book speaks of a different approach to the world: poetic contemplation, scientific inquiry, comparative analysis, critical investigation, aesthetic appreciation.

This book also aims at celebrating geography's visual and material richness. Textual sources, ranging from ancient Greek literature and patristic theological writings to Romantic poetry and twentieth-century popular science and geography books, are thus set in dialogue with graphic representations and objects, including Byzantine textiles, medieval and Renaissance maps, modern artworks, and geological diagrams. In weaving all these threads together, this book aims at shedding light on continuities and discontinuities in the history of Western geographical imagination and doing the job of a "good" metaphor: opening up new ways of seeing a rapidly changing world.

PART I

Clothing Creation

Mythical Cloaks

A man drives across a vast yellow plain swept by the wind. He follows a straight asphalted road for miles and miles in what appears to be a journey toward the infinite. There are no landmarks or signs, only dry grass and the slate-tinted sky. Ahead of him are just the road and the flat line of the horizon. Exasperated by the monotony of the landscape, the man stops the car and gets off to consult his map. Where is he? For how many more miles shall he endure this desert of dullness? As he spreads the map on the car's hood, he accidentally ripples it. Suddenly something extraordinary happens. A mountain emerges on the horizon. Its shape uncannily resembles the ripple the man just produced on the map. The man smiles. He knows that the map is no longer there to be looked at, but to be acted upon. As he impetuously crumples it up, new reliefs dramatically arise from the depths of the earth, one after the other. Each movement of the hand translates into an earthquake; each wrinkle on the map becomes a new relief in the landscape. Chthonic forces are violently released from unseen profundities to give the land a new shape. Geological time is compressed at the speed of a hand's gesture. Flatness is morphed into a bedraggled sequence of heights and depths, of rugged peaks and steep ravines, of sharp pinnacles and dark crevasses. The straight line of the road now bends and zigzags. It ramifies into endless winding paths. Satisfied, the man gets back in his car ready for new adventures.

This scene, a TV commercial for a French SUV, not only encapsulates the aspirations of a new type of customer dreaming a world centered around him; it also evokes the transformative power of maps.[1] It is after all a reminder that by acting on the map, we also act on the world; that every project on territory—from town planning to boundary setting and road building—starts from a map. It is a reminder that mapping is a creative act and a physical act;

that maps are more than visual representations, they are also material objects we grasp and handle, roll and fold, ripple and crumple. In the hands of the French driver, the map is not a simple printed sheet of paper; it becomes a malleable substance akin to clay.

Maps transform the surface of the earth, but they are in turn surfaces liable to transformation. They are part of the "stuff" of the world they portray. Hence, the commercial presents us with a paradox: how is that the man is embedded in the very landscape he holds in his hands? How is that the landscape starts to take shape and make sense to him only when he creatively engages with the map? What is that mysterious force that binds the map and the land together?

The same paradox haunts the very origins of Western cartography. The first Greek image of the inhabited world, we are told, was crafted by the Milesian philosopher Anaximander in the sixth century BCE.[2] At approximately the same time, Pherecydes, a mythographer from the small island of Syros in the Aegean, recounted the extraordinary story of the mystical wedding between Zeus and Chthonia in one of the earliest fragments of Greek prose that has survived to us. On the third day of marriage, Zeus, the ruler of the gods, places a mantle he has embroidered with the shape of the land and the ocean upon Chthonia, the still formless earth. In other words, he covers his bride with a giant map made of fabric. Thanks to the molding power of the mantle, cartographic order is imprinted on chaos and Chthonia suddenly takes shape; she becomes Gē (or Gaia), Mother Earth.[3] As with the driver in the French commercial, Zeus is at once inside and outside of the scene: he holds the map in his hands and, through that very map, he transforms the reality of which he is part. The act of creation is ascribed to the power of the divine mantle; signification to cartographic inscription. Chthonia, the amorphous primordial matter, the monstrous abyss, can only be gazed at through the mediation of the cartographic mantle.

Extraordinary as it might sound, Pherecydes's story is no isolated tale. The ancient world is shrouded in fabric. The Babylonians envisaged the sky as a mantle.[4] Egyptians devised a "living mantle" in the shape of the goddess Nut, whose elongated body formed an arch, literally wrapping and sheltering the earth.[5] Indian mythology similarly refers to a hill goddess who lifted herself from the ground, acting as a protective mantle for the people and animals of her region when it was attacked by the rage of Indra, the divinity of the thunderbolt. Myths from Central Africa describe the earth being "rolled out like a mat," while the Quran likens it to a carpet spread out by Allah and held in place by "firm mountains" that serve as weights or pegs.[6] In the Maori tradition, the Nga Uri cloak is symbolic of care and protection for the earth, while Native American mythology ascribes the birth of summer to a "new cloak of green" unfolded by the Shining One upon Mother Earth, "beautiful with all her flowers and birds," and to a soft cloak of dark blue subsequently spread over the sky, in which many a star sparkled and twinkled.[7] Why is this the case? Why are cosmogonic mantles common features across such a diverse range of cultures?

According to art historian Ewa Kuryluk, the universal appeal of textile metaphors comes from our symbiotic relationship with fabric: humans are born naked, but are wrapped in cloth as soon as they emerge from their mothers' bodies. At the moment of death, humans are likewise wrapped in cloth, before being wrapped by the dark matter of the earth. We need garments in order to survive, and to die too. The rich symbolism of thread and fabric, Kuryluk argues, "resonates in everyone because of textiles' omnipresence in swaddling clothes, garments, bedsheets, towels, blankets, bridal veils, burial shrouds, bandages, tablecloths, curtains, tents, sails, flags, banners, canvases, screens, scrolls, sacks, bags, rugs, and other textiles that provide us with comfort and pleasure."[8]

In the Greco-Roman world, however, mantles, cloaks, and textile metaphors in general seem to hold a special degree of complexity and resilience. Long before Pherecydes, Homeric heroines were embroidering mythical narratives on mantles and nuptial blankets, while goddesses and nymphs perpetually spun the threads of human destiny on their laps, and poets wove their hymns of praise. Nearly half a millennium after Pherecydes, Strabo (64 BCE–20 CE) described the earth as a large island shaped as a *chlamys*, the short cloak worn by Macedonian warriors, which also characterized the outline of Alexandria. Romans provided the metaphor with material texture. As early as 39 CE, we are told, Queen Kypros of Judea sent Emperor Gaius a linen or wool cloth depicting Earth and Ocean, accompanied by the following lines: "Modelling all with shuttle on the loom [Kypros] made me [the textile], a perfect copy of the harvest-bearing earth, and all that the land-encircling ocean girdles, obedient to great Caesar, and the grey sea too."[9] Extending *ad fines orbis terrarum* (to the edges of the earthly globe), the Roman empire had become Zeus's mantle. Gaia belonged to Gaius.

Why were textile metaphors so pervasive in the ancient Mediterranean? Why were mantles so deeply interlaced with the earth and its images? What have metaphors to do with myth? And what have myths to do with metaphor? This chapter explores the rich fabrics of the mantle metaphor and its manifold ramifications in the Greco-Roman world. In particular, taking its point of departure from Pherecydes's myth, it illustrates two aspects of the earth's mantle that will run through the rest of the book: first, the ability of the mantle to conceal and reveal, as a metaphor for geographical knowledge; and second, its textured nature, as a metaphor for interconnectedness and thus for the harmony of the cosmos and humankind.

Surfaces and Depths

Various commentators have noted the close relationship between the stories of Anaximander and Pherecydes, that is, between the first Greek map and the mythical cartographic mantle. As with Pherecydes, Anaximander was interested in the origins of creation. He postulated the existence of an eternal and unchanging primordial element, the *apeiron*, out of which an ordered world had come into being through the separation of opposites generated by its perpetual motion (we can see here parallels with Pherecydes's evolutionary narrative of the transformation of primordial Chthonia into ordered Gē).[10]

Much of Anaximander's fame, however, is tied to cartography. Probably inspired by maps he might have seen in Egypt and in the Near East, Anaximander crafted his own map, which he circulated together with a book titled *Periodos Gēs* ("Circuit of the Earth"). Pherecydes, as the classicist Alex Purves notes, is also believed to have seen maps, but, unlike Anaximander, he uses cartography in a metaphorical sense. And yet, the outcome is the same: *geography*, the depiction and description of the surface of the earth.[11] It is to Anaximander that the origins of the expression *graphein tēn gaian* ("to write the earth") are ascribed, for he literally "wrote" the earth on a tablet, *en pinaki*.[12] In other words, the Milesian philosopher was the first Greek who translated the complexity of the earth onto a flat surface, thus literally putting the world in the hands of the beholder. This translation, however, had already happened in Pherecydes's tale: by embroidering his mantle (*pharos*)[13] with the land and the ocean, Zeus also crafted a map and made the earth legible to the human eye.

The tablet and the mantle are nonetheless different kinds of crafts. Unlike the *pinax*, the *pharos* is not a rigid surface. It bends and adapts to the mass and shape of what it covers. Moreover, while the tablet can be imagined in isolation, the mantle evokes the mysterious presence of what it hides beneath its fabric; it fulfills its function, its *raison d'être*, precisely in the symbiosis with this invisible body. Examples of *pharoi* in ancient Greek literature range from the long white robes wrapping "the sweet forms" of Aidos and Nemesis (Hesiod, *Op.* 198) and the beautiful body of Calypso (the nymph who held Odysseus captive, *Od.* 5.230), to the sail of Odysseus's ship woven by the same nymph and molded by the invisible breath of the breeze (*Od.* 5.258). Death itself is cloaked in a *pharos*; a notable example is the shroud pitifully covering Patroclus's body amid women's funerary lamentations (*Il.* 18.353).

Concealment and death are intrinsically related. Their bond is at once topographical (Hades, the realm of the dead, was believed to be located in the underground, away from the human gaze) and semantic, for the verb *kalyptein* (to hide) can also signify "to bury." Indeed, Calypso is both "she

who hides" and the goddess of death. This meaning, Hadot observes, corresponds to the representation of both "the earth that hides the bodies of
the dead" and "the veil with which the heads of the dead are covered."[14] In
Homer's verses, the thick vegetation surrounding Calypso's dark cave (the
place for hiding *par excellence*) is but an extension of the veil that covers the
head of the nymph, or of the fabrics she weaves with her golden shuttle (*Od.*
5.90). Elsewhere, Homer writes that death is a band (*telos*) wrapping the head
and especially the eyes of the dying person (*Il.* 16.502). The endless violence of
battle scenes in the *Iliad* repeats over and over in Homer's phrase "and darkness veiled his eyes" (*Il.* 4.446; 6.1; 13.487), as a mournful refrain. In Euripides's
Hippolytus an agonizing Phaedra, consumed by passion, begs her nurse to
hide her "blond head" under a *pharos*. The nurse obeys but adds, "I veil you,
but when will death cover my body?" (*Hip.* 133). Death looms over humans as
a shroud, a cloud, a mystery.[15]

Inspired by classical authors, Italian Renaissance artists used the iconography of the head covered by the shroud in personifications of the Nile,
whose mysterious sources had remained concealed to geographers and explorers since antiquity. The Latin poet Tibullus (55–19 BCE) wondered where
Father Nile had concealed his head, and Ovid (43 BCE—18 CE) replied that
the river had submerged its head in the ends of the earth to escape the heat
generated by Phaeton's loss of control of the sun's chariot. Indeed, in the first
century BCE, the Latin phrase *caput Nili quaerere* ("to seek the Nile's head")
became a common expression indicating an impossible task.[16] The *pharos* is
therefore a threshold between life and death, between appearance and disappearance, between the known and the unknown.

What lies beneath the *pharos* Zeus donated to Chthonia? Pherecydes,
Purves observes, starts from the premise that the earth can be approached
in two ways: vertically and horizontally. In the beginning the earth is chthonic,
an epithet ancients Greeks used as a surname for infernal divinities and which
we still use to indicate something subterraneous, obscure, deep, invisible, in
other words, "in or beneath the earth."[17] Chthonia is a monstrous creature,
because she cannot be seen; because she has no shape. She is not *omorphē*
("beautiful, well-formed"), but *a-morphē* ("amorphous, without form"). After
her marriage to Zeus, however, Chthonia takes on a new name. She becomes
Gē, the Greek word for "land," or the substance that covers the surface of
the earth. Gē is the Gaia of the Latins, she who smiles and shines. While the
chthonic quality of the earth is traditionally understood to be hidden, Gē
encompasses "that which lay on the surface of the earth and which would be
spread out for view."[18] Darkness is replaced with transparency and clearness
of vision; verticality is replaced with horizontality.

FIGURE 1.1 Scene from the ceremony of the *anakalyptria*, delightfully represented on the east frieze of the Parthenon (ca. 440 BCE). The ritual consisted in the substitution of the bride's veil with a cloak offered by the groom. Here Hera, sitting next to Zeus, graciously lifts her veil.
British Museum. Photo by the author.

Matter, Bachelard argues, is the unconscious of form.[19] Chthonia, the dark matter of the earth, we are told by Pherecydes, has always been.[20] By contrast, Gē, the surface, the form, is a creature of time; she comes into being. How does this transformation happen? Pherecydes explains that the wedding between Zeus and Chthonia is the first sacred marriage: "From this event the *nomos* [custom] of the *anakalyptria* [the unveiling of the bride] originated among gods and men."[21] The ritual consists of the substitution of the primordial veil of the bride with a cloak donated to her by the groom (fig. 1.1). As she receives the new mantle, the bride also changes her name; as it still happens today, she adds the name of the husband to her own.[22] Hence Chthonia becomes Gē. She becomes an object to be looked at.

The climax, however, is reached not at the conclusion of the ceremony, but at the moment of the unveiling in the wedding chamber, when Chthonia takes her primordial veil off, so that Zeus can replace it with his geographical mantle. The bride remains naked only for a moment. According to the Italian geographer Franco Farinelli, it is only in this very brief moment that truth can be grasped:

Truth shows itself in the twinkle of an eye. But it is just a twinkle of an eye; the gift of the groom immediately covers the body of the bride again; the abyss can be gazed at only for a moment. And in that moment we cannot know anything because sky and earth are undistinguishable, duplicity becomes undifferentiated unity. Only through Zeus' mantle are we able to know something, but what? By the end of the wedding, all we see is the image of the mountains, of the rivers, of the palaces, in other words the earth not as Chthonia, but simply as Gē, as a face; what we see is not the thing, but the image of the thing.[23]

We are only allowed to know the appearance of things. As Democritus later wrote, "we know nothing in reality, for the truth is within a deep abyss."[24] Our knowledge of the world is by necessity horizontal, superficial—geographical in the literal sense of the word. According to Heraclitus, one of the earliest great thinkers of antiquity who lived in Ephesus just after Pherecydes, the nature of things demanded to be hidden. For Heraclitus, *physis* (what we call nature) indicated the springing-forth of things, the process of self-realization, of growth, especially vegetal growth (modern Greeks still call plants *phyta*, while the English language has inherited the prefix *phyto-*).[25] Gradually, however, a power was imagined to produce this manifestation, and by the first century BCE nature had become a personification, a goddess:

> Principle and origin of all things
> Ancient mother of the world
> Night, darkness and silence.[26]

The goddess was identified with Isis, a monstrous female creature covered with breasts. Like Chthonia, Isis was frightening; she could not be looked at; and like Chthonia, she was wrapped in a veil. In the words of Porphyry (234–305 CE), one of Plotinus's pupils, "Nature hates to expose herself uncovered and naked in the view of all. Since she has concealed the knowledge of her being from mankind's coarse senses, by hiding beneath the vestments and envelopes of things, likewise, she has wished that sages discuss her mysteries only under the veil of mythic narratives."[27] Myths thus served a function akin to that of metaphors; they were interfaces necessary for comprehending experience; they allowed the ancients to approach the unknown by way of the known.

Weaving Myths

Myths are similar to maps. They are nothing but models of reality. As such, they owe much of their effectiveness to the plainness of their plot: the simpler their structure, the stronger their appeal and resilience. Edmund Leach has drawn attention to the markedly binary patterns of mythical narrative; to their opposing categories and mirrorlike correspondences. The reason, he argues, is experiential: "Humans delight in symmetry and duality, probably because of their bodies' symmetrical nature, the division into two sexes, the duplicating character of procreation, and their sense of separation between body and mind."[28]

Juxtaposing Chthonia and Gē, groom and bride, seen and unseen, the myth of Zeus's mantle is paradigmatic in this respect. The plot sketches a move from primordial disorder to cartographic order. It is about making sense of the earth; it is about mapping as a cognitive process. The binary pattern, however, is embedded not only in the narrative plot of the myth, but in the very fabric of Zeus's mantle and its making. Weaving is more than a spatial act; it is also a conjugal vocation. To weave is to unite, to interlace, to bind. More specifically, as John Scheid and Jesper Svenbro suggest, to weave is to unite opposites: the vertical warp (*stemon*, or *mitos*, both gendered in the masculine) and the horizontal weft (*krokē*, gendered in the feminine). Weaving thus offers a simple model to the mind seeking order. "The original disorder, the raw wool is replaced by an organized fabric in which each fiber is in place."[29] Like mapping, "to weave is really to give order to a great tangle of matter, in order to put each matter in its proper place."[30] Yet, to weave is also to "give birth" to a new tissue. In Scheid and Svenbro's words:

> During the wedding ceremony, Zeus transforms himself into Eros in order to be able to create the opposing elements of the entire cosmos. This is an interlacing that is essentially weaving. It allows Zeus to create not only the wedding cloak but also the union of the bride and the groom, beginning with his own union with Chthonia, henceforth elevated to the rank of goddess of marriage. . . . Both literally and figuratively, Zeus is a weaver.[31]

In classical literature matrimony, the union of opposites, the coming together of two distinct life threads, is symbolized by another cloak, the Greek *chlaina*. This was the matrimonial blanket that covered the couple at the moment of consummation. Formed by the tight knotting of weft and warp, the *chlaina* was a metaphor for union—what the Romans called *conubium*.[32]

In bringing Zeus and Chthonia together in the same mythical account, Pherecydes himself wove a fabric—a textual fabric. His writings are contemporary not only with the first Greek map, but also with the birth of prose—as if geographical description, *graphein tēn gē*, demanded a new writing style.[33] In the ancient world, the image of weaving was used extensively in connection with "weaving narratives," hence, literary and visual descriptions of the earth were neatly intertwined.[34] Through the creative act of writing, both geographers and mythographers "interlaced" words and crafted narrative patterns.[35] The very word "text" is related to the Latin *textere* ("to weave"). Today the metaphor endures in expressions such as "to spin a tale," "to dress up in words," "narrative thread," "texture of speech," "web of history," while in Italian the word for narrative plot (*trama*) is the same as the word for weft.

In classical mythology the metaphor takes visual shape—and, once again, through the mantle. Poets from Homer to Ovid and Catullus tell us about beautiful *pharoi* and *chlainai* embroidered with mythical narratives: from the battles of the Trojans and the Achaeans woven by Helen into the cloak she gave Telemachus as a farewell gift (*Iliad* 3.123–38), to the gigantomachy skillfully crafted by Arachne in her weaving competition with Minerva (Ov., *Met.* 6), and the scenes of Ariadne's tragic story decorating Peleus and Thetis's nuptial blanket. These cloaks were fine art crafts manufactured by goddesses and heroines, as well as powerful narrative devices in the hands of the writer. In the *Iliad*, Helen depicts the same battles that Homer himself is in the process of telling, as a "spatial tableau of the poem."[36] In Catullus's description of Peleus and Thetis's wedding, the Fates spin the destiny of Achilles, the son who will be conceived beneath their nuptial blanket.[37] Similarly, the transmutations of gods into animals woven by Arachne foreshadow the tragedy of her own transformation into a spider.

Diverse as they are, these tapestries offer a complete view of the plot that is to be unfolded.[38] They operate as maps in that they empower the reader with the omniscient gaze of the gods. The poem is conceived of as a craft that can be seen and grasped in its totality, like the picture of the cosmos famously engraved on Achilles's shield (*Il.* 18.478–608).[39] The weaving metaphor conveys both a sense of space and a sense of time to the poems: patiently woven by their makers, the scenes come into being one after the other; they are always in the process of being crafted, of being unfolded by the poet.

Gē's mantle in Pherecydes's tale is the ultimate map. As with Achilles's shield, it offers a view of the entire earth at a glance, with Ocean embroidered on its edge.[40] But the tale is itself a map of creation. It is a dynamic map that brings together history's sequentiality (the account of the wedding) and geography's simultaneity (the cartographic mantle). The term Pherecydes uses to designate the act of embroidering is *poikillein*. This verb has the same root as the word *poikilia*, meaning "variety," especially of colors. It was used in the context of embroidery and painting, but it was also applied to literary and musical styles in the sense of intricacy and ornamentation,[41] as well as to the description of different aspects of creation, from a starry sky to a "dappled" earth bursting forth with flowers in the spring.[42]

Poikillein evokes the splendor of diversity. Pherecydes informs his readers that Gē's mantle was both vast (or *megas*, like the epic tapestry woven by

Helen) and beautiful (*kalon*). In this sense, the mantle is reminiscent of Aphrodite's divine cloak (*peplos ambrosios*) crafted by the Graces and the Hours and "dyed with the flowers of spring—such flowers as the Seasons bring—in crocus and hyacinth and flourishing violet and the rose's lovely bloom, so sweet and delicious, and heavenly buds, the flowers of the narcissus and lily."[43] Gē's mantle is also reminiscent of the "flowery robe" that Persephone was weaving, as described in the Orphic Rhapsodies, when she was abducted by Hades. Persephone's robe, Purves comments, has a certain cosmic significance: "her weaving of the flowers into the fabric suggests that it should be imagined as the soil of the earth upon whose surface flowers and crops emerge with the cycling of the seasons. The emergence of the plants on the surface of the cloth is reminiscent of a simultaneous movement in creation myths detailing the progression from unformed clod of earth to a fully articulated, manufactured cosmos."[44]

Pherecydes's image of the intricately woven robe bears also an important philosophical connotation. It signifies that the earth is invested with order, design, and perfection, and is therefore the handiwork of a rational creator. "The robe is the product of a carefully employed skill, a *technē*; on the same level, the earth is the result of a rational and purposeful act of creation. As in the weaving of a beautiful garment, so in the making of the world there is nothing random."[45] The creator remains external to his creation: Zeus, the weaver and embroiderer of the cosmic mantle, remains physically detached from it in the same way an artisan remains detached from his handicraft. In this sense, the myth foreshadows and casts the ground for the idea of a separate cause to creation that underpins later Greek philosophy, from Aristotle's efficient cause to Plato's demiurge.[46]

If for mythographers and geographers the mantle conveyed a poetic quality, and for philosophers a creative one, for the historian it could turn into an instrument of parody. Herodotus (484–425 BCE) exploited the beauty and concealing quality of the mantle in one of his numerous caricatures of Xerxes, the king of the Persians and stereotype of foolish barbarian. Having already revealed his childish *hubris* in cutting the isthmus of Mount Athos and "punishing" the sea for having interfered with his military plans, Xerxes displayed his stupidity also in personal affairs.[47] Showing off in a marvelous gaily colored *pharos* just woven by his wife Amestris, he promised Artaynte, his son's future bride, whatever she wanted in exchange for her favors. The girl asked him to swear:

> He promised this, supposing that she would ask anything but that; when he had sworn, she asked boldly for his mantle. Xerxes tried to refuse her, for no reason except that he feared that Amestris might have clear proof of his doing

what she already guessed. He accordingly offered her cities instead and gold in abundance and an army for none but herself to command. Armies are the most suitable of gifts in Persia. But as he could not move her, he gave her the mantle; and she, rejoicing greatly in the gift, went flaunting her finery.[48]

Here the mantle's qualities are turned upside down. The *pharos* woven by the wife, a Greek symbol of conjugal devotion (think of Penelope at her loom as she patiently waits for Odysseus), becomes the tangible proof of infidelity. It becomes a metaphor for barbarian immorality. Rather than uniting, it becomes the expedient for potential separation between husband and wife. Rather than covering, the mantle unveils Xerxes's depravity; it lays bare his own vanity and stupidity. *Poikilia* betrays him.

Alexandria and the Fabric of the World

The extravagant figure of Xerxes endured through the centuries as a literary *topos* and counterpoint to another ruler, Alexander the Great (356–323 BCE)—and to his mantle. Succeeding his father Philip of Macedon at the age of twenty, Alexander spent most of his ruling years on an unprecedented military campaign through northeast Africa and Asia. His military successes led to the defeat of the Persians and the creation of one of the largest empires of the ancient world, stretching from Greece all the way to India. Over the centuries, the figure of the Macedonian king-warrior crystallized into one of the most pliant heroes ever spawned by Western culture.[49] It was, however, when Greece became a province of the Roman empire that Alexander, predictably, achieved his almost mythical status.

Four centuries after his death, Alexander's brutal conquest of the known world was idealized by the Romans as "the quest of a youthful heroic and semi-divine figure towards the edges of space and time" and was used as a "pedigree" for their expanding global empire. The Alexander myth, with its rhetoric of benevolent empire attributed to the cosmopolitan Macedonian king, was indeed in large measure a literary product following Augustus's victory over Mark Antony and Cleopatra at Actium (31 BCE) and Rome's worldwide domination.[50]

Much of the success of the Alexander myth in the Roman empire is owing to Plutarch (45–125 CE), a Greek biographer and essayst who lived under Roman rule. Elevated to a paradigm of wisdom, in his *Moralia*, the figure of Alexander is systematically contrasted to Xerxes's barbarian foolishness and arrogance. For example, while the Persian king "punished" the sea and cut the isthmus of Mount Athos, Alexander is said to have judiciously rejected his architect's grandiose plans to carve the mountain into a colossal statue of himself holding the city of Alexandria in his hand. The reason for the refusal, we are told by Plutarch, was purely logistical: how could the mountain's rough topography have allowed the cultivation of grain and thus the sustainment of its own inhabitants? How could such a large city have been self-sufficient?[51] Instead, Plutarch tells us, the Macedonian opted for a totally different setting and layout. Alexandria was to be built on the flat Mediterranean shores of Egypt, on the fertile delta of the Nile—and, intriguingly, in the shape of a *chlamys*.[52]

The *chlamys* was a multipurpose cloak woven from wool and worn by men. It consisted of a single piece of fabric, characteristically rounded at the edges.[53] Traditionally associated with travelers and hunters, the *chlamys* was part of Alexander the Great's everyday outfit. It was also worn by his elite cavalry, and perhaps also by his *bēmatistai* (or military surveyors).[54] With the progressive expansion of the empire, the cloak turned into a fashionable garment and status symbol that spread across the newly annexed territories; it was enthusiastically adopted by Greek and non-Greek soldiers alike, and by those who wanted to imitate the dashing fashions of those adventurous young men. The *chlamys* was draped over the left shoulder and pinned together over the right one. This left the right arm free to hold a sword or a horse's reins, while covering most of the rest of the upper body. It could also be used as a blanket when camping overnight.[55] As opposed to Xerxes's "effeminate" *pharos* appropriated by Artaynte, Alexander's *chlamys* was an expression of Greek heroic masculinity. But why give a city the shape of a cloak?

In the Greco-Roman world, the act of weaving was associated with peacemaking and social cohesion. Indeed, Scheid and Svenbro observe, of all Greek representations, "weaving seems to *fabricate* society more than any other."[56] Weaving essentially means interlacing different threads (the weft and the warp) to create a new fabric. In his *Statesman*, Plato (428–348 BCE) thus uses weaving as a metaphor for the union of opposites and as a paradigm for the art of politics. The good statesman, he argues, makes a tightly woven fabric out of citizens with different characters and dispositions; he "weaves all into the unified fabric [of the *polis*] and with perfect skill."[57]

The Platonic weaving metaphor was physically "enacted" through the ritual manufacturing of textiles by women from cities previously at war against each other. In his archaeological survey, Pausanias (110–80 CE), for example, writes of women from sixteen different cities coming together to weave a *peplos* for Hera in the temple of Olympia in the Peloponnese, while Plutarch associates the foundation of Athens with the institution of a festival centered on a large piece of fabric destined to cover Athena Polias in the Acropolis, symbolizing the unity of all the inhabitants of Attica.[58] Just as Zeus's cartographic mantle transformed amorphous Chthonia into a legible ordered surface, ritual weaving dispelled the chaos of war and social unrest, and brought instead order and harmony. Every time a cloak was woven to Hera (or Athena), a cosmogony was thus reenacted, if only at the micro-scale of the *polis*.[59]

In selecting the *chlamys* as the outline for his imperial city, Alexander extended the metaphor to a totally new scale. Alexandria was different from Plato's Athens and the other *poleis* described by Pausanias and Plutarch in that it interwove different ethnic and religious cultures in the same urban fabric. The origins of its "threads" were not to be sought within its walls, nor in the surrounding region, but rather all around the *oikoumenē* (the inhabited world).[60] The reason is that, according to Plutarch, Alexander always sought to bring people together, "uniting and mixing in one great loving cup as it were, men's lives, their characters, their marriages, their very habits of life."[61] In other words, he was a cosmic weaver—like Zeus.

After the death of its founder, not only did Alexandria pursue its cosmopolitan vocation, but it also thrived as the capital of the Ptolemaic dynasty that ruled Egypt for the following three centuries.[62] And so did the *chlamys*. Appropriated by the Ptolemies as their dynastic emblem, Alexander's cloak became a consistent symbol of royal and military power in Greco-Egyptian iconography and, by extension, of the Greek influence that "wrapped" the city and the rest of the civilized world. The *chlamys*-shaped city reflected the glory of the dynasty and its global ambitions.[63] Alexandria's cloak encompassed the ethnic diversity of the entire *oikoumenē*: Greeks and Egyptians, and also Syrians, Jews, and people from all the states of Asia Minor, North Africa, Europe, as well as people from beyond the Pillars of Heracles, and far-off Indus. A thriving port and major commercial hub, the city of the Ptolemies gained the fame of *pantrophos* ("universal nurse"), that is, a microcosm embodying the greater world beyond.[64]

Ptolemaic Alexandria was also a powerful magnet for intellectuals from across the Mediterranean, attracted as they were by its famous library. The Bibliotheca was a global archive that gathered all the texts ever written in the Greek world and beyond; it was also a "centre of calculation" in which threads of knowledge from across the *oikoumenē* converged and intertwined, giving birth to new texts and textures.[65] Intriguingly, city and archive were metaphors of one another. The urban space was organized according to the same principle as the universal library—what Giorgio Mangani called "*memoria locativa*," or knowledge retrieval through a topographical classification system. Like the Bibliotheca, the city was a self-contained cartographic space, bounded as it was by the edges of the *chlamys*. Its fabric was divided into different ethnic quarters identified by letters of the Greek alphabet forming the signature of its creator and master: "King Alexander, descendant of Zeus, founded it."[66]

In a sense, the city was a map of itself. It is therefore appropriate that the first scientific world map was produced in Alexandria and that it took the very shape of the city—the *chlamys*. It is also fitting that its author was the head librarian of the Bibliotheca under the third Ptolemy. This librarian and polymath, Eratosthenes (276–194 BCE), is generally remembered for the remarkable accuracy with which he measured the circumference of the earth. Among his other achievements, however, are the development of a map based on his new measurements, the adoption of a plane projection system with meridians and parallels, and, not least, the conception of the *oikoumenē* as a vast island surrounded by Ocean and located in the northern hemisphere, yet occupying less than half its surface.[67]

The *oikoumenē*, Eratosthenes believed, was approximately twice as long as its width, and it was shaped like a *chlamys*, that is, "like a rectangle rounded at its two ends and punctuated by various seas and gulfs."[68] As Christian Jacob comments, "if Eratosthenes chose to give the earth the shape of the city, the political symbolism was obvious: there was an analogical link between Alexandria and the earth, the microcosm and the macrocosm." In extending the *chlamys* from the city to the *oikoumenē*, "the map could illustrate the pretensions of the dynasty to rule over the whole world through symbolic mediations. The map, like the thousands of papyrus rolls archived in the Library, condensed the whole earth into the king's palace itself."[69]

Eratosthenes's original writings perished with the library, but fragments of his work survive through later authors, along with his *chlamys*-shaped *oikoumenē*.[70] Strabo, for example, justified this outline on geometrical grounds: the hybrid form of the *chlamys* was the only one that could accommodate an insular *oikoumenē* matching half of the northern hemisphere.[71] The Amaseian, who was nevertheless more interested in providing a vivid account of places and people than in creating a mathematical model of the earth, also used the *chlamys* to bring Alexandria's shape before his readers' eyes.[72] Just as the earthly *chlamys* was a canvas on which he sketched different regions and places, each with their own *idiōmata* ("peculiarities"), so was the urban mantle a container of memorable wonders and wonderful memorials: the magnificent royal palaces, the Museum, the burial places of Alexander and King Ptolemy with their sarcophagi made of glass and gold, the imposing Pharos, the manmade Paneium hill, the famous Gymnasium and Hippodrome, and so on.[73] In Strabo's account, the world mantle and the urban mantle were both narrative containers in which each *topos*, each embroidered feature, became the potential springboard for a story—what Pierre Nora called a "memory place."[74]

Strabo's *chlamys* nonetheless maintained its symbolic value: at the urban scale, it acted as an iconic reminder of Alexander's legacy of the city; at the global scale, it implicitly extended his military power and political aspirations to the Roman empire, of which Strabo was a citizen. Indeed, under the Romans the *chlamys* continued to be an element of imperial costume and to commemorate the universalist political ideas attributed to Alexander. Known as *paludamentum*, it was worn by emperors and senior officers. We regularly find it in depictions of this category of dignitaries and gods, thrown over their shoulder as part of their military uniforms.[75] The *chlamys* was therefore a reflection of the political nature of Strabo's *Geography*, whose utility, he believed, lay in informing "the activity of statesmen and commanders."[76]

Grounded in vivid topographic description, Strabo's *Geography* has been contrasted to the other great *Geography* of the Greco-Roman world, that of Claudius Ptolemy (100–168 CE). This Ptolemy was an astronomer, geographer, and mathematician from Alexandria, and he is one of the most influential figures in the history of Western cartography. As opposed to Strabo's descriptive account, Ptolemy's *Geography* was essentially a scientific treatise ultimately aimed at the construction of a mathematical model of the earth.[77] Ironically, while this text lays the foundations of modern Western cartography, no original map survives to us; indeed, it is uncertain whether it even contained maps, the first surviving examples being reconstructions by medieval Byzantine scholars.

Alongside a theoretical discussion on the nature of geography and a list of coordinates of the part of the world known to the Roman empire during his time, Ptolemy provided detailed instructions for projecting the sphere of the earthly globe on a flat surface. He devised two projections (fig. 1.2): the first, which is still used to our days, was obtained by fitting the northern hemisphere in an imaginary cone, whereby 180 straight meridians radiated from the North Pole and crossed 80 curved parallels; in the second projection, by contrast, only the prime meridian was straight, and the others bent around it, thus mimicking the terrestrial curvature.[78] The resulting shape was that described by his predecessors: the *chlamys*, that is, the iconic layout that shaped both the foundations of Alexandria and Eratosthenes's *oikoumenē*.

Ptolemy's geography was intrinsically linked to cosmography, as it relied on the use of celestial coordinates to create a grid (latitude and longitude) upon which could be plotted locations on the surface of the globe. In this new grid—the same we use today—parallels and meridians crisscrossed in the

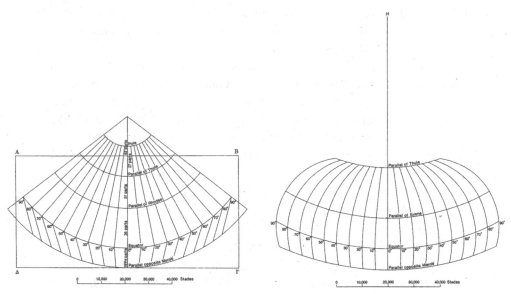

FIGURE 1.2 Ptolemy's first and second projections, which visually render the inhabited world in the shape of a mantle. From *The History of Cartography*, ed. Harley and Woodward, 1:187.

same way as the weft intersects with the warp. Like the loom, Ptolemy's grid provided the framework that made possible weaving the contours of islands and continents. At the same time, it connected the earthly globe with the concentric spheres of the heavens, in line with the ancient Greek belief in the "affinity of the stars with mankind and that our souls are a part of the heavens."[79] Ultimately, the Alexandrian scientist challenged humans to turn their gaze upward, to the starry sky, and, like Zeus, embroider their own mantle of the earth.

*

In antiquity, mythical narratives helped explain natural phenomena. As with metaphors, they were necessary to make sense of the world—and this is why they are found in almost every culture. Indeed, for Lakoff and Johnson, "no culture could function without myths, much in the same way we cannot function without metaphors."[80] Like textiles and texts, myths and metaphors are "woven" crafts. Myths interlace the visible with the invisible. Metaphors bring two distinct images together to produce something new. The creative power of the latter was understood by Aristotle, Alexander the Great's tutor, when he claimed that "the greatest thing by far is to be a master of metaphor."

Ordinary words, the philosopher argued, "convey only what we know already; it is from metaphors that we can best get hold on something fresh."[81]

In its various incarnations, from Zeus's *pharos* to Alexander's *chlamys*, the ancient mantle of the earth was both a myth and a metaphor. As a myth, it explained processes of creation (of the earth, of a city, of a dynasty). As a metaphor, it fulfilled the yearning for purpose and order of classical civilization, while acknowledging that reality is a complex *symplokē*—a song and fabric patiently woven by the nymphs.[82]

Weaving the *pharos* for his bride, Zeus created bonds of harmony between opposites; but so did Alexander and his successors in weaving Alexandria's cosmopolitan *chlamys*, the geographers who intertwined her urban cloak with that of the *oikoumenē*, and Ptolemy, the Alexandrian astronomer who, through his mathematical calculations, wove the mantle of the heavens into the earth's mantle. Just as Zeus's mantle imposed order upon Chthonia and allowed humans to look at her face, so did Ptolemy's scientific methods make the world comprehensible through the imposition of geometrical order onto its chaotic surface, while retaining a sense of wonder at its infinite variety.[83]

Under the Ptolemies, and especially under the Romans, the *chlamys* was such an effective metaphor not only because of its synthetic qualities, but also thanks to its power to symbolically project military and political power on a global scale.[84] For Strabo, a Roman citizen influenced by Stoic thought, natural features explained human things; they concealed deeper and more complex meanings than their simple appearance.[85] His *chlamys*-shaped *oikoumenē* was both a social fabric and a political concept, or in Christiaan van Passen's words, "a spatial entity of mutually related and interdependent peoples, . . . within which they have a place and a function."[86] The geography of the "*chlamys*," argued Strabo, was a description not only of its shape and size, but also of the structural relationship between its various parts.[87] His encyclopedic work was therefore akin to the construction of a statue, or what he called "*kolossourgia*," "which sets out major matters, how they came about, and things as whole."[88]

With its emphasis on universal sympathy and on the interrelations and affinities among things in the whole creation, Stoicism encouraged cosmopolitanism. Indeed, Plutarch's famous claim that "Alexander desired that all men be subject to one law of reason and one form of government and . . . reveal all men as one people" reflected the thoughts expressed by Zeno (333–264 BCE), the founder of Stoic philosophy. Stoicism implicitly justified the discourse of a universal Roman empire encompassing global destiny and extending *ad termini orbis terrarum*—that is, enveloping the world like a vast mantle.[89]

Characteristically, Marcus Aurelius (121–80 CE) likened the entire world

to a single "great city" and stage, which he invited his readers to view from
above. Taking a distance, "as from a lofty watchtower," the Stoic emperor-
philosopher argued, one would realize that we are part of the "skein" and
"web" of the "one whole," and that our place in society is assigned and woven
in our particular web. "Willingly give thyself up to Clotho, one of the Fates,
allowing her to spin thy thread into whatever she pleases. . . . Constantly regard the universe as one living being, having one substance and one soul; . . .
observe too the continuous spinning of the thread and the contexture of
the web."[90]

By way of such textile metaphors, Stoics like Plutarch and Marcus Aurelius
thus stressed the strong bonds between the macrocosm and the microcosm
that is the human being.[91] It is therefore fitting that in Plutarch's account—
and only in his account—Pherecydes is put to death by the Spartans and his
skin is preserved by their kings, like a worn-out *pharos* once wrapping the
microcosm of his creative soul.[92]

Biblical and Byzantine Garments

Praise the Lord, O my soul.
O Lord my God, you are very great;
you are clothed with splendour and majesty.
He wraps himself in light as with a garment;
he stretches out the heavens like a tent
and lays the beams of his upper chambers on their waters.
He makes the clouds his chariot
and rides on the wings of the wind.
He makes winds his messengers,
flames of fire his servants.
He set the earth on its foundations;
it can never be moved.
You covered it with the deep as with a garment;
the waters stood above the mountains.

(PS. 104:1–6)

The scalar tension between Pherecydes's cosmic mantle and Plutarch's account of the mythographer's tragic end—that is, between the fabric of creation and the fabric of the human body—mysteriously echoes through the pages of the Bible and throughout the history of Judeo-Christianity. Scripture is shrouded in mantles, veils, and garments of all sorts. The heavens cover the earth "like a tent" (Ps. 104:2). Kings, high priests, and scribes are clothed in long robes (Luke 20:46). Charismatic prophets wander through the silence of the wilderness wrapped in dusty cloaks and raiments of camel's hair (Matt. 3:4).[1] Colored mantles joyously pave Christ's way to Jerusalem (Matt. 21:8; Mark 11:8; Luke 19:36), and the simple touch of his *himation* is said to have cured a woman affected by an issue of blood, the same woman who, according to tradition, on the way to Golgotha offered Christ the linen upon which his face was miraculously impressed (Matt. 9:20).

Cloth, clothing, and their various synonyms metaphorize qualities of divinity, humanity, and the universe: "God is 'robed in majesty' and 'girded with strength' (Ps. 93:1). Humans either 'put on' righteousness and 'wear' justice 'like a robe and a turban' (Job 29:14), or they are 'clothed with shame and dishonor' (Ps. 35:26) and 'wrapped in cursing' (Ps. 109:18–19) and 'despair' (Ezek. 7:27)," while "God threatens humanity that he will withdraw light and 'clothe the heavens with blackness, and make sackcloth their covering' (Isa. 50:3)."[2]

Read through the spiritual eyes of the Church Fathers, the entire drama of human salvation, as Orthodox theologian Nicholas Constas observed, can be expounded in terms of nudity, clothing, and dress. For example, in the narrative deployed by Proclus, archbishop of Constantinople in the fifth century CE, Adam and Eve are said to "have been originally clothed in 'garments of glory.' Stripped of those garments at the time of their transgression, they are subsequently clothed in 'garments of skin' (Gen. 3:22). To remedy the nudity of the fallen Adam, Christ 'clothed himself' with Adam,"[3] that is, by entering the fleshy garment of a human body. Finally, thanks to Christ's self-sacrifice, the lost garments of glory are restored to the faithful through the mystery of baptism, for "all who have been baptized in Christ have been clothed in Christ" (Gal. 3:27). God's self-disclosure can only be experienced through the multiple veils in which He wrapped himself, from Christ's fleshy garment to the garment of creation and its universe of symbols. In other words, divine presence can only be experienced through concealment—through the mantle. How to explain this paradox? And why the mantle?

Symbols reveal by veiling. During his terrestrial ministry, Christ himself spoke in parables, that is, through metaphors. The Church Fathers followed the course. Weaving the threads of Scripture and classical education, they developed a complex understanding of space and time, which they expressed through the veiled language of prophecy, typology, and song, and which lies at the heart of Eastern Christian spiritual culture.[4] With their rich visual import, tactile qualities, and cultural heritage, textile metaphors offered the Fathers privileged media for expressing the mystery of Incarnation, that is, the appearance of the invisible God in the flesh.[5] Textile metaphors, nonetheless, were not confined to the human body; they also encompassed the entirety of creation, and extended to global narratives of decay and renewal. Within this symbolic framework, the ancient poetic figure of the mantle of the earth was revived both in word and in image. But was it the same as Pherecydes's mantle?

The Church Fathers, many of whom had been schooled in the classics, found the mantle appealing for the same reasons as their ancient predecessors, that is, for the aesthetic pleasure and the sense of harmonious interconnectedness it had the power to evoke, as well as for its ability to conceal. At the same time, however, they emphasized two new qualities: its liability to be consumed by time, and its ability to operate as an interface with the Creator. The mantle of biblical creation was different from the eternal mantles woven by the Olympian gods. Its beauty was ephemeral, it was bound to vanish.

The contemplation of the mantle of the earth was therefore not meant to exhaust in its texture, nor was it stationed on its surface; it was rather meant to redirect the inner eye of the faithful to the invisible presence of its celestial weaver. After all, in shifting their attention from mundane to spiritual affairs, Christ himself had warned his disciples about the transient nature of the earthly mantle: "Consider the lilies, how they grow: they toil not, they spin not; and yet I say unto you that Solomon in all his glory was not arrayed like one of these. If then God so clothes the grass, which is today in the field and tomorrow is cast into the oven, how much more will he clothe you, O ye of little faith?" (Luke 12:27–28).[6]

"The classic spirit," writes literary critic Alfred Biese, "seemed to shudder before the eternity of the individual, before the unfathomable depths which opened up for mankind with this religion of the soul, which can find no rest in itself, no peace in the world, unless it be at one with God in self-forgetting devotion and surrender."[7] This chapter navigates through the depths and folds of the biblical mantle and its various incarnations in the Byzantine empire, which was the continuation of the Roman empire in the East during late antiquity and the Middle Ages, and home to the Greek Church Fathers.[8] In particular, it explores the mantle metaphor at the different yet interlocked scales of the earth, the heavens, and the human body. It surveys the polychrome surfaces and delicate textures of the earthly, celestial, and fleshy garments found in the Bible and follows their symbolic threads through Byzantine understandings of creation.

The Garment of Creation

Byzantine textiles hold an irresistible aesthetic appeal. While their fine designs and bright colors delight the eye, their fabric invites the hand to gently caress their surface; it compels the finger to follow its regular geometric patterns, to retrace the contours of intricate naturalistic embroideries, to pause on the fine detail, or simply to feel the texture. It is no surprise that from the early days of the empire, textiles featured as highly desirable luxury items.[9] Patterned and figured, monochrome or multihued, embroidered with silk or gold, precious textiles were made into imperial garments and vestments for powerful clergy, they served as wrappers for relics, and were draped over the remnants of saints in shrines and tombs.[10] Sometimes their decorative patterns exceeded ostentation. One can only sympathize with the anxieties of Asterios, bishop of Amaseia (350–410 CE), over the "vain and curious broidery which, by means of the interweaving of warp and woof, imitates the quality of painting" and that ended up clothing certain wealthy Christians.[11]

Some of them, as Asterios tells us, went about cloaked not only in sumptuous vegetable or animal patterns, but even in scenes from the Gospels. These "iconic clothes," the bishop feared, were nothing but glittering surfaces masking a lack of spiritual depth.[12]

Transferred from the human body to the earthly globe by way of metaphor, Byzantine garments simply grew in splendor. Mantle metaphors are found in the encomiastic writings of early Christian authors praising both beautiful sceneries and emperors. For example, Gregory of Nyssa (335–95 CE) writes of his friend Adelphios's country estate at Vanota, in central Anatolia: "Vines, spread out over the slopes, and swellings, and hollows at the mountain's base cover with their color, like a green mantle, all the lower ground. The season at this time even added to their beauty, displaying its grape-clusters wonderful to behold."[13] Gregory's poetic pen makes the reader almost touch those gentle slopes, their velvety texture embroidered with gemlike grape clusters.

No less poetically, in his encomium to Constantine (the first Byzantine emperor), Eusebius of Caesarea (263–339 CE) praises the Creator who "clothed the previously shapeless eternity with beautiful colors and fresh flowers," "encircled the entire heaven like a great cloak [*megan peplon*] with the manifold colors of a painting," and encompassed the earth "with the ocean as with a beautiful azure vesture [*peribolē*]."[14] Eusebius's panegyric unfolds between these earthly and cosmic mantles and the imperial mantle—the purple *chlamys*.[15] A reflection of Christ on earth, the emperor is said to rule over the world in direct imitation of the Ruler of the cosmos.[16] Just as the divine Logos holds together the green mantle of the earth, the great cloak of heavens, and the azure vesture of the ocean, so does Constantine "unite all the nations in one harmonious whole," in a "universal empire."[17]

The Stoic motif of Roman law bringing different peoples under the same *chlamys* here is extended to divine law. The nations of the empire are united not only by the same civil code and regulations, but also, and more significantly, "in the pursuit of a godly life, in praising the one Supreme God, in acknowledging his only begotten Son their Savior as the source of every blessing, and our emperor as the one ruler on the earth, together with his pious sons."[18] Wrapped in textile metaphors, Eusebius's cosmographic panegyric thus follows the same dynastic thread as Alexander the Great's *chlamys*-shaped *oikoumenē* and the maplike cloth with which Queen Kypros flattered Emperor Gaius.

Christian earth's mantles were nonetheless much more than fine-looking aesthetic coverings, rhetorical devices, or emblems of power; they were surfaces that embedded deep mystical meanings. On two fourth-century Roman sarcophagi, Christ triumphs over the material world by sitting not on a sphere, but on a textile arch (the allegory of Cosmos), which a man holds

FIGURE 2.1 Detail from the sarcophagus of Junius Bassus (fourth century CE) with Christ triumphing over Kosmos, the allegory of the material world, here covered by a textile arch.
Museum of Saint Peter's Basilica. Photo: United Archives GmbH / Alamy Foto Stock.

above his head (fig. 2.1). The iconography of the representation has been linked to Roman triumphal arches, to the perishable nature of fabric, to the tablecloth of the altar upon which the Eucharist is prepared, and therefore to Christ's victory over death.[19]

The most profound and compelling descriptions of earthly mantles (or garments, as they were more frequently called), however, are found in patristic commentaries to Genesis (*hexaemera*). In his commentary, for example, Basil of Caesarea (330–79 CE) explains how, at God's command, "the cold and sterile earth travailed and hastened to bring forth its fruit, as it cast away its sad and dismal covering to clothe itself in a more brilliant robe [*peribolē*], proud of its proper adornment and displaying the infinite variety of plants."[20]

Here, as in Eusebius's encomium, one can easily trace immediate parallels between the replacement of the earth's garment of mourning with her colorful attire and the substitution of Chthonia's primordial veil with Zeus's "cartographic" mantle. Unlike Pherecydes, however, Basil immediately directs the attention of the reader from the charms of the visible scene to the voice of the invisible Creator. The earth's mantle does not simply cover up hidden underground abysses, it shrouds (and reveals) God himself. Unlike Zeus "the weaver," Aristotle's efficient cause, or Plato's demiurge, the Christian God is not external to creation, but inhabits it: "He was in the world, and the world was made by him" (John 1:10). In other words, God created the earth not from outside, but from within; he did not set it spinning and backed away, but, through his incarnation, "he remained mystically at the centre of it, keeping it going."[21] In Eusebius's words, God's presence "pervades the depths of unexplored and secret wisdom."[22] The Christian mantle of the earth is therefore first of all an interface between humans and the Creator. It is through the sight of visible and sensible things, Basil observes, that "the mind is led, as by a hand, to the contemplation of invisible things." As to the form of them, he argues, "we also content ourselves with the language of the prophet, when praising God that stretches out the heavens as a curtain and spreads them out as a tent to dwell in."[23]

Nowhere is the relationship between the mantle and the divine Logos more powerfully expressed than in John Chrysostom's homilies on Genesis:

> Who could fail to be absolutely astonished at the thought of how the word uttered by the Lord, "Let the earth put forth a crop of vegetation," penetrated to the very bowels of the earth and, as though with some magnificent veil [peplō tini thaumastō], adorned the face of the earth with a variety of flowers? In an instant you could see the earth, which just before had been shapeless and unkempt, take on such beauty as almost to defy comparison with heaven.[24]

In his second theological oration, Gregory of Nazianzus (329–89 CE) likewise praises the beauty of the earth, including the variety and lavish abundance of its fruits, the grace and qualities of its colors, and the brilliant transparency of precious stones, as reminders of God's merciful glory. He then asks the reader to

> [t]raverse the length and breadth of earth, the common mother of all, and the gulfs of the sea bound together with one another and with the land, and the beautiful forests, and the rivers and springs abundant and perennial, not only of waters cold and fit for drinking, and on the surface of the earth; but also such as running beneath the earth, and flowing under caverns. . . . Tell me how and whence are these things? *What is this great web unwrought by art [ti to mega touto kai atechnon hyphasma]*? These things are no less worthy of admiration, in respect of their mutual relations than when considered separately.[25]

Gregory's use of the word *hyphasma* is worthy of note. Every *hyphasma* ("woven robe") or *hyphos* ("web") is made of horizontal and vertical threads neatly interlaced by the weaver. For the Greek Fathers creation was not a romanticized object of aesthetic contemplation; its beauty lay not in its forms (that is, in the individual "threads"), but in their function, in their being part of the same great fabric, or *hyphasma*. Through the garment metaphor, Gregory thus compels his audience to see creation not through human eyes, but through the eyes of God—to see it, that is, from the end for which it was created for the union of all things in the person of his incarnate Son.[26] God's earthly *hyphasma* is a palpable manifestation of his creative power.[27] Like the fleshy garment of the incarnated Christ woven in the womb of the Virgin, the visibility of the earthly garment clothes the invisibility of the Creator. The mantle of creation allows humans to physically experience God's presence. It at once conceals and reveals.

Hyphasma as a metaphor for the mystical interlocking of the elements of creation in a single harmonious whole adds to a wider patristic inventory. Creation, for example, is often also compared to a statue made of different parts and to the human body formed by its various organs (the same metaphors employed by Strabo and Ptolemy when describing the scope of geography).[28] It is also likened to a many-stringed lyre or lute composed of chords of different pitch tuned to produce a single perfect melody,[29] or to a book "whose creatures are like letters proclaiming in loud voices to their divine master and creator the order and harmony of things."[30] The textile metaphor, however, holds a unique appeal in that it survives not only in words, but also as a material object.

Between the fifth and seventh centuries, the metaphor found tangible expression in a series of textiles featuring allegories of Gē—a sort of continuation of the tradition inaugurated by Queen Kypros. Produced and traded in Egypt, these Byzantine "earth's mantles" usually portrayed Gē as a richly dressed or bejeweled, fruit-bearing female figure, in line with Eusebius's and Basil's characterizations of the "adorned" earthly mantle (fig. 2.2). The relatively small sizes of the surviving weavings give weight to the hypothesis that some of them might have been in turn embedded in garments (possibly of the sort criticized by Asterios).[31] The same iconography graces the mosaic floors of Byzantine basilicas of the same period in Jordan—perhaps, as Byzantine art historian Henri Maguire suggested, as an invocation and auspice of bounty in those arid regions.[32]

Occasionally featuring Ocean and marine creatures surrounding Gē, the precious textiles skillfully woven by the hands of Queen Kypros and her Byzantine successors hold a cartographic quality: they are consciously works of

FIGURE 2.2 Byzantine tapestry-woven panel featuring a personification of bountiful Gaia (Mother Earth) surrounded by crosses (fifth century CE), a reminder of the sacredness of the earth's mantle. Paris, Musée du Louvre. Photo © Musée du Louvre, Dist. RMN-Grand Palais / Georges Poncet.

art and miniature replicas of the world. As such, they bear a double significance. As we have seen, in the ancient world, the image of weaving was used extensively in connection with "weaving narratives," and hence literary and visual descriptions of the world are neatly intertwined.[33]

While such artifacts are predominantly a phenomenon of the early Byzantine period, texts, textiles, and world images intertwined again in the Middle Ages, with the rediscovery of Ptolemy's work.[34] The first attested reconstruction of the Ptolemaic world map is ascribed to the polymath monk Maximos Planudes (1255–1305), and today it survives in various manuscript copies.[35] These maps feature the part of the world known to the Alexandrian geographer, generally in the first projection (fig. 2.3).[36]

The world's regions are usually depicted in bright colors starkly set against a dark blue ocean, in line with Ptolemy's and Strabo's conception of geography as the sum of different parts. The result is an opening patchworked mantle woven in a loom of parallels and meridians. Its surface is invisibly traversed by winds blown through the pipes of male allegorical figures set on the edges of the map—an evocation of the apocalyptic image of the angels releasing the winds from the four corners of the earth at the end of times (Rev. 7:1).[37] Figures of the zodiac are embroidered on the eastern extremity of the cartographic mantle, as if to remind the viewer of the interlocking of earth and heaven stressed by Ptolemy and his Byzantine successors.

Planudes crowned his achievement in verse. He composed two poems in honor of the Alexandrian astronomer and his work, which he claimed to have brought to light after centuries of oblivion. The first poem, *Praise to Ptolemy*, is in effect an elaborate dedication to his patron Andronicus Palaiologos II, whereby the rediscovery of Ptolemy's *Geography* is woven into a rhetoric of imperial expansion. The second poem, *Heroic Verses on Ptolemy's Geography*, is a shorter geographic *ekphrasis* in which the map speaks to the reader in the

FIGURE 2.3 Late Byzantine world map drawn in Ptolemy's first projection (fifteenth century). The map illustrates a manuscript of Ptolemy's *Geography* followed by Maximos Planudes's *Heroic Verses*. © The British Library Board, Burney Ms. 111.

first person. Intriguingly, both poems contain textile metaphors. Following Strabo (though rejecting his view of an *oikoumenē* entirely surrounded by the ocean), in the *Heroic Verses* the learned monk recuperates the ancient cloak metaphor: "I am not a sling; you look at me in the shape of a *chlamys*," the map says to the reader.[38] Yet the mantle metaphor stretches beyond the mere shape of the earth. In the same poem, Planudes builds up a cosmogonic drama that culminates with the newly rediscovered Ptolemy's *Geography* triumphantly "covering" previous scientific accomplishments:

> When the sun emerged, the light of the stars was hidden
> And the moon drew off rapidly her own ray
> And as the very geography just appeared
> She placed a dark cover [*zoferēn kalyptrēn*] on all previous ones.[39]

Illuminated by Planudes's pen, in the *Praise* the "dark cover" reveals its fine texture and design. Ptolemy's map is evocatively compared to Athena's splendid polychrome "*peplos*":

> It is a great earthly miracle how Ptolemy in his wisdom
> led the round earth before our eyes, as if someone had
> depicted a small town on maps. I myself have never seen
> such a *peplos* of Athena bearing such an elaborate,
> multicolored and well-designed decoration as this.[40]

Planudes's description of the cartographic *peplos* skillfully interlaces Ptolemaic science with early patristic tropes. Like his predecessors, the monk is keen to celebrate the harmonious integration of geographical elements: "the mouth of rivers," "the paths over the snow-clad mountains," "the nations that inhabited the earth," "the sea with its islands."[41] Like Basil's and Gregory's descriptions of creation, Ptolemy's map, Planudes suggests, has the power to bring all these features before one's eyes without losing sight of the whole. The joy the viewer feels in front of the cartographic mantle's "orderly arrangement" and its beautiful variety, the author fittingly concludes, is comparable to another mantle: "a meadow carpeted with spring flowers."[42]

Delightful as they are to look at, Byzantine Ptolemaic cartographic mantles nonetheless blend the aesthetic appreciation of the world with the apocalyptic evocation of its destruction. Fixed around the hem of the *chlamys*, the personifications of the winds mark the edges of space and time.[43] They remind the pious viewer of the ephemeral nature of creation—and of mantles. The Old Testament repeatedly describes God's creation as a garment that shall "wax old," be "fretted away," and perish, as opposed to the eternal Logos (Ps. 102: 26; Is. 51:6; Heb. 1:11).[44] Likewise, biblical heavens are shaped "like a tent" (Is. 40:22; Ps. 104:5), "worn out," and rolled up "like a mantle," or a "scroll," at the end of times (Is. 34:4; Rev. 6:14; Heb. 1:12).[45] Beguilingly, one can see the prophecy fulfilled in the Ptolemaic maps themselves and in their bright surfaces consumed by time, as colors and substance fade, as folia are eaten up by bookworms, as manuscripts are torn apart and silently disappear in the deep folds of history.

The Veil of Heaven and the Skin of the Firmament

The transiency of earthly and heavenly mantles was a recurrent motif in early patristic commentaries and homilies. Textile metaphors were used by the early Church Fathers to stress the corruptible and mutable nature of creation. For example, Tertullian (155–222 CE), who dedicated an entire oration to his *pallium* (the cloak associated with the philosophers), writes about the earth "clothing herself in different clothes,"[46] while John Chrysostom (347–407 CE) extends the metaphor to the sky and to the entire cosmos. The sun, he writes, adorns the heaven, as it were, "with a saffron-colored veil"; it makes the clouds "like roses," running unimpeded all the day. "Beholdest thou, then, its beauty? Beholdest thou its greatness?" Chrysostom asks his readers. But he also warns them: "[L]ook also at the proof of its weakness!"[47] Clustered in a thick mantle, these very same clouds can block its golden beams. In the same way, he explains, God fashioned creation "beautiful and vast, but on the other hand corruptible."

> And both these points the Scriptures teach, for one in treating of the beauty of the heavens thus speaks: "The heavens declare the glory of God." And again, "Who has placed the sky as a vault, and spread it out as a tent over the earth" (Is. 40:22). . . . But another writer, showing that although the world would be great and fair, it is yet corruptible, thus speaks: "You, Lord, in the beginning hast laid the foundation of the earth, and the heavens are the work of Your hands. They shall perish, but You remain, and they shall wax as does an old garment [*imation*], and as a vesture [*peribolaion*] shall Thou fold them up, and they shall be changed" (Heb. 1:11; Ps. 102:25–27).[48]

These apocalyptic verses find their perhaps most dramatic visual expression in the church of the Chōra, the monastery at the outskirts of Constantinople, where, incidentally, Planudes taught and where he rediscovered Ptolemy's writings.[49] The decoration of the church was part of a large-scale renovation of the monastery started a decade after Planudes's death, and it revolves around the themes of the Incarnation and the Resurrection.[50] As with Byzantine Ptolemaic maps, it embeds different temporal scales. Biblical past, present, and apocalyptic future are invoked simultaneously in the same three-dimensional architectural space, as the visitor's gaze progresses downward from heaven to the dark regions of the Last Judgment, to culminate in Christ's descent to Hades with the rescue of Adam and Eve and the promise of eternal life.

One of the most striking features of the scene is the image of the scroll of heaven literally rolled up by an angel at the end of time (pl. 1).[51] Fittingly painted on the domical "vault" (Is. 40:22), the *peribolaion* features as a beautiful transparent veil encrusted with golden stars, the sun, and the moon. The heavenly veil spins toward the center of the celestial vault like a whirlpool, drawing the surrounding space into its centripetal movement (fig. 2.4). Floating choirs of hierarchs, martyrs, and holy men and women orbit around the heavenly scroll in "spatial containers," so that the whole scene becomes a chorographic map of the future and a choreographic chart of salvation.

Scripture repeatedly contrasts the perishable garment of creation with the eternal Logos. On the Chōra fresco, the transient heavenly *peribolaion* is visually balanced by the timeless figure of Christ set underneath, "wrapped in light as in a garment" (Ps. 104:4). The Lord is embedded in five concentric circles painted in different gradations of blue, which symbolize the human experience of the uncreated light and the "divine darkness" inside which God dwells (*phōs oikōn aprositon*).[52] God's garment of light is transfigured into the Aristotelian (and Ptolemaic) cosmos, made, as it was, of concentric planetary spheres revolving around the earth and encompassed by the outer sphere of the fixed stars, or firmament.[53] Here the center of the universe is none other than Christ, who sits on an orange ribbon after Isaiah's verse "the heaven is my throne and the earth my footstool" (Is. 66:1; Matt. 5:35).[54]

Byzantine cosmographic imagination was dominated by two models: the spherical Aristotelian model, and the model of Hebrew cosmology, which ascribed to heaven the form of a vault.[55] Visual representations of the former model are found in the beautiful translucent orbs featuring in the hands of Christ Pantocrator ("the Ruler of All Things"), of angels, and of emperors, as symbols of universal dominion inherited from ancient Roman iconography. The latter model, by contrast, is best expressed by the vaulted, tabernacle-shaped cosmos devised by Kosmas Indikopleustēs in the sixth century (fig. 2.5), an imaginative alternative to what he rejected as dangerous "pagan" models.

The Greek word *kosmos* means both "universe" and "ornament."[56] In ascribing to the cosmos a spherical shape, the philosophers of antiquity emphasized its beauty and mathematical perfection. Kosmas's vaulted shape, by contrast, emphasized literal biblical exegesis. The archetypal structure Yahweh revealed to Moses on Mount Sinai (Ex. 24–25), Kosmas believed, was

FIGURE 2.4 Detail of the scene of the Second Coming with the heaven "departed as a scroll when it is rolled" (Rev. 6:14), Chōra monastery (Kariye Djami), Istanbul, fourteenth century.
Photo by the author.

not only that of the tabernacle of the Temple of Jerusalem, but that of the entire universe. Kosmas's main problem with classical spherical models was the notion that the whole universe was eternal, an idea contrary to biblical narrative. Throughout antiquity time was understood as intrinsically connected to space. The temporal (or timeless) dimension of the universe was thus reflected in its very shape. In particular, Kosmas argued that "whatever

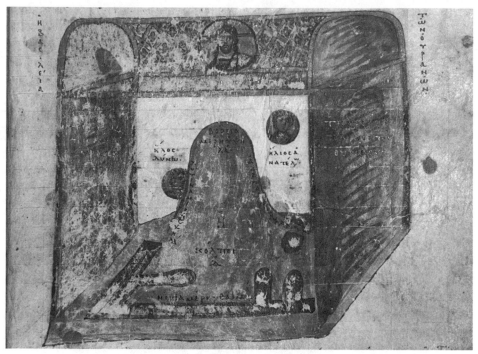

FIGURE 2.5 Kosmas Indikopleustēs's tabernacle-shaped cosmos featuring a diapered gold cloth decorated with fleurs-de-lis in the upper vault, a celestial threshold between God's eternity and the terrestrial world inhabited by humans. Sin. Gr. 1186, fol. 69r, eleventh century.
By permission of Saint Catherine's Monastery, Sinai, Egypt.

is spherical and perpetually rotating must necessarily be eternal and because the universe cannot be eternal, therefore it is not spherical and does not rotate."[57] In his *Topographia Christiana* Kosmas therefore devised a model that combined transiency and eternity in distinct yet interlocked regions of the cosmos: a terrestrial prism inhabited by humans at the bottom, and the celestial vault inhabited by God at the top.

In spite of their radical differences, both models conceptualize the cosmos as a closed, self-contained system made of interlocking parts and defined by an outer boundary. In both models this boundary is the firmament (*stereoma*), the "solid" (*stereon*) perceptible limit of the visible creation, behind which is concealed the uncreated God.[58] In the Aristotelian model, the firmament is the outer sphere of fixed stars encompassing the mobile planetary spheres, which, in its Christian rendering, is in turn encompassed by God. In Kosmas's three-dimensional diagram, by contrast, the celestial boundary features as a blue veil interposed between the prism and the vault of the cosmic tabernacle. In illustrations of the Second Coming from medieval manuscripts of the same treatise, the firmament is indicated either as a broad horizontal line, or as a diapered gold cloth decorated with fleurs-de-lis, an allusion to the tabernacle curtain.[59] But what is the firmament, exactly? And how is it different from heaven?

Basil of Caesarea made a basic distinction between the two. While the heavens were believed to be made of a "light substance," the word "firmament" suggested to him something stronger than the heavens that had been created previously; it was composed of some resistant matter, such as condensed air.[60] In patristic literature and liturgical texts, the perceptible nature of the *stereoma* is thus conveyed through a series of textile metaphors ranging from "a skin" (*derrin*)[61] to a "veil" or "curtain" (*katapetasma* or *parapetasma*). The first word, *derrin*, emphasizes organic integration with the rest of the cosmos. By contrast, the other two words are both derived from the verb *petannymi*, which denotes the opening of doors and portals, and more generally a "spreading out" and "opening outward."[62]

In both the Old and New Testament the heavens (and by implication the firmament) are said to "open" and "close." For example, Christ speaks of the famine in Elijah's time in which "the heaven is closed for three years and six months" (Lk. 4:25), implying a lack of both rain and God's care for the sinful people of Israel. Conversely, the "opening heaven" signals theophanic eruptions; it indicates that the intercourse between God and humans is taking place. The heavens open at the moment of Christ's baptism (Lk. 3:21), during Stephen's martyrdom (Acts 7:56) and, most intriguingly of all, when Peter has a vision of "a giant sheet" containing the bounty of creation let down from heaven by its four corners (Acts 10:11)—a giant map bound to vanish together with Peter's brief ecstatic experience.[63]

The firmament and the heavens are therefore veils that at once separate and unite; they are boundaries and thresholds between upper and lower worlds, "between the visible world and the invisible, between being and becoming."[64] Building analogies between the earthly ruler and the celestial ruler (that is, Constantine and God) and the spaces they inhabit, Eusebius thus metaphorically compares "the vast expanse of the heaven" to

> An azure veil [*parapetasma*], interposed between those without, and those who inhabit his royal mansions: while round this expanse the sun and moon, with the rest of the heavenly luminaries (like torch-bearers around the entrance of the imperial palace), perform, in honor of their sovereign, their appointed courses; holding forth, at the word of his command, an ever-burning light to those whose lot is cast in the darker regions without the pale of heaven.[65]

More specifically, for Byzantine commentators the veil of the firmament was symbolized by the curtain of the Temple of Jerusalem. The curtain is described in Exodus, as Moses enters the dark veil of smoke hanging on the top of Mount Sinai to encounter God and behold the "pattern of the heavenly tabernacle" (Ex. 25:9). Here the prophet is instructed to "make a veil of blue and purple and scarlet woven, and fine linen spun" (Ex. 26:31).[66] According to the twelfth-century homilist Iakobos Kokkinobaphos, "the curtain of the temple is a veil, because it shrouds in mystery the secrets of holiness. And the sky above us is also a veil, for the heavenly azure conceals the depths of the universe."[67] The veil of the temple, Kokkinobaphos concludes, was intended by Moses to symbolize the veil of heaven, and both these veils prefigured the flesh of Christ, which at once enfolded and concealed his divinity.[68]

The Garment of Flesh

Byzantine tradition intertwines the weaving of the veil of the Temple with the weaving of Christ's fleshy garment.[69] Together with other virgins brought up in the temple, Mary, Kokkinobaphos explains, was responsible for the manufacturing of the veil of the sanctuary. When the archangel Gabriel greeted her, he found the girl spinning the skein of purple wool out of which the veil was to be crafted. Spinning therefore mystically coincided with the moment of incarnation. The production of the red fabric prefigured the royal purple of the world's future king and the scarlet robe in which "the King of the Jews" was dressed by the Roman soldiers. More subtly, spinning red thread and weaving red cloth were metaphors for blood coagulating and solidifying into flesh: "the weaving of the 'red thread' into the 'red veil' conveys transition from fluidity to solidity, from time to space, from a dot (sign, letter, word) to surface and texture, and from the immediacy of conception to the emergence of a 'true' image."[70] By contrast, Christ's undressing on the top of Golgotha and the "unmaking" of his seamless tunic (that is, its partition amid the soldiers) foreshadowed death (John 19:23–24).

The image of the Virgin spinning the red thread of life resonates with the Davidic verses "For it was you who formed my inward parts; you wove me together in my mother's womb" (Ps. 139:13). It also resonates with the rich repertoire of ancient Greek textile metaphors, from the Moirai "spinning" human destiny and Empedocles's belief that reincarnated souls were dressed by nature in new "mantles of flesh" (*sarkōn chitōni*) to metaphors that are still current in everyday speech, such as "the thread of life," "organic tissues" and "fibers," or the very word "histology" (from *histos*, or "loom").[71] More specifically, the biblical scene echoes in a commentary on the cave of the nymphs in the *Odyssey*, written by the neo-Platonist philosopher Porphyry of Tyre (234–305 CE). The Homeric verses describe the nymphs weaving beautiful purple webs in a cavern, which, "under the veil of allegory,"[72] Porphyry argues, symbolizes the mysterious moment of conception, when souls descend to earth and are clothed in the garment of the body.[73]

The metaphor of corporeal weaving, however, acquired special currency among Christian commentators from the early centuries of the Byzantine empire. Proclus, for example, compares the womb of the Virgin to a "spider net,"[74] to a "workshop" containing "the awesome loom of divine economy," or simply to a "loom" intertwining the two natures of Christ.[75] Indeed, the very word "conception" (*syllēpsis*) denotes the weaving together of different threads—of flesh and soul or, in this case, of humanity and the divinity, "a paradoxical union without confusion."[76] Epiphanios of Salamis (310–403 CE) situates the making of Christ's fleshy mantle within the broader narrative of Salvation history. Eve, he argues, wove garments to cover the visible nakedness of human flesh; by contrast, the Virgin "clothed all the faithful with garments of incorruptibility," freeing them from their invisible nakedness.[77]

In Scripture, the fleshy garment of the human body and the mantle of the earth are made of the same corruptible fabric. The one is but the reflection of the other. "My flesh is clothed with worms and clods of dust; my skin is broken and has become loathsome," cries Job (Job 7:5). "The moth shall eat [mortals] up like a garment, and the worm shall eat them like wool," warns Isaiah (Is. 51:8).[78] As with the transiency of the mantle of the earth, the ephemeral nature of the fleshy garment was largely expounded and commented upon by the Church Fathers.[79] Gregory of Nyssa, for example, contrasts the stationary nature of the human soul with the mutable nature of the fleshy mantle wrapping it, "for the body is on the one hand altered by way of growth and diminution, changing, like garments, the vesture of successive statures, while the form, on the other hand, remains in itself unaltered through every change."[80]

More intriguingly, Augustine of Hippo (354–430 CE) draws a direct link between the micro-scale of the human body and the macro-scale of Creation, comparing the spirit to heaven and the fleshy mantle to the earth's mantle.[81] Three centuries later, these two constituents of human nature were literally mapped on different physical regions of the cosmos, and, once again, by way of textile metaphors: "the life in the body is spent in the most divine and lovely region, while the life in the soul is passed in a place far more sublime and of more surpassing beauty, where God makes his home, and where he wraps man about as with a glorious garment, and robes him in his grace, and delights and sustains him with . . . the contemplation of himself," writes John of Damascus (676–749 CE).[82]

FIGURE 2.6 The vision of Saint Peter of Alexandria. Christ's garment is torn by the heretic Arius (who is shown bent in the darkness under Christ's feet). Dionysiou Monastery, Mount Athos, 1547.
Photo: Father Apolló of Docheiariou.

Through the centuries, matter and metaphor intertwined in ever richer and more complex textures that continued to unfold in different spaces and at different scales. For example, the seamless tunic of Christ (John 19:24) was taken as a symbol of the body that was virginally made, but it was also taken to symbolize the seamless unity of the divine and human natures of Christ, as well as the unity of the Church threatened by heresies, such as Arianism (which denied Christ's double nature). Since the thirteenth century the seamless garment torn by Arius featured in the liminal spaces of Orthodox church buildings connected to the mystery of the Eucharist, whereby the body of Christ is physically consumed by the communicant (fig. 2.6).[83]

Likewise, the purple robe of the Virgin (*maphorion*) came to be associated with her physical protection. Brought from Palestine to Constantinople in the fifth century, the sacred garment was taken on procession around the city's walls during revolts and sieges, such as the Avar attack of 626 and the Arab siege of 717. The protection provided by the *maphorion* during the Russian attack of 860 is evocatively described by Patriarch Photios: "the city put [the robe] around itself and bedecked itself with it."[84] The Mother of God's mantle, in other words, embraced the entire city, and it also embraced the entire universe, for by making her fleshy mantle a "container of the uncontainable" (*chōra tou achōrētou*), the Virgin had made herself "wider than heaven" (*platytera tōn ouranōn*).[85]

As the "veil of the Logos," the Virgin was also identified with the veil of the Temple of Jerusalem, which was in turn identified with creation. The veil could only be passed through by the high priest, and only on the Day of Atonement, the holiest day of the year in Judaism. On that day the high priest wore special vestments that were fashioned after the veil and represented the fabric of creation, "For upon his long robe the whole world was depicted" (Wis. 18:24). For the Hellenistic Jewish philosopher and exegete Philo of Alexandria (25 BCE—50 CE), the priest's robe was connected to the idea of weaving as a "manner of making the world":

> I fix my eyes upon the section of the earth upon the spheres of heaven, the many different kinds of animals and plants, and that vast variegated piece of embroidery, this world of ours. For I am straightway compelled to think of the artificer of all this texture as the inventor of the variegator's science, and I do homage to the inventor, I prize the invention.[86]

In Philo's symbolic framework, the high priest was a figure of the heavenly priest (the divine Logos), who similarly passed through the veil not in ascending into the sanctuary, but in descending from the divine throne to earth (Wis. 18:14–16). As the Logos descended through the veil of the heavens, it took form and became visible, clothing itself in the garment of creation.[87]

The veil of the Logos, which revealed its presence precisely by concealing it, passed into Christian usage as an expression of the idea of the Incarnation. In the words of Elder Aimilianos of Simonopetra, the Nativity of Christ created a new veil: "through the incarnation of the Word, divinity was concealed, so that its light would not blind and burn humankind."[88] The Virgin was therefore both the weaver of the veil and, as the chief mediator and intercessor before God, the veil itself—a giant veil spread between earth and heaven.[89]

This paradox found, and continues to find, both visual and material expression in the central doors and curtain of the icon screens of Orthodox churches, traditionally graced, as they have been, with representations of the

FIGURE 2.7 Central curtain of icon screen featuring the scene of the Annunciation. The Virgin spins the skein of purple wool for the veil of the Temple as she is about to weave Christ's veil of flesh. Philotheou Monastery, Mount Athos.

Photo: Father Apolló of Docheiariou.

mystical dialogue between the Virgin and the archangel (fig. 2.7). The screen separates the space of the nave occupied by the congregation from the space of the sanctuary, which is restricted to the clergy.

Conceptualizing the church building as a microcosm, the late Byzantine commentator Symeon of Thessaloniki (1375–1430) identified these two spaces with heaven and the regions beyond heaven, respectively.[90] In this cosmographic scheme, the icon screen therefore marks the limits that oppose and distinguish two worlds (earth/heaven, microcosm/macrocosm, sensible/intelligible), while at the same time constituting "the paradoxical place where those two worlds communicate."[91] The curtain of the icon screen is explicitly identified by Symeon with the flesh wrapping and concealing the divine Logos, while the veil covering the altar table is taken to symbolize the veil of the heavens with "the immaterial tabernacle around God . . . by which he himself is concealed, 'clothing himself with light as with a garment.' "[92]

The screens and veils of Orthodox liturgical performance, however, are not necessarily material, or at least not in the same way as the icon screen. Intangible boundaries are fabricated by ephemeral curtains of incense, or by the sonic curtain of chants "veiling" the mystical prayers recited by the priest behind the iconostasis at given moments of the liturgy. As symbolic boundaries between the sensible and the intelligible, these veils of fabric, smoke, and sound effectively realize "the unique Byzantine understanding of the incarnation as a paradoxical dialectic of revelation and concealment"[93]—the same mystery that lies at the heart of the cosmos.

*

Textile metaphors weave in and out of biblical narrative. Through the veiled language of Scripture, the Church Fathers saw, and poetically expounded, the perfect harmony of creation. Their exegetical commentaries and homilies wove splendid tapestries in which the divine Logos welded together opposed elements. As with Pherecydes's mantle of Gē, the biblical "cosmographic mantle" was a manifestation of God's creative power. In Byzantium its richness inspired the most splendid human crafts: from maps to music, from poems and paintings to embroidered textiles and silky liturgical curtains. At the same time, however, the Church Fathers warned, the garment of creation was also a synonym for changeable, ephemeral, and superficial beauty—no matter what its scale was.

The biblical mantle was a liminal space. Its surface was a point of tension, or rather, of contact, between the visible and the invisible, between the transient and the eternal. If it separated, it was also "the very thing that enabl[ed] contact, disclosing or revealing precisely to the same degree that it conceal[ed]."[94] In Mary's weaving of the veil of the Temple, Kokkinobaphos saw a prefiguration of the Incarnation: "Christ clothed himself in the royal robe of the flesh woven from the body of the Virgin."[95] As with the mantle of the heavens, Christ's mantle of flesh both enfolded and concealed his divinity. At the same time, however, the mantle made that divinity accessible to humankind. In Constas's words, Christ's fleshy garment was "a veil which stood between creation and the Holy of the Holies, like an iridescent silk in which the contrasting colors of divinity and humanity alternately shimmer and play. It was the glorious clothing of the naked God which covered the shame of humanity and granted access to the heavenly sanctuary 'by the new and living way that Christ opened for us through the veil of his flesh' (Heb. 10:20)."[96]

The analogy is made explicit in the poetic verses of the following early Byzantine Easter hymn (to be dated between the late fourth and early fifth centuries):

> Today is hung upon the wood He,
> who hanged the earth upon the waters.
> A crown of thorns is put upon Him,
> who is King of the angels.
> A fake crimson robe surrounds Him
> who surrounds the sky with clouds.
> A slap on the face suffers He,
> who freed Adam in the Jordan.
> With nails was He detained,
> the Bridegroom of the Church.
> With a spear was He pierced,
> the Son of the Virgin.[97]

Here the macro-scale of the cosmographic mantle is mapped on the micro-scale of the surface of the suffering body of Christ. As his skin is pierced by a spear, the veil of the Temple is severed, the sun eclipsed, Chthonia's darkness revealed (Matt. 27: 51). Christ's mantle of flesh and the mantle of the earth are woven into one another. They are barriers, and yet at the same time they are gateways to what humans cannot otherwise access because of their bodily limits. As such, both mantles follow a cartographic logic: they are textures of symbols through which humans make sense of the world. The following chapter turns to the reception of the patristic earth's mantle in the Latin West.

3

Medieval Vernicles

While I drink in the sweetness of the flowers, the thought occurs to my mind of the fragrance of the clothing of the Patriarch Jacob, which the Scripture compares to the odour which mounts from a fruitful field. When I delight my eyes with the bright colours of the flowers, I am reminded that this beauty is far above that of the purple robe of Solomon, who in all his glory, could not equal the beauty of the lilies of the field, although to him there was wanting neither richness of material, nor wisdom and taste in arrangement. In this way, while I am charmed without by the sweet influence of the beauty of the country, I have not less delight within in reflecting on the myster ies which are hidden beneath it.[1]

In the Middle Ages the warm embrace of the patristic mantle encompassed not only the Byzantine empire, but also the politically fragmented Latin West. Its mighty fabric spread from the jagged coasts of Ireland to the snow-clad Italian Apennine; it unfolded over the dark forests of Germany and the verdant valleys of the Loire and northern France; it covered ancient roads, hamlets, and abbeys; it alighted upon the soft landforms of the Roman countryside. The quiet contemplation of the polychrome mantle of nature, intimated Bernard of Clairvaux (1091–1153), soothes the weary mind and edifies the soul, for it recalls "the thought of the heavenly sweetness towards which [it] aspires."[2] Not only did the fragrance and smiling countenance of the landscape surrounding his abbey in northeastern France evoke splendid biblical garments, but they also revealed God's presence and redirected the spiritual eye to the mysteries hidden beneath the flowery mantle.

With their promise of freshness, or simply by virtue of their stunning beauty, poetic mantles and garments brought messages of hope and spiritual renewal. In a famous passage from about the millennial year 1000, the French monk and chronicler Rodulfus Glaber (985–1047) explained how after "a mist of blindness" had darkened the eyes of the Roman Catholic Church, "it was as though the very world had shaken herself and cast off her old age, and were clothing herself everywhere in a white garment of churches."[3] This garment was a macrocosmic echo of the garment of renewal wrapping the soul of the newly baptized Christian (Gal. 3:27). It was also a reenactment of God's creative act expounded by Genesis and so prolifically commented upon by the early Church Fathers in their *hexaemera*. After Augustine, Thomas Aquinas

(1225–74) saw in the green garment of the earth the last phase of Creation, that is, God's adornment of the world. In the beginning, comments the theologian, the earth, invisible, void, shapeless, empty, is "without the comeliness which it owes to the plants that clothe it, as it were, with a garment."[4]

The motif of the garment as a metaphor for the splendor of creation and, in turn, creation as a visible manifestation of God's mercy was prominent in the medieval West, as it was in the Byzantine East. In spite of obvious continuities, however, Eastern and Western garments presented compelling differences in their manufacture. To echo Glacken's elegant phrasing, if the symbolic world of the Church Fathers and of the Byzantines was "a great hemispheric tent" (Is. 40:22), the world of their Western medieval counterparts continued to be a tent, yet one replete with "busyness, practicality, interest in the immediate."[5]

While the early patristic and Byzantine view of nature was essentially poetic, rather than scientific, in the Latin West, by the early thirteenth century, natural theology was developing into an effort "to understand God's mind by discovering how his creation operates."[6] If the Greek Fathers insisted on human limits before God's mystery, for Aquinas it was reason's task to clarify "in logical terms, as far as possible, the deliverance of revelation."[7] As the poetic language of song and prophecy started to give way to scholasticism and rationalistic inquiry, Glacken comments, "the world remained a symbol all right, but within it the humble artisan portrayed branches and blossoms realistically, and did not neglect self-advertisement."[8]

The tension between symbol and scholastic exegesis set the precondition for the development of one of the most characteristic visual expressions of Western medieval culture: the so-called *mappae mundi*, or Christianized images of the earth. While Byzantine cosmographic diagrams, such as that of Kosmas Indikopleustēs, presented the viewer with an utterly symbolic interpretation of creation, and later Ptolemaic maps offered a scientific rendering of the *oikoumenē*, Western *mappae mundi*, which literally translates as "world cloths," were in essence "visual encyclopedias" exalting the marvel and multiplicity of creation and the place of the Creator within it. Through the cartographic medium, *mappae mundi* implicitly addressed questions that were at once theological and geographical. These included the precise location of the garden of Eden, or the shape of the earth and of the cosmos (questions that Greek Fathers like John of Damascus deemed beyond human remit and spiritual self-interest).[9] In other words, *mappae mundi* accomplished in drawing what contemporary theological *summae* and universal encyclopedias accomplished in words.

Mappae mundi form part of a broader set of no less compelling metaphorical and material textiles, ranging from elaborate cosmographic mantles for princely use to veronicas (Western copies of the cloth on which Christ miraculously imprinted his face, also known as "vernicles"). The biographies of these objects intersect with the lore of ancient traditions and with the pragmatics of geopolitics and the silk trade, with old prototypes and new devotional cults, with local piety and international diplomacy. Poetic and material crafts invested with powerful theological and political meanings, medieval cosmographic mantles, *mappae mundi*, and vernicles all give expression to a specific understanding of creation. By exploring these three types of objects, the following pages show how some of the classical and patristic ideas expounded in the previous chapters were interpreted and given visual, textual, and material form in the Latin West. In particular, this chapter shows the continuing potency of the earth's mantle metaphor in fostering a perception of the world and of the cosmos as a perfectly integrated and self-enclosed space inhabited and governed by the Creator. The last part of the chapter shows how this vision was translated onto the "mantle" of a physical landscape increasingly modified by human action.

The Politics and Poetics of Cosmographic Mantles

Sometimes in 1018, as winter was about to encroach, a nobleman from Bari traversed the Italian peninsula and reached the palace of the Ottonian emperor Henry II (973–1024) in Bamberg, southern Germany. He arrived a broken man. Melus of Bari, this was his name, was the leader of a group of insurgent local aristocrats who had repeatedly resisted the political leadership of the Capetan, the representative of the Byzantine emperor in Apulia. Supported by the Norman and Lombard troops and by the pope, who envisaged the Byzantine presence in southern Italy as a threat to the Vatican territories, Melus organized a series of revolts against the Byzantines. The last one, in October 1018, however, turned into a disastrous defeat. Melus therefore fled to Bamberg in search of imperial support. To maximize his chances of diplomatic success, the nobleman presented the king with an extraordinary gift: a cosmographic mantle (fig. 3.1).

Made of precious blue silk finely embroidered with gold, the mantle confronted the king with the totality of the celestial hemisphere—with a "*descripcio tocius orbis*," as stated in one of its various inscriptions. The mantle featured thirty-two constellations depicted in mythologized, figural form, including the signs of the zodiac, and exquisite depictions of the northern and southern hemispheres.[10] Amid these heavenly shapes, venerable silhouettes of saints and evangelists emerged in roundels alongside seraphs, cherubs, and the graceful figure of the Virgin (appropriately labeled "*stella*

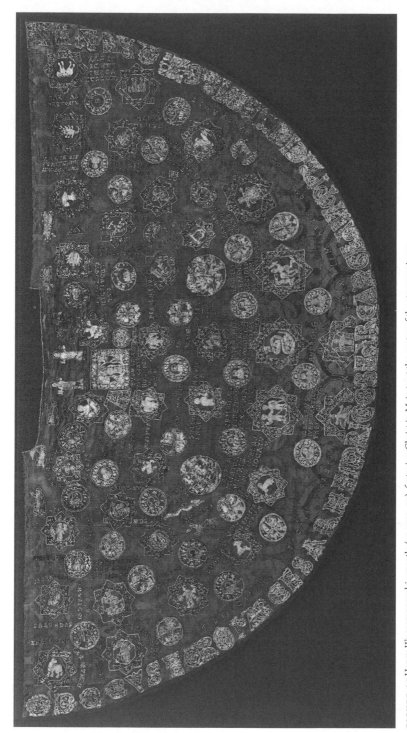

FIGURE 3.1 Henry II's cosmographic mantle (ca. 1019–20), featuring Christ in Majesty at the center of the starry universe. Diözesanmuseum Bamberg. Photo: Uwe Gaasch.

maris"). This universe of celestial and holy bodies mystically orbited around a triumphant Christ in Majesty, the focus of the entire composition.

Along the hem of the cosmic mantle ran Melus's dedication to Henry in exquisitely elaborate golden characters: "O ornament of Europe, Emperor Henry, you are blessed. May the king who rules forever increase your realm."[11] The king was thus connected to the cosmos in both senses of the word: as an "ornament" (which was in turn reflected in the splendor of the mantle he was supposed to wear) and as the universe ruled by God, of whom the king deemed himself a direct emanation. The mantle was therefore a surface mediating between the terrestrial and the celestial emperor, and by extension, between earth and heaven: "by wearing it, Henry would put himself under the direct protection of Christ."[12]

The fineness and sheer size of this amazing textile (approximately 3 meters wide) must have struck Henry, as might have the attention paid to the small flattering detail (for example, the distinctive position of Gemini, the king's zodiacal sign, is contiguous with the king's name in the dedication). The gift clearly had the desired effect: in spite of his defeat, Melus was invested by Henry with the highest honors, including the prestigious title of duke of Apulia. And yet, like all mantles, the mantle of glory Melus received in exchange for the cosmographic mantle was an ephemeral craft. The man did not live long enough to cherish it: shortly after his arrival in Bamberg, death took him by surprise at the imperial palace. The celebrations of Holy Easter solemnly presided over by Pope Benedict VIII had just been completed.[13] The cosmographic mantle nonetheless survives (in restored form) as a testament to Melus's political efforts and to a well-established tradition of "silk diplomacy" that he inherited from his enemy, the Byzantines.[14] More intriguingly, the mantle bears witness to the ideology of the Holy Roman Emperors (likewise mimicked from the Byzantines), which set a direct linkage between the terrestrial ruler and Christ, the Ruler of the cosmos.[15]

Henry II's mantle is part of a long political and poetic tradition of cosmographic garments and stellar ceremonial mantles. Art historian Eliza Garrison traces the custom of clothing the emperor in the cosmos back to Roman antiquity. The Capitoline temple of Jupiter housed a collection of star cloaks that were presented to the emperor for the celebration of military victories.[16] After the fall of the Western Roman empire in 467 CE, the popes consciously adopted imperial ritual and rhetoric. Their consecration of feudal monarchy from the time of Pepin and the Frankish kings "sustained in the fragmented West of Europe an illusion of sacred imperial continuity."[17] Melus's gift to Henry, Garrison argues, was therefore nothing but a conscious "revival of Roman tradition in east Frankish territory" with respect to Byzantium.[18]

Indeed, the Byzantines too used ceremonial star cloaks. The *De ceremoniis aula imperatoris* (1030), a book of the protocol at the imperial court in Constantinople almost contemporary with Henry's star cloak, presents the evocative description of an imperial mantle styled after the high priest's cosmic robe of the Old Testament (Wis. 18:24).[19] The book reports the Byzantine emperor's "golden mantle" being decorated with "a zodiac made of gold and pearls and precious stones" and embellished with a fringe of "three hundred and sixty-five golden bells shaped like orange blossoms and just as many oranges" (compare Ex. 28:23). As Muthesius comments, "the Emperor vested in priestly robes with cosmic designs was almost a symbol of the amalgamation of the earthly to the heavenly bodies: the kingdom of earth and heaven."[20] In evoking the high priest's mantle of the Old Testament, the *descripcio tocius orbis* embroidered on Henry's mantle similarly emphasized the Frankish emperor's role as *rex et sacerdos*. Yet, was the cosmographic mantle a regal garment only? Was it a mere Byzantine import, or can we trace a distinctive Western development of this extraordinary cloth?

In medieval Europe the cosmographic mantle was first of all a literary phenomenon. In the Latin West, as in the Byzantine East, metaphorical and allegorical mantles wrapped the heavens from late antiquity throughout the Middle Ages. The Latin writer Martianus Capella (360–428 CE), for example, dressed several of his celestial characters in starry garb, including Geometry, whose rules govern planetary motions.[21] His contemporary Claudianus (370–404 CE) likewise inscribed the dress of Proserpina with an account of the birth and infancy of the sun and moon, while, five centuries later, the Carolingian Benedictine monk and scholar Remigius d'Auxerre (841–908) likened night to "a starry peplum."[22] These mantles foreshadowed the magnificent cosmographic and magic garments of twelfth-century Frankish poetry, which reflected a renewed interest in astronomy and astrology.

The most striking parallels between such garments and Henry II's star mantle are found in Chrétien de Troie's famous romance novel *Erec et Enide*, composed around 1170. As Prince Erec, the hero of the novel, is about to step into his new role of "philosopher king," Chrétien dresses him in a shiny ceremonial robe featuring the four mathematical arts of the *quadrivium* embroidered by the four fairies. The fairy who embroidered Astronomy, writes Chrétien, took counsel directly from the Moon and the Sun.[23] At Erec's marriage, his bride Enide prays at the altar of the Virgin in the church of Carnant, asking that the couple might be granted a heir, and offers "a green silk cloak, like of which had not been seen before and a large chasuble embroidered all over with pure gold."[24]

Similarities between Henry's star mantle and the poetic garments crafted by Chrétien are striking not only in terms of their regal status and materialities but also in the very way in which they come into being, are performed, and move around. Enide's pious offering was originally a garment made of silk from Almeria embroidered in the Val Perilleus by fairy Morgan, who had intended it as a gift for her lover. Through a subterfuge, Queen Guinevere, Erec's mother, obtained the garment and transformed it into a chasuble for her private chapel before passing it on to her future daughter-in-law. As it moves across western Europe, the embroidered textile progresses through different religious contexts: "from Islamic—the belief of the Almerian creators of the silk—to pagan in the hands of Morgan the fairy, to Christian in Guinevere's care."[25] On the Virgin's altar in the church of Carnant, the magic garment completes its process of Christianization and turns into a sacred treasure. The same garment that was originally tailored by a fairy as a token of physical love morphs into a symbol of love for God through Enide's pious offering. Marking the progress from corporeal to spiritual love, the precious textile becomes an interface, a mediator between earth and heaven.

Their liability to be moved around and transformed made textiles particularly suitable objects for reuse in different settings. As with Enide, Henry II did not keep Melus's cosmographic mantle for himself, but deposited it in the treasury of Bamberg Cathedral soon after he received it. The inscriptions on the mantle mark the passage between different hands: Melus, the original donor, had his name inscribed along the hem (*pax Ismaheli qui ordinavit*); Henry, the recipient and new donor, set his dedication to God under the figure of Christ in Majesty, at the center of the mantle (*superne usye sit gratum hoc Caesaris donum*) (fig. 3.2).[26] Like Enide, Henry turned a precious possession into a pious donation, placing himself under the direct protection of the Ruler of the Cosmos.[27]

The most elaborate cosmographic mantles were crafted by the eclectic French theologian and poet Alain de Lille (1128–1203). His work is populated by a cast of mythical allegorical figures vested in magnificent garments. For Alain and his neo-Platonic contemporaries, allegory was itself a garment.[28] The *auctores* of antiquity, whose authority permeated medieval scholarship, were deemed to have used *involucra* "to veil their most profound utterances from the eyes of the ignorant and profane."[29] In this, medieval scholars believed, the ancients followed the practice of Nature herself, who, as Macrobius famously wrote, "just as she withheld an understanding of herself from the uncouth senses of men by enveloping herself in variegated garments, has also desired to have her secrets handled by more prudent individuals through fabulous narratives."[30]

Like Martianus, Alain clothed the seven maidens of his *Anticlaudianus* in shiny cosmographic garments. Most amazing of all, however, is the

FIGURE 3.2 Detail from Henry II's cosmographic mantle with Christ in Majesty.
Diözesanmuseum Bamberg. Photo: Uwe Gaasch.

magnificent cosmic robe he wove for Natura in his *De planctu* (written in the
late 1160s). In this work, a mix of prose and verse, Alain illustrates how humanity, through sexual perversion, has defiled itself from both nature and God. As
the poet-theologian mourns this sad state of affairs, Nature makes her appearance in the form of a beautiful virgin wearing a starry diadem and a cosmic
garment, to whose description the author devotes several pages. As in Homeric
literature, here the fine embroideries of Nature's robe serve as narrative devices
unfolding entire worlds in motion. Unlike Helen's cloak, or Peleus and Thetis's
nuptial blanket, however, Nature's garment is not a mere tableau animated by
the pen of the poet, but a fluid surface in a continuous state of transformation.
It is "a garment, woven from silky wool and covered with many colors . . . its
appearance perpetually chang[ing] with many a different color and manifold
hue."[31] To each color corresponds a different element of creation:

> At first [the garment] startled the sight with the white radiance of the lily.
> Next, as if its simplicity had been thrown aside and it were striving for some-
> thing better, it glowed with rosy life. Then, reaching the height of perfection,

it gladdened the sight with the greenness of the emerald. Moreover, spun exceedingly fine so as to escape the scrutiny of the eye, it was so delicate of substance that you would think it and the air of the same nature. On it, as a picture fancied to sight, was being held a parliament of the living creation.[32]

After birds of all sorts have paraded under the writer's ecstatic gaze, Alain turns to Nature's marine mantle, a precious craft fashioned in "fine linen with its white shaded into green, which the maiden . . . had woven without a seam." Silk ripples change into sea waves, as the mantle's many intricate folds reveal "the color of water, and on it a graphic picture told of the nature of the watery creation, as divided into numerous species": sea dogs, herrings, plaices, salmons, mullets, dolphins, and other creatures likely to have populated contemporary *bestiaria*, including "a fish with the lower members of a siren, and with the face of a man." Having exhausted the long catalogue of marine species, Alain finally turns to the land, "a damask tunic, also, pictured with embroidered work . . . starred with many colors, and massed into a thicker material approaching the appearance of the terrestrial element."[33]

The changing nature of the garment reflects the ancient motif of the mutability of creation expounded by the early Church Fathers and, by implication, its transiency. The bounty and variety embroidered on Nature's mantle is a fleeting thing. It is akin to the "giant sheet" full of animals that God let down from heaven by its four corners and which was nevertheless destined to vanish together with Peter's brief vision (Acts 10:11).[34]

More crucially, Alain employs the garment to give expression to the old patristic understanding of the order of creation and its neatly woven fabric, "united in unbroken elegance." "Offspring of God" and "mother of all things," Nature joins "all things in firmness with the knot of concord," and "with the bond of peace marr[ies] heaven to earth," and "cloaking matter with form, shape[s] the cloak of form with [her] finger."[35] In turning Nature into allegory, Alain removes her, as in neo-Platonism, from the One or from God. He apostrophizes her as a creation of God, as his faithful deputy setting patterns of harmony in the cosmos. The ordered design of Nature's garment sets a stark contrast with the tragedy of its defilement. Alain's repeated use of textile metaphors adds further pathos to the drama. "Stripped of *the cloak of decency*, man overthrows natural law with lust and depravation."[36] The stars "*clothe* [the firmament] with their splendors," serving Nature's majesty; by contrast, Nature laments,

Many men have taken arms against their mother in evil and violence. . . . [They] lay me the hands of outrage, and themselves tear apart my garments piece by piece, and, as in them, force me, stripped of dress, whom they ought to clothe with reverential honor, to come to shame like a harlot. This tunic, then, is made with this rent, since by the unlawful assaults of man alone the garments of my modesty suffer disgrace and division.[37]

Alain's reproach to mankind is a moral one: of all created beings, man alone in his depravity fails to obey the laws of Nature. He parts her garment, just as the centurions parted Christ's tunic (Matt. 27:35; Luke 23:34; John 19:24). As her garment is torn apart, a dreadful Chthonian abyss is opened. Vice is laid bare:

> Now guile does not seek the robe of hypocrisy, nor does the foul odor of vice look for the balsams of the virtues to furnish a mantle for its stench. The nettle, indeed, does cloak its poverty with roses, sea-weed with hyacinths, dross with silver, rouge-paint with a true glow, that thus, for a time, as appearance may make amends for evil. But crime puts off all ornaments. . . . For vice strips itself openly; falsehood becomes the tongue of its own madness.[38]

The stark contrast between the beautiful mantle of Nature and the ugly nakedness of human vice and corruption calls to mind the crude verses of the Cistercian monk Hélinand of Froidmont (1150–c. 1229):

> A well-nourished body, a delicate flesh
> is nothing but a garment of worms and fire.[39]

Conflating the appealing image of a soft skin with the appalling worms of the cemetery and the fire of hell, Hélinand turns the mantle inside out; he reminds his readers of the other side of their garments of flesh.

Mappae Mundi and Veronica's Veil

Deeply intertwined in the fabric of Alain de Lille and in Henry II's mantles is a profoundly theological understanding of creation. In Alain's De planctu, Nature is called to assist the Creator, once he "had clothed all things with the forms for their natures and had wedded them in marriage."[40] Likewise, the two registers that govern the graphic composition of Henry's star mantle (secular constellations and Christian figures) are fabricated into one another. The interpenetration of cosmological and biblical motifs gives visual expression to Christ the Logos as the force that harmoniously binds creation together—a concept repeatedly emphasized by the early Church Fathers and by the Irish theologian John Scotus Eriugena (815–77), who, in a striking image, called Scripture and the cosmos "the two garments of Christ."[41]

Christ as the ordering principle of creation was a common motif in Western medieval art, and especially in representations of the Crucifixion. In an ivory relief on a book cover from Reichenau (Germany; fig. 3.3) contemporary

FIGURE 3.3 Detail from ivory relief on book cover featuring the scene of the Crucifixion with allegories of the four elements. At bottom, Ocean faces breast-feeding Gaia, who is covered by a veil. Reichenau, Germany, mid-ninth century.
Inv.-Nr. MA 160, Photo Nr. D49620 © Bayerisches Nationalmuseum München. Photo: Krack, Bastian.

with Eriugena, the crucified Christ features as the center and structure of the cosmos.

The four Aristotelian elements are represented by allegorical figures. At the top of the composition are personifications of the sun and the moon (which stand for fire and air, respectively). At the foot of the Cross, Oceanus, who embodies the primeval element of water, sits astride a sea dragon. Opposite him on the right sits Gaia holding a cornucopia, her traditional attribute, and breast-feeding two children under her protective veil (a distant echo of the arched textile covering Kosmos on early Christian Roman sarcophagi;

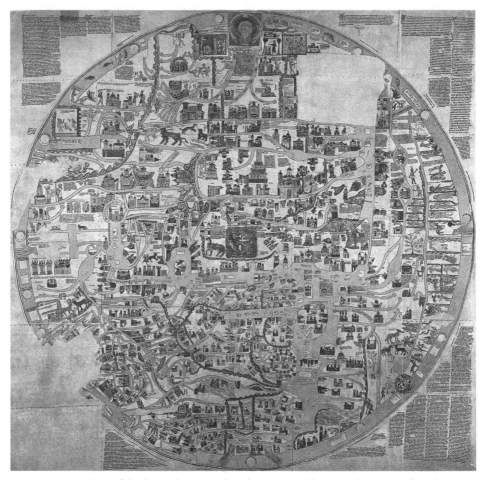

FIGURE 3.4 Copy of the thirteenth-century Ebstorf *mappa mundi* featuring Christ mystically embedded in the fabric of the world, with his head, hands, and feet marking the four cardinal directions. As the napkin on which Christ's face is imprinted suggests, the world was a giant vernicle itself imprinted with the vestiges of the Creator.
Leuphana Universität Lüneburg.

see fig. 2.1). Mother Earth is conflated with Eve; the Holy Cross with the Tree of Life (twined by the snake). The crucified Christ is a *cosmocrator*, the axis around which the entire creation revolves; at the same time, he is also the axis of human salvation, bridging, through his sacrifice, the dark recesses of Hades with the upper chambers of the firmament.[42]

On a thirteenth-century *mappa mundi* manufactured for the nuns of a Benedictine convent in Ebstorf, Gaia, her veil, and the body of the crucified Christ all merge in a single image (fig. 3.4).[43] On this giant "world cloth" (at

3.6 meters square, the largest we know of), Mother Earth's protective veil shifts into a variegated fabric embroidered with a true encyclopedia of creation. As with Nature's cosmographic garment in de Lille's *De planctu*, medieval *mappae mundi* were deemed to contain not only places and geographical features, but also plants, birds, fish, terrestrial animals, people, and monstrous creatures, among other things—in other words, the totality of the created world. Inheriting from earlier exemplars their circular form, tripartite scheme, and orientation with East at the top, thirteenth-century monumental *mappae mundi* like the Ebstorf example are striking for their complexity. They served as visible testimonies of the variety and wonder of the created world; as proofs of Aquinas's equation between multiplicity and perfection.[44] In Evelyn Edson's words, they "pulled together the world view of the high Middle Ages and presented it in a spatial format, incorporating history, geography, botany, zoology, ethnology, and theology into one harmonious and dazzling whole."[45]

Thirteenth-century *mappae mundi* are also similar to de Lille's mantle of Nature in that they portray not a static image, but a whole world in motion—in this case, nothing less than a universal history of the human race unfolding from Creation to the Second Coming. *Mappae mundi* were inevitably affected by Augustine, the most influential figure in Latin Christianity. In particular, his attempt to explain Creation and the shift from eternity to time fostered a keen interest in geography among medieval theologians; for them, space was a product of time, and geography an outcome of history.[46] Unlike modern maps, which usually offer a snapshot in time, *mappae mundi* showed present, past, and future converging on the same image, from the Creation and Fall of Adam and Eve in the garden of Eden to the vanished empires of the ancient world, Christ's Resurrection in Jerusalem, and his Second Coming.

As in the Reichenau Crucifixion (and on Henry II's cosmographic mantle), on the Ebstorf *mappa mundi* Christ mystically remains at the center of creation, structuring both space and time. Here, however, the body of the crucified Christ is literally woven into and made one with the *orbis terrarum*.[47] Head, hands, and feet pierce the landmass at the cardinal points and mark the limits of the *oikoumenē*. Christ's feet are at Gibraltar, the bottom of the circle and western exit of the Mediterranean where Hercules erected the pillars with the inscription "*non plus ultra*" ("there is nothing beyond").[48] Right opposite, at the top of the circle (that is, at the farthest east), Christ's face is clearly visible next to the garden of Eden. His gaze directly meets the gaze of the reader, inviting meditation; his body is "a universal figure that dominates all knowledge on the map, a grounding principle of the encyclopedia, the origin of the infinite diversity of the world. Seeing the map means seeing Christ and being seen by him."[49]

A similar pattern is found in the Lambeth Palace *mappae mundi* (ca. 1300). Here the world embraced by a figure of Christ is presented in a diagrammatic T-O structure. The T shape composed by the Nile, the Don, and the Mediterranean forms the arms and lower part of the cross, the "backbone" of creation, while, inscribed in a square handkerchief, the circle of the *orbis* holds the parts together in a coherent, self-enclosed whole. The world features as an agglomerate of names of provinces, cities, and islands, rather than as pictorial signs. The image calls to mind Isidore of Seville's (560–636 CE) belief that names were the key to the nature of things and that secular knowledge, including geographical knowledge, was necessary for a good understanding of Scripture.[50]

Contrived for meditative and educational purposes, these maps give expression to God's immanence and omnipresence, while at the same time shaping a discourse of place. The result is, in Marcia Kupfer's words, "a vision of universal participation in the Church as Christ's body."[51] Through the map, the world literally becomes the eucharistic body of Christ, a giant wafer, preserving the ontological distinction between Logos and creation. Underlying this "cosmic host" is the ancient Greek theory of microcosm and macrocosm, according to which each human being embodied the world, while the world took the form of a gigantic man.[52] On both maps this man is none other than Christ. On the Ebstorf map, however, his representation is twofold. His former manifestation, large and immediately visible, encompasses the entire earthly landmass; the latter is spatially located on the *umbilicus mundi*. It consists of a small vignette that shows Christ's body emerging from the tomb in Jerusalem and bringing the message of the Resurrection to humankind.[53]

The encyclopedic *mappa mundi* was a mnemonic image recalling the diversity of God's creation. It was, in Christian Jacob's words, "a rhetorical figure, a visual hyperbole commensurate with the power of the Creator."[54] The author of the Ebstorf map defined *mappa* as a "*forma*," and therefore the *mappa mundi* as a "*forma mundi*," a giant spatial container encompassing the beautiful totality of the created world.[55] Its power and significance nonetheless lie not only in what the map showed, but also in *what it was*. Like the Greek *pinax* (or the Latin *tabula*), the word *mappa* indicates the physical support on which a graphic representation is inscribed or drawn, rather than the graphic representation itself—a reminder that maps are weavings of a mental image through a material medium, "the materialization of an abstract intellectual order extracted from the empirical universe."[56] *Mappa* originally designated a "tablecloth," as well as "a piece of cloth that was thrown in the Roman circus to signal the beginning of public games."[57] The term is thus concerned with a textile, with a material, with a surface. In the case of the Ebstorf map, this surface was made of thirty goatskins sewn together. Although it was not crafted on a loom, the map nonetheless shared the assemblage-like quality of fabrics and the materiality of a garment—a garment of skin, to be sure. And what better medium than a skin to portray the incarnate body of Christ?

Thus understood, the Ebstorf map was nothing but a replica of Veronica's veil: the latter displayed the impression of Christ's face on a cloth; the former, the Creator's imprint on the world and in nature—an argument expounded by the early Church Fathers (Greek and Latin alike) and found among virtually all theologians of the Middle Ages.[58] If the created world was a huge *mappa* painted by God, the map was, in Kupfer's words, "a simulacrum of the fabric of creation in which the *vestigia dei* (God's footprints) are immanent" as *loci* to be imprinted on the surface of memory. By analogy with the veronica, "the Ebstorf map thus likens the world to a cloth on which God impressed the *forma* of his Eucharistic body."[59] The motif is echoed in the very iconography of the map: the T-O shape is that of the wafer used in the Latin Eucharist, while the face of Christ is impressed on a veil in a fashion reminiscent of medieval veronicas.

While the sacred veil was a powerful relic venerated in Byzantium, the diffusion and popularity of the cult reached its apogee in the West, and precisely at the time of the production of the Ebstorf *mappa mundi*. The cult of the veil had originated in Edessa and moved to Constantinople in 944, where the veil became the palladium of the city. After the Latin sack of the Byzantine capital in 1204, the cloth was taken to Saint Peter's in Rome.[60] Four years thereafter, Pope Innocent III, the man responsible for the development and spread of the cult in the West, initiated an annual custom of processing with the veronica from St. Peter's to the hospital of the Holy Ghost. Intriguingly, one of the earliest accounts of this procession is provided by Gervase of Tilbury, likely to be the author of the Ebstorf map, who was in Rome for the coronation of Otto IV the following year and left a description of the veronica.[61]

The cult of the veronica was connected with the mystery of the Eucharist and, more specifically, with the doctrine of the transubstantiation of bread and wine into the body of Christ, which the Fourth Lateran Council elevated to dogma in 1215.[62] Like the Eucharist, the veronica had the ability to be infinitely reproduced and to spread Christ's presence throughout the *oikoumenē*. The hidden yet true presence of the body of Christ in the veil of the host was visually expressed in the iconography of the holy veil and in the Ebstorf map, which likewise mystically embedded the presence of Christ. As in Byzantine iconography, Western representations of veronicas typically featured a white cloth imprinted with the face of Christ with his neck truncated, as if by the border of an invisible tunic. Sometimes the face was surrounded by a tripartite halo, as on the *veronica* at the top of the Ebstorf *mappa mundi*. Here the face of Christ thus becomes a sort of mirror image of the T-O map in which it is inscribed.[63]

The peculiarity of both Byzantine and Latin representations of the Holy Veil is that the face of Christ does not lie within the folds of the cloth, but seems to be projected, or floating on the top of it, as if to stress the paradoxical nature of the image: instead of hiding, the cloth makes the face of God visible. The cloth "is not a transparent membrane that would permit an unclear looking through to the spiritual world beyond, but it is rather marked by an image projected from our side."[64] As with the fleshy body of the incarnated Christ, it fulfills the function of the Veil of the Temple. In this, *sudarium* and *mappa* are cognates: by virtue of their materiality, they enable the viewer to see the invisible. They are also cognates in that "the denotative sense 'handkerchief' or 'napkin' endures even as the connoted representation takes precedence and functions independently: a defining image, replicated in other media, subsumes the originary cloth support."[65]

The word "cloth" comes from the Germanic root *kli-*, which means "to stick," "to cling" (as in the words "clay" or "cleave"). Cloth is thus "that which clings to the body, or that which is pressed or felted together."[66] Made of vegetal fibers or animal skins, cloths embed an organic quality; they form a continuum between the cosmos and the human body, a "second skin" necessary for human survival. On Veronica's original *mandylion*, cloth and skin blended in the same texture. Soaked with bodily fluids, the holy veil offered spiritual presence while remaining bound to the body.[67] The linen was a sponge impregnated with and exuding divine Grace. It was a material vessel inhabited by Christ. By inscribing the world with the body of Christ "on a skin," the Ebstorf *mappa* gave visual expression to this blend. It conveyed the holiness of creation.

During the fourteenth century, the veronica became a central icon and universal symbol of the "*Ecclesia Romana, id est universalis*."[68] As the faithful flocked to the *vera icon*, the demand for copies reached such proportions that by the late Middle Ages a guild for the purpose was founded in Rome. Just as the divine power exuded through Christ's garment healed Saint Veronica by the mere touch of its hem, so was the veronica's power transmitted to its replicas through physical contact with its surface. These copies were treated as "contact relics" sharing in the grace of the original. Churches and cloisters throughout Europe thus acquired their own veronicas and organized their own cults of the Holy Face.[69] The cult of the veronica gained such significance that, in the words of art historian Neil MacGregor, from the fourteenth century on, "wherever the Roman Church went, the veronica would go with it."[70]

The Ebstorf eucharistic map was similarly a symbol of the universal Roman Church and of the Holy Roman Emperor. Its topography placed emphasis on Jerusalem, the target of the Crusades and center of the map, and on Otto's connections and allies, while its textual contents largely drew on a text itself dedicated to the emperor (Gervase of Tilbury's *Otia imperialia*).[71] Displayed on the wall of a monastic foundation under Otto's tutelage, the *mappa mundi* would have fulfilled its spiritual goals while at the same time featuring as a proud symbol of terrestrial lordship.

Landscape as a Veronica

Cosmographic mantles and *mappae mundi* empowered the beholder with a totalizing and moralizing view of the universe and of human history. They offered a synoptic picture of a God-inhabited space and time. Yet, how did this picture translate to the mantle of the physical landscape? How did it translate to Western medieval perceptions of place? Did these perceptions change over time?

The French historian Jacques Le Goff has likened the face of medieval Christian Europe to "a great cloak of forests and moorlands perforated by relatively fertile cultivated clearings," a sort of "photographic negative of the Muslim East." While in the Near East, observes Le Goff, oases were "islands" in the desert and trees meant civilization, in western Europe timber was plentiful and islands of civilization were carved out of it. "Any progress in medieval western Europe meant clearings, struggle, and victory over brushwood and bushes."[72] From the eleventh through the thirteenth centuries, the landscape of western Europe underwent dramatic transformations, including large-scale deforestation. Such a substantial and unprecedented environmental change was supported by the belief that humans were God's helpers in finishing the creation. Ironically, one of many epithets applied to wilderness was *loci non vestiti* ("unclothed places"), as though cultivation, and thus the clearing of the green mantle, was a necessary precondition for "clothing" God's work.[73]

Hence in Bernard's description of Clairvaux, landscape is transformed from untamed wilderness into a charming ordered setting in perfect harmony with the monks' activities—into a beautiful "*vestitus*." Landscape is made productive and pleasant through human action; "it is given meaning because humans impose an order upon it."[74] Grain and vineyards both "offer to the eye a beautiful sight and a needful support for the inmates"; on the summit of mountains the brethren collect and burn dry branches, grub up the brushwood that "disfigures the ground," dig the soil. In the valley, fruit trees are enclosed by a wall in a fashion redolent of the garden of Eden, while the water of the nearby river, diverted by the monks, flows through workshops, boilers, and mills, relieving the brotherhood from heavy labor.[75] For Bernard and his like, human action is necessary to soften and "polish" the mantle of the earth; it constitutes the ornament, the fine embroidery of the precious cloak unfolded by God upon humankind. A beautiful landscape is thus a tangible expression of the collaboration between humans and the Creator, but first of all, of God's care for the world. It is a mirror of his benevolent face. Like a veronica, it is a skin impregnated with divinity, an intermediary between humans and their celestial Father.[76]

That beautiful nature was first of all cultivated nature is an idea with ancient roots. Eden was beautiful, it was the ornament of the earth, because it was ordered and self-enclosed. It was not a forest or a jungle, but a garden. The natural beauty praised by the Church Fathers and their successors was therefore almost invariably farmland, parkland, cultivated land, in other words, bounded land that was qualitatively different from the threatening and confused wilderness that spread outside of it—the desert of the East, and the *silva* of the West. Thus, in a letter to Valentinian Augustus dated 384 CE, Ambrose noted how "[f]ormerly, the earth did not know how to be worked for her fruits. Later when the careful farmer began to rule the fields and to clothe the shapeless soil with vines, she put away her wild dispositions, being softened by domestic cultivation."[77]

This perception of the earth as a mantle waiting to be "softened" by human action is deeply rooted in the classical world and, in particular, in Vergil's works. The Latin poet delighted himself in the variety of the physical landscape; he took pleasure in examining different kinds of soils and considering what could best be grown on each. Indeed, he linked the development of civilization to subsequent transformations of the land.[78] In the golden age, Vergil tells us, the unbidden earth "gave more freely of all her store, in that none asked her bounty."[79] Jove stopped this bounty, fashioning nature as we know it now, but Ceres then showed men how to cultivate the soil. Vergil thus calls upon the farmer to tend to his land, to cosset the plants, to know the soils, to weave hedges to keep the flocks from the vines:

> And tame with culture the wild fruits, lest earth
> Lie idle. O blithe to make all Ismarus
> One forest of the wine-god, and to *clothe*
> With olives huge Tabernus![80]

Vergil's gaze is not bound to the surface of the earth, but oscillates between its variegated mantle and the mantle of the heavens. It pays attention to the

FIGURE 3.5 Frontispiece of Petrarch's copy of *Commento di Servio a Virgilio* illustrated by Simone Martini, ca. 1336. Vergil lies under a tree behind a thin curtain. Petrarch believed allegory to be "the very warp and weft of all poetry." Like allegory, the thin curtain at once conceals and reveals the mystery of creation. GL Archive / Alamy Foto Stock.

seasons, to atmospheric phenomena, to the constellations. As Bernardus Silvestris wrote in his *Cosmographia*, "In the stars Vergil composes with grace."[81]

On the exquisite frontispiece of Petrarch's copy of Vergil's works illustrated by the Italian artist Simone Martini (ca. 1336), we see the poet reclining on the soft grass under a tree as he follows the course of the stars with his pen, seeking inspiration (fig. 3.5). His figure is revealed behind a thin curtain drawn aside by the fourth-century grammarian Servius, the author of an influential commentary on Vergil. By drawing back the curtain, Servius symbolically "unveils" the poet to posterity, while the protagonists of the *Aeneid, Georgics,* and *Eclogues* undertake their tasks in the idyllic quiet of a serene starry night: Aeneas escorts Severius, the peasant prunes the vine, the shepherd tends to his flock.[82]

Martini's frontispiece offers the viewer a visual summary not only of Vergil's work, but of medieval society at large; a tripartite society made of laborers, warriors, and priests (the last here being incarnated by Vergil, the "high priest" of Latin poetry and secular prophet of Christianity).[83] As with the Reichenau crucifix and *mappae mundi*, Martini's composition presents a perfectly integrated microcosm in which the earth forms a continuum with heaven; it is a hierarchically arranged world where the *vita activa* unfolding in the lower part of the composition is balanced by the poet's *vita contemplativa* in the upper part.[84]

The fabric of the curtain speaks of the allegorical nature of the composition with its implicit Christian undertones (the shepherd, the vine).[85] Petrarch, who cherished both classical scholarship and his Christian faith, and wished his literary life to parallel that of Vergil, called theology "poetry concerning God" and allegory "the very *warp and weft* of all poetry."[86] Compelling the gaze to penetrate the depths of heaven, the opening curtain reminds the viewer that "although the world of Christianity was enclosed and shut off on this earth below, it was wide open above."[87]

The light curtain is a thin threshold; it is a veil that at once conceals and reveals the mystery of creation. Its opening calls to mind Giotto's representation of the heavens "being rolled up like scroll" in his fresco of the Last Judgment in the Cappella degli Scrovegni in Padua (1305; pl. 2). The realistic, if not illusionistic, three-dimensional representation of the angels literally "peeling off" the skin of the firmament (which continues through the window on the wall) sets a compelling contrast with the stylized symbolic rendering of its Byzantine counterpart in the Chōra fresco examined in the previous chapter (see fig. 2.4). The two images are good illustrations of contemporary attitudes toward divinity and creation: while the prominent Byzantine theologian Gregory Palamas (1296–1359) insisted upon the possibility given to humans to behold the hem of God's uncreated garment of light, his English contemporary William of Ockham (1285–1347), in the wan northern light of Oxford, championed with equal forcefulness his belief that "when it comes to nature, what your eyeballs see is precisely what you get."[88]

In the fourteenth century, as Western sacred art grew more and more realistic, shrouds veiling God the Father or the Holy Trinity became a common feature in representations of the Crucifixion (fig. 3.6). Used as intermediate backgrounds to stress monumental drama and highlight the fleshiness of Christ's body, these textiles were akin in thickness to the "skin of the firmament" in Giotto's fresco. As Kuryluk notes, in such representations nothing

FIGURE 3.6 Central panel of Luca di Tommè's triptych, *The Trinity and the Crucifixion*, with scenes from the life of Christ, ca. 1355, tempera on panel, 56.9 × 54.4 cm (22 3/8 × 21 3/8 in.). Shrouds veiling God the Father (or the Holy Trinity) became popular in representations of the Crucifixion from the fourteenth century, as Western sacred art became more and more realistic.
Putnam Foundation, Timken Museum of Art, San Diego.

is more important than the accurate treatment of skin and the cloth and the paradoxical relationship between them: "as skin dies, cloth comes alive and replaces the body."[89] Or rather, the entire body of Christ becomes a vernicle—a threshold space between the viewer and the invisible God.

*

In its threefold artifactual incarnation (the cosmographic mantle, the *mappa mundi*, and the vernicle), the earth's mantle, as this chapter has shown, endured in the Latin West as a powerful metaphor for a sanctified cosmos held together by the Creator and imprinted with his own vestiges. This chapter has also shown how the dramatic transformation of the physical landscape in medieval Europe did not contradict, or compromise, this vision. Western Europeans envisaged themselves as God's collaborators; they helped him polish and beautify the earth's mantle through their work—be it clearing a forest, planting a new crop, or reclaiming a marsh.

While in the West the mantle continued to operate as an interface between humans and God, it nonetheless progressively lost the symbolic and mystical quality it continued to hold in the Byzantine East. The appearance of curtains and veils in fourteenth-century Western art marks the passage to a new sensibility toward creation, of which Petrarch has been traditionally identified as the forerunner.[90] Compelling the eye to push beyond them, veils start to open up a third dimension, which would find its fulfillment in the Renaissance with the theorization of linear perspective, as we shall see in the next chapter. While Byzantine icons "wrapped" the viewer, making him or her the vanishing point of the composition, linear perspective sets the vanishing point within or beyond the canvas, creating the illusion of three-dimensionality. Viewers are set at a distance, outside of the composition; they become external to landscape. Made visible as a landscape ruled by linear perspective, nature is thus reified and deprived of her veil. Landscape is no longer a veronica; the veronica is just an object within the landscape.

On a panel of Hans Memling's *Crucifixion* featuring Saint Veronica (1470; pl. 3), the holy woman and her cloth are set against a verdant landscape, the stereotypical idyllic landscape of chivalric poems, with the *silva* and the fortified walls of a castle looming in the distance. A transparent veil covers the forefront of Veronica, the mystical bride. A white turban crowns her head, reminding viewers of her Eastern origins. Her red and blue garment, the colors of the Virgin's outfit, contrasts with nature's green bloom. The fabric is smooth, silky, soft; it invites touch. Piously suspended over her womb, the holy *mandylion* celebrates the mystery of incarnation. Yet, unlike in the Ebstorf and Lambeth Palace *mappae mundi*, the veronica is now a humble part of the cosmos, rather than being intertwined in its fabric. Nature is now put into perspective, has receded to the background. She has lost her veil.

The chapters in Part II discuss this transformed perception of space and nature in early modern Europe by way of "unveiling" metaphors and, eventually, the Romantic return to a transcendental, yet now largely secularized, attitude toward creation.

Unveiling Space

4

Renaissance Stage Curtains

In times of great discoveries, a poetic image can be the seed of a world.[1]

In modern literature, Patricia Oster observes, "the metaphor of the veil appears every time when borders between truth and appearance, illusiveness and disillusion, a real world and a world of fiction are being touched."[2] In the history of Western cartography and of Western science at large, the metaphor seems to have followed a similar course. Veils, mantles, and stage curtains start to appear on maps and on frontispieces of atlases in the sixteenth century and proliferate over the following two hundred years, as the boundaries between geography and the geographical imagination were being questioned and redefined through the great voyages of discovery.

As European vessels crisscrossed the Atlantic and eventually the Pacific, and Jesuit missionaries ventured to the Far East, an expanding world was unveiled and celebrated behind opening stage curtains and drapes lifted by allegories of Divine Providence and figures of classical mythology. At the same time, theatrical unveilings of Isis, the frightening goddess-personification of nature, became a standard subject on the Baroque frontispieces of anatomy and natural philosophy books. Flying veils and opening stage curtains are nonetheless a metaphor for a more profound change: a change in human consciousness within an expanding world or, in the words of the nineteenth-century historian Jacob Burckhardt, "the discovery of the world and of man."[3]

The Renaissance can be defined as "the Age of Opening"—the opening of time (through the systematic rediscovery of the classical past), as well as the opening of space and of nature.[4] The great geographical discoveries opened up both new spatial and intellectual horizons: fifteenth-century Portuguese navigators sailing down the coasts of Africa disproved the Aristotelian belief in an impassable torrid zone around the equator; Columbus's voyages rebutted the classical notion of its uninhabitability; the appearance of a new continent forever shattered the ancient world island. In the words of Dan O'Sullivan, "the voyages of discovery were in a way large-scale experiments, proving or disproving the Renaissance concepts inherited from the ancient world."[5] The answers to questions such as whether or not there was a southern landmass, or the earth was flat, or the Atlantic navigable were not to be found in a yet more attentive rereading of Strabo or Aristotle; they could only be answered through direct experience. In this, geography was similar to experimental science: in both cases, real-world experience eventually triumphed over received authority. It is therefore not so surprising that the term "discovery" soon came to be used to describe both the progress of geographical exploration and scientific "breakthroughs."[6]

The "opening of space" indeed happened at different scales and went hand in hand with the "opening of nature." In the sixteenth century, the progressive unveiling of global space was paralleled by the opening of the fleshy mantle and the unveiling of the micro-spaces of the human body prompted by the pioneering of human anatomy. At the same time, the implementation of linear perspective in Western painting opened up a third spatial dimension: depth. It invited the eye of the beholder to penetrate into and beyond the canvas, to "pierce" the veil of landscape and peer into the mysteries of nature. As Catherine Smith noted, one of the most pervasive features of early Renaissance culture was indeed the belief that "nothing is truly known unless as it is seen."[7] Creation was no longer a veronica, an unviolated sacred cloth; it was rather a thick skin to be cut and penetrated by the human eye through scientific observation. As the body of Christ became disentangled from the map, "man ceased to look at nature as a child looks at her mother, taking her as a model; he wanted to conquer her, and become her master and possessor."[8]

The morphing of the earth's mantle from veronica into opening mantle, or stage curtain, gives visual expression to this transformation. As to why such images were more widespread in the Renaissance than they were before, we need only turn to the etymology of the word "dis*covery*." As Jerry Brotton observed, the word became common currency in the English language in the sixteenth century, that is, at the height of the Age of Discovery.[9] In Portuguese (one of the first languages to record seaborne discoveries) "to discover" was regularly used as a synonym for "to explore" (but also "to find by chance"), and in Dutch it embedded the meaning of "finding the truth," or "detecting a mistake." In all these languages its primary and literal meaning was to "uncover," that is, to "lift up" the veil. This chapter charts the transformation of this veil in the Renaissance as a visual metaphor for the opening of space happening at different scales: from the global scale of geographical exploration and sixteenth-century Ptolemaic world maps to the micro-scales of landscape painting, the theater, and the dissected human body.

The Opening of the *Mappa Mundi*

> I kept a diary of noteworthy things, that if sometime I am granted leisure, I may bring together these singular and wonderful things and write a book of geography or cosmography, that my memory may live with posterity and that the immense work of Almighty God, partly unknown to the ancients, but known to us, may be understood.[10]

The opening of global space that took place between the fifteenth and seventeenth centuries, that is, during the so-called Age of Discovery, forever transformed Western perceptions of space and the image of the earth itself. Yet, how was this change visualized? How did it translate onto the map?

The self-enclosed medieval *mappa mundi*, with its rigid schemas, was clearly not suitable to represent the new expanding world—at least, not Amerigo Vespucci's world and the "singular and wonderful things" continuously unfolding before his eyes. Conceived of as a tool for learning and meditation, this type of map was not designed to accommodate new landmasses, or the unprecedented flow of new geographical data streaming from exploration. Likewise, while the portolan charts produced and used by sailors provided extremely detailed coastal renderings, they did not take into account terrestrial curvature. Attempts at integrating the two genres and reconciling biblical and secular knowledge can be read on late *mappae mundi*. Fra' Mauro's famous map (1459), for example, contains no fewer than 3,000 inscriptions, including a wealth of locations reported in Marco Polo's travels to the Far East, and portrays the Mediterranean coastlines in the scalloped portolan chart fashion. The map seems to gesture toward "the growing status attributed to observation over contemplation."[11] Yet, so densely packed with information is its circular world island that it seems on the verge of exploding. And at some point it did indeed explode. When and how did this happen?

The rupture of the closed eucharistic order of the *mappa mundi* was paralleled by the development of Ptolemaic cartographic science in fifteenth-century Italy. Unknown in the West for more than twelve centuries, Ptolemy's *Geography* was brought to Florence in 1397 by the Byzantine scholar and diplomat Manuel Chrysoloras. By 1410 it was translated into Latin. Ptolemy's mathematical projections and grid system produced a new type of cartographic representation: one that, unlike the *mappa mundi*, allowed the mapping of an unlimited set of data onto a mathematically predetermined surface, and was thus particularly well suited to an age of geographical discovery.[12] Unlike the medieval *mappa mundi*, Ptolemaic maps were ruled by a "logic of expansion";[13] they were *open systems*. The map ceased to be the body of Christ and a closed container of place-events; instead, it became an expanding archive of locations to be mapped on an orthonormal geometrical space. Within the loom of the ancient Ptolemaic grid, new cities, islands, and mountains could be easily added, preexisting locations amended, coastlines rectified, and even new continents accommodated.

Although it was no longer a veronica, the map nonetheless did not cease to be a mantle—the mantle of the earth. Ptolemy's projections visually rendered the metaphor, as they portrayed the earth in the shape of the ancient Macedonian *chlamys* referred to by Eratosthenes and Strabo (see fig. 1.2). Indeed, some Italian humanists called Ptolemy's first conical projection *mantellino*, or "little mantle," on account of its shape.[14] In the sixteenth century, Ptolemy's second projection grew especially popular, for it created the illusion of sphericity and elongation, thus conveying the unfolding vastness of an ever expanding world. The most extraordinary example is the Waldseemüller map (1507; fig. 4.1). This was the first Ptolemaic map to stretch across the earth's full 360 degrees and the first-ever map to portray the Americas as a distinct continent.

Like a vast opening mantle, the Waldseemüller map revealed a totally new world order to its viewers. As the cartographic *chlamys* opened, the ancient round world island of the medieval *mappa mundi* broke into four continents and fragmented into myriad islands. The sense of fragmentation is accentuated in color versions of the map, where continents painted in different tints and polychrome insular "crumbs" starkly contrast with the uniform dark blue ocean. On the top of the cartographic mantle is no longer the face of Christ (as on the Ebstorf *mappa mundi*), but the mirroring portraits of Ptolemy and Amerigo Vespucci, the geographer of the Old World and the discoverer of the New. The former holds a quadrant, the symbol of terrestrial and astronomical measurement; the latter holds a pair of compasses, "a more modern and practical insignia of his modern navigational method." The geographer and the explorer look down to the hemisphere with which their discovery is associated. "The two men's eyes meet as they gaze across each other's respective spheres of influence, a suitable look of mutual admiration."[15]

At the same time, the gazes of the wind heads framing the map silently traverse its surface, weaving an invisible texture of longitudinal and latitudinal lines. As Jacob notes, these gazes circumscribing the world uphold "a dialectical, if not contradictory, relationship with the single totalizing effect yielded by the extent of the terrestrial surface, unless the winds are a metaphor of a unique gaze, God's gaze, which cannot be represented." The wind

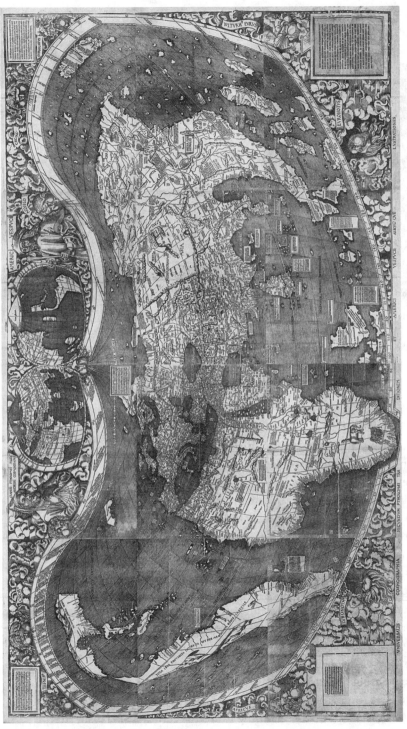

FIGURE 4.1 The opening of global space: Martin Waldseemüller, *Universalis cosmographiae* (1507), the first Ptolemaic map to stretch, like an opening mantle, across the earth's full 360 degrees and to portray the Americas as a distinct continent.

Niday Picture Library / Alamy Foto Stock.

FIGURE 4.2 Domenico Ghirlandaio, fresco showing the Madonna della Misericordia protecting members of the Vespucci family under her celestial mantle. Vespuccis' private chapel, Florence, ca. 1472. From Andreas Quermann, *Ghirlandaio*, Maestri dell'arte italiana (Cologne: Könemann, 1998).

heads float on the margins of the map, in a space utterly different from that of the human world: "the turbulence of the winds, clouds, and tempests exist only to better emphasize the machinery of the cosmos, this order surviving human ephemerality. Permanence in the midst of a swirl."[16]

Claudio Piani and Diego Baratono connect the shape of the map to representations of Marian mantles in contemporary votive paintings of the Madonna della Misericordia (Virgin of Mercy). These images typically showed the Virgin sheltering a group of people under the folds of her outspread celestial mantle.[17] In particular, Piani and Baratono link the Waldseemüller map to a fresco by Ghirlandaio commissioned by the Vespuccis for their private chapel in Florence (ca. 1472; fig. 4.2). Here the Virgin protects the family under her blue cloak, whose shape bears a striking resemblance to that of Waldseemüller's map. Intriguingly, as with the map, the fresco includes a portrait of young Amerigo among the family members. The inscription "The earth is full of God's mercy" (under the Virgin's feet) reinforces her benevolent gesture, while at the same time prefiguring the missionary aspect underpinning oceanic discoveries.[18]

These sentiments are best expressed in Alejo Fernandez's famous *Virgin of the Navigators* (pl. 4). The painting forms the central panel of an altarpiece commissioned by the Casa de Contratación for the chapel in its building in Seville sometime before 1536. The earliest known painting whose subject is the discovery of the Americas, the *Virgin of the Navigators* explicitly connects the Marian mantle with oceanic exploration. The Madonna is depicted spreading her dark blue mantle over the protagonists of the great voyages and their patrons. In the foreground are Ferdinand II of Aragon and Emperor Charles V (cloaked in red), alongside Christopher Columbus, Amerigo Vespucci, and one of the Pinzón brothers, shown kneeling. The Virgin's mantle straddles the seas; it bridges the continents; it protects the ships and their crews as they embark on the perilous Atlantic crossing. Around the Virgin, in the semidarkness of the background, gather the indigenous peoples of the Americas converted from their original faiths by the navigators who have set sail in her name.[19] A threshold between the sea and the sky, the dark mantle at once protects, conceals, and reveals.

The discovery of America, Piani and Baratono observe, seems to be a privileged *topos* in which the sacred act of unveiling assumes increasingly Christian tones. On Johannes Stradanus's engraved frontispiece of *Americae retectio* (fig. 4.3), a picture atlas designed to celebrate the first centenary of the discovery of the New World, a mantle is lifted by the dove of the Holy Spirit, thus revealing the new cartographic order: a globe centered on the Atlantic and crossed by European vessels bound to the New Continent. The configuration of the mantle is, once again, reminiscent of the Waldseemüller map and the Marian cloak, while also evoking the shape of the biblical "tent of heavens" mentioned in the Book of Psalms (Ps. 104:5).

Stradanus's image fuses biblical allusions with classical mythology. Complicit with the dove in unveiling the new world order are the mythical figures of Janus and Flora, which Stradanus associates with Genoa and Florence, the native cities of Christopher Columbus and Amerigo Vespucci. As a whole, the image evokes Pherecydes's ancient myth, while at the same time perpetuating the salvational axis found on early Christian sarcophagi (see fig. 2.1) and Carolingian crucifixes (see fig. 3.3). Above the dark chthonic figure of Neptune supporting the globe is the luminous dove of the Holy Spirit; above the untamed seas are the portraits and instruments of Christopher Columbus and Amerigo Vespucci. The "unveiling" of the New World is only the most recent act of the unfolding drama of salvation history.

The act of opening is intrinsically temporal; it embeds a sense of imminence; it pre-announces what is about to come, the "unfolding" of events. Stradanus's mantle acts as the connective tissue between classical mythology, biblical past, and the oceanic present. Yet, in its unfolding motion, the mantle also connects past and present to the future, to that part of the globe that

FIGURE 4.3 Johannes Stradanus, frontispiece to the commemorative atlas *Americae retectio* (1592). A dove (the Holy Spirit) unveils the new global order produced by Christopher Columbus's and Amerigo Vespucci's discoveries.
Old World Auctions.

remains still hidden to the viewer. Inherent to the mantle is thus a dynamic quality, a quality that is also found on the Waldseemüller map and other Ptolemaic "cartographic cloaks," such as the map of Palazzo Besta in Valtellina (northwest Italy) (fig. 4.4).

Part of a cycle of frescoes featuring scenes from Genesis, this map represents God's creation. As such, it provides a powerful visual link between ancient biblical visions and the new horizons of geographical discovery.[20] It expresses the tension between ancient and new world orders. Framed by a golden ribbon, the opening cartographic mantle brings together the New World and vestiges of old medieval *mappae mundi* (for example, the Red Sea painted in red). The map is embedded in a bucolic verdant landscape, as if to convey the beauty of God's newly crafted handiwork. Taken as a whole, the fresco combines geography and topography, cartography and landscape painting. Most intriguingly, it combines the two ways of seeing and representing space typical of the Renaissance: the Ptolemaic grid and linear perspective, that is, the horizontal expansion of space and the opening of spatial depth.

The Opening of Depth

The development of Ptolemaic cartography and linear perspective are both responses to and manifestations of the "opening of space." Even before being representational techniques, linear perspective and Ptolemaic projections are conceptual tools for rationalizing and grasping the world. Both of them are concerned with the mastery of space through the transposition of three-dimensional realities on two-dimensional surfaces: Ptolemy's projections enabled the cartographer to transfer the sphere of the globe onto the flat surface of the map; linear perspective enabled the artist to "enframe space" and recreate the illusion of three-dimensionality on the surface of the canvas (or the wall).

Svetlana Alpers and other scholars have stressed the distinct emergence, development, and characteristics of these two types of spatial organization. Perspectivalism appeared in fifteenth-century Italy and became the organizing principle for seeing and representing the landscape.[21] Ruled by the theoretical principles exposed in Leon Battista Alberti's treatise *De pictura* (1435), it

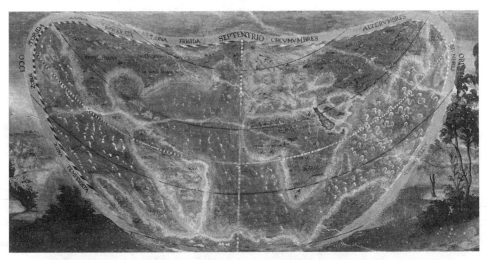

FIGURE 4.4 Frescoed scene of Creation combining Caspar Vopell's map and a landscape view. Drawn according to Ptolemy's second projection, the newly created world unfolds like a *chlamys*. The fresco is located in the Creation Room of Palazzo Besta, Valtellina (late sixteenth century).
Study © Claudio Piani. By kind permission of Ministero per i Beni e le Attività Culturali e per il Turismo—Polo Museale della Lombardia. Photo: Ivan Previsdomini © 2007.

presupposed a single, immobile viewer perceiving the image through a window that exactly located the height and distance of their point of view. Everything was centered on the sovereign eye of the beholder, "like the beam from a lighthouse, only instead of traveling outward, appearances travelled in"—what Alberti called "visual pyramid."[22]

Through linear perspective, the artist established the arrangement of the composition and the "point of view" to be taken by the observer, while controlling through framing the scope of reality revealed.[23] The subject was reduced to a disembodied "winged eye," which Alberti chose as his own emblem, as if to underline the quasi-divine and detached act of rational seeing.[24] This eye was understood to be static and unblinking, fixed and absolute, rather than dynamic. In Martin Jay's words, "it followed the logic of the Gaze rather than the Glance."[25] It arrested the restless flux of life in an "eternal moment of disclosed presence."[26]

Projectionism, which Alpers links to the distance point method used in seventeenth-century Dutch painting, by contrast, "de-emphasized the centre and stressed instead the spreading of the grid in all directions from the perimeter."[27] In Alpers's words, "whereas the Albertian perspective posits a viewer at a certain distance looking through a framed window at a putative substitute world, Ptolemy and distance point perspective conceived of the picture as a flat working surface, unframed, on which the world is inscribed."[28] The aim of Dutch map makers and painters was "to capture on a surface a great range of knowledge and information about the world. Theirs was not a window, as in the Italian model of art, but rather . . . a surface which was laid out as an assemblage of the world."[29] It was a panoptic space articulated through multiple points of view—those of the wind heads framing Waldseemüller's cartographic mantle (see fig. 4.1).[30]

These fundamental differences notwithstanding, Ptolemaic cartographic science and perspectival painting both rested on the conceptualization of Euclidean space as an absolute dimension preexisting its contents, as an objective entity that could be mastered and controlled through geometry. In other words, they both normalized space by turning it from a closed ensemble of objects (or place-events, as on the medieval *mappa mundi*) into an open system of spatial relations regulated by mathematical laws; into a uniform, infinite, isotropic space.[31] As such, the two systems (or elements of them) were often used by the same practitioners. Leonardo, for example, experimented with geographical projections and created his own world map, while Alberti was so obsessed with the cartographic grid as a means for organizing space that he advised painters to set out a veil, or curtain, "loosely woven of fine thread . . . divided by thicker threads into as many parallel squares as you like, and stretched on a frame." The veil was to be positioned "between the eye and the object to be represented, so that the visual pyramid passes through the loose weave of the veil," enabling the viewer to better observe and copy the world behind it.[32]

If the Ptolemaic mantle gave visual expression to the unfolding world in its entirety, the Albertian veil was the precondition for representing the immediate environment in a realistic, three-dimensional fashion. Not only did the *velum* make it easier for the painter to transfer details onto the canvas, but, more important, "it trained both the artist and the viewer to see the underlying geometry of nature, the truth of visual reality established by God at the creation."[33] By a strange paradox, nature was to lose her veil thanks to the mediation of another veil.

Alberti's *velum* was much more than a valuable artistic device. It was also a powerful metaphor that underpinned human vision and perception of the world. Intriguingly, in the *De pictura* the Italian architect repeatedly uses textile metaphors to speak both of nature's hidden geometry and of the functioning of vision itself. For example, when drawing human and animal bodies, the artist is to imagine first their essential, hidden structure (that is, the skeleton) and then "dress it with flesh, in the same way we first draw a naked man and then wrap him in clothes."[34] Surfaces, argues Alberti, are formed by multiple lines "akin to the threads of a fabric,"[35] while the rays of the visual pyramid emanating from the eye are similar to "the thinnest threads [*fili sottilissimi*] tied very tightly inside of the eye, as in a fabric [*mappa*]."[36] For Alberti, the *velum* was therefore nothing but the natural continuation of the myriad invisible threads and geometrical lines structuring the fine fabric of the world.

Alberti's technique is illustrated in a famous woodcut by Albrecht Dürer printed in 1525, almost a century after the publication of the *De pictura* (fig. 4.5). The woodcut shows the artist as he views his model through the gridded *velum* and reproduces her on paper through a system of coordinates reminiscent of the Ptolemaic grid. The image has been used to illustrate the power of perspective and the strong division of gender roles in early modern Europe: the male artist actively looks out, while the female model is passively looked at; the man sits upright, while the body of the woman lies horizontally on the table as it is being visually "dissected."[37] In between the two figures is the framed *velum*.

At the same time, the figures of the artist and of the model are framed for the viewer. Their bodies are set against windows revealing glimpses of landscape. Gillian Rose noted the close association between landscape and the female figure: here, as in many paintings—from Giorgione to Thomas Gainsborough—Woman and Nature share the same horizontal topography

FIGURE 4.5 Albrecht Dürer, woodcut showing an artist drawing a nude with the *velum*, a perspective device described by the German artist in his *Painter's Manual*, 1525. Metropolitan Museum of Art, New York.

of passivity and stillness. "The viewer's eye can move over the canvas at will, just as it can wander across a landscape painting, with the same kind of sensual pleasure."[38]

As in Alberti's text, in Dürer's illustration, fabric underpins the dialectics of the composition. The rational, masculine, objectifying gaze of the artist is filtered through the vertical gridded screen, through a perfectly flat, geometrical veil. The rectilinear space mapped out by the artist (and by Renaissance episteme) becomes, in Paul Carter's words, "a machine for firing eyesight into space." By conceiving of sight metaphorically as a "spear," this gridded space pierces "the chaotic maze of appearances and marks as it targets the vanishing point where all comes together and the enigma of the environment's multiplicity is commanded and reduced to order."[39] Yet, in trespassing the *velum*, the artist's gaze encounters a second, this time obstructing, veil: the rippled blanket covering the most intimate parts of the horizontal female body, as if it were the woman's (that is, Nature's) last defense.

If the model's blanket half-conceals, the gridded screen is the precondition for "seeing" and reproducing space. Paradoxically, its very flatness is the prerequisite for opening up depth, that is, for enabling the artist to create the illusion of three-dimensionality. The result is a staged, theatrical space. Staging and theatricality were the hallmark of Renaissance landscape in its various manifestations—from the paintings and illusionistic frescoes decorating the walls of Andrea Palladio's villas to the physical landscape of Italian piazzas and of the Venetian countryside, tamed, staged, and framed as it was through porches, windows, and colonnades for the visual consumption of patrician landowners.[40]

Appropriated by the British architect Inigo Jones, who had himself spent study periods in Venice and become acquainted with Palladio's work, this perspectival "way of seeing" informed the design of his scenes for Ben Jonson's London court performance *Masque of Blackness* (1605). The play is thought to be the first to make use of the Albertian central-point perspective on the British stage and thus create the illusion of depth. The lines of perspective converged in the eye of the king, who sat on his throne in a privileged elevated position opposite the stage upon which a cast of beautiful court ladies and the queen herself performed, masked in elaborate dresses and makeup. Significantly, the play also features the earliest use of the word "landscape" in the scenic sense, as well as its unusually early identification of landscape scenery with nature.[41] In the play's sequel, *The Masque of Beauty* (1608), Germinatio, a female mask representing the fertility that Nature brings to the earth, is clothed in the colors of spring. According to Kenneth Olwig, like the landscape scenery, Germinatio

> provided a means of portraying the generative powers of nature. Natura was traditionally depicted as a modest goddess who did not like being gazed upon or having men tear the fabric of her garment (the physical universe). In the masques, on the other hand, nature becomes the object of a gaze that verges on the pornographic.[42]

In Jonson's play, landscape was the mantle of Natura, a motif we explicitly encounter in contemporary Venetian theater writing. In a comedy dating five years before *The Masque of Blackness*, for example, the goal of art is said to be the innocent, playful "imitation of nature," including "painting the mantle of the earth [*il manto della terra*] with charming and beautiful flowers."[43] In Jonson's play, nonetheless, landscape scenery embeds an inherently political quality: it is the charming mask of Britain's body politic commanded by the king's gaze. It is a cover, a mantle veiling divergences and conflict under its harmonious uniform appearance.

The concept is visually rendered on the famous frontispiece of Thomas Hobbes's *Leviathan* (1651), one of the most influential works on statecraft in western European history (fig. 4.6). Written during the English Civil War, the *Leviathan* argues for a strong, undivided government ruled by an absolute sovereign. On the frontispiece, the giant crowned figure of the king, formed of over three hundred tiny persons, emerges from the landscape holding a sword and a crozier, the symbols of temporal and religious power, which are reflected in the pairings of emblems in the lower part of the composition (the castle and the church, the crown and the mitre, the cannon and excommunication, the weapons and logic, the battlefield and the religious courts).

FIGURE 4.6 Frontispiece to Thomas Hobbes's *Leviathan* by Abraham Bosse, with input from Hobbes (1651). The king surveys the landscape from his height, while a white drape theatrically announces the title and author of the book.

Classic Image / Alamy Foto Stock.

The giant holds the symbols of both sides, reflecting the union of secular and spiritual in the figure of the sovereign.

The vertical posture of the king's torso contrasts with the horizontality of the rolling hills and town forming the landscape underneath. Landscape veils the lower part of the king's body, like the white drape on which the title of the work is inscribed. Landscape is a mantle akin to the cloaks wrapping the anonymous figurines that form the body of the king. At the same time, landscape is the "means for making visible the abstract power of the state."[44] Captured from a low angle, landscape is the stage the king gazes upon. And the king, in turn, is the vanishing point of the landscape. Journeying inward into the composition, crossing drape, town, and nature, the inquisitive gaze of the viewer eventually converges with the penetrating gaze of the sovereign.[45]

The Opening of the Human Body

In Renaissance Europe, the opening of geographical space and of spatial depth were paralleled by the opening of the surface of the human body and the unveiling of its thus far hidden interior. This became possible thanks to the penetrating perspectival gaze and to the study of human anatomy pioneered by Andreas Vesalius (1514–64), professor at the University of Padua and later imperial physician at the court of Emperor Charles V.

Brought to the West by Byzantine scholars roughly at the same time as Ptolemy's *Geography*, Greek medical manuscripts stimulated the study of anatomy, which opened up a largely uncharted territory.[46] However, while Greek anatomical science rested mainly on the dissection of animals, Vesalius and his followers worked with human corpses. Like the oceanic explorers of his time, the physician set out to challenge the authority of the ancients through direct experience, rather than inference. Just as Christopher Columbus upon his return from his first voyage observed that "although there was much talk and writing about [the Indies], all was conjectural, without ocular evidence," so did Vesalius stress the primacy of eyewitness in his exploration of the human body.[47]

Heroic anatomist-explorers dissected human bodies in public anatomical theaters, whose structure often mimicked the concentric model of the Aristotelian cosmos, as if to stress the relationship between micro- and macrocosm, that is, between interior and exterior worlds.[48] Surrounded by large crowds of spectators, they dramatically launched themselves on voyages of discovery of the human body, often borrowing the narrative tropes of oceanic exploration. In the words of the Italian humanist Giovanni Ciampoli (1589–1643), "just as a Florentine sailor ensured the eternity of his name with an America that was unknown to the ancients, a Modenese doctor, with the muscle of Fallopius, inscribed his name within every human being, as the discoverer of this previously unobserved part."[49]

In this pioneering phase of human anatomy, which Jonathan Sawday maps roughly between 1540 and 1640, the interior of the body, like the world's continents, began to be charted and to take its modern shape and features: "Fallopius mapped the female reproductive organs, Eustachius the ear, Realdus Columbus and Fabricius of Aquapendente the venous system, and Michael Servetus the pulmonary transit of the blood." Like the Columbian explorers, Sawday observes, "these early discoverers dotted their names, like place-names on a map, over the terrain which they encountered."[50] These early anatomists' mapping enterprise presented challenges similar to mapping the globe, such as the rendering of a three-dimensional object on a flat surface and its partitioning. Cartographers dissected the earth in continents and regions, and, thanks to Ptolemy's projections, they flattened it on atlas tables. Likewise, anatomists cut up the human body in pieces on the surface of the dissection table and flattened them on the tables of the anatomical atlas (fig. 4.7). In other words, they made a world hidden beneath the mantle of skin accessible through its transposition on a flat surface. What did this paradox hide? What did the mantle of skin hide?

"Skin is everything," Kuryluk observes. Skin holds the different parts of the human body together. Skin wraps and contains the body's physical substance, as well as the soul.[51] Skin is the precondition for human existence. In the Renaissance, piercing the mantle of skin was more than a scientific enterprise: it was a gesture imbued with a deep moral meaning. Penetrating the new, unexplored inner landscapes of flesh somehow meant also penetrating the depths and folds of the human soul; it meant getting to "know thyself." To open the body was a form of self-analysis, and the anatomical performance was ultimately "a lesson in human mortality." It reminded the public that, after all, the body was but a temporary, corruptible garment of the soul.[52] Just as the opening of the Ptolemaic mantle of the earth confronted the self with its tiny place in the face of geographical immensity, so did the anatomist's opening mantle of human skin confront the self with its ephemerality in the face of eternity.

On the title page of his *De humanis corporis "fabrica"* (1543; fig. 4.8), the first anatomical atlas, Vesalius points at the womb of a dissected female corpse lying on the anatomist's table.[53] The atlas was published the same year as Copernicus's *De revolutionibus orbium coelestium*, a notable coincidence of the history of the opening of microcosm and macrocosm. What is depicted on Vesalius's title page is nothing less than "a demonstration of the structural

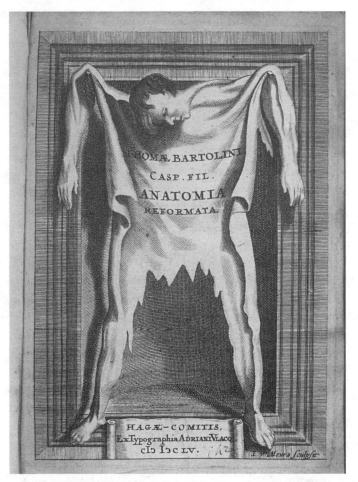

FIGURE 4.7 Frontispiece of Thomas Bartholin's *Anatomia reformata* (1666). The fleshy mantle of the dissected human body lies flat, like the skin of the globe projected on paper by Renaissance map makers. Courtesy of Princeton University Library, Rare Book Division, Department of Books and Special Collections.

coherence of the universe itself, whose central component—the principle of life concealed within the womb—Vesalius is about to open to our gaze."[54] Vesalius's deictic opening gesture is rhetorical; it is directed to the spectator; it invites the gaze into the dark space of the human microcosm, as it is being revealed for the first time.[55] Like the perspectival gaze of Dürer's artist (see fig. 4.5), it invites the eye to penetrate space, to pierce the fleshy *velum* of the female body.[56]

Vesalius's gesture, however, is also admonitory: hidden within the scene, yet in the midst of it, a skeleton hovers over the open abdomen of the dissected

FIGURE 4.8 The opening of the human body: title-page of Andreas Vesalius's *De humani corporis fabrica* (1543), the first anatomical atlas. With his deictic opening gesture, the pioneer of human anatomy reveals the unexplored space of the human body's interior.

Courtesy of Linda Hall Library of Science, Engineering & Technology.

woman. Bone is juxtaposed to flesh; the womb, that is, the origin of life, is brutally juxtaposed to the end of life. The iconography is reminiscent of the scene of the chasing of Adam and Eve from Paradise in the Genesis cycle of the Palazzo Besta, in which the progenitors of humanity are haunted by the skeleton of their own mortality. On Vesalius's frontispiece, the woman's procreative organs are half-hidden yet in the process of being "unveiled"; by contrast, the specter of death faces the spectator in open view, a reminder of the only certitude in human life.

While on Vesalius's title page the dissected body lies passively on the anatomist's table, the following pages are replete with upright, self-demonstrating figures lifting up and holding their own skins. These curious figures multiplied over the following century. Unlike the horizontal female corpse on the title page of the *Fabrica*, these self-dissecting vertical bodies take on a life of their own. They are no longer unveiled by a discoverer, but stand up on their feet unveiling themselves, opening up their bodies with their own hands (fig. 4.9). They become active agents disclosing their flesh for the spectator, in theatrical self-revelatory gestures comparable to Vesalius's deictic hand. Sometimes the opening of their fleshy garments is echoed and reinforced through the presence of mantles and flying veils (fig. 4.10). What was the purpose of these images? Why animate cadavers? Why represent them as active living beings?

Collaborating with the anatomist, Sawday suggests, the self-opening body of the *écorché* served a double function: it at once proclaimed the truth of anatomical study and acted as a powerful *memento mori*. It was "an instrument of propaganda on behalf of this new discovery of the human world," as well as a reminder of its own fragility.[57] According to Raphael Cuir and Yves Hersant, these discursive, rhetorical figures held an erotic and disturbing dimension particularly perceptible when a female body was offered to the male gaze. They also held a Christian spiritual dimension for, in inviting meditation over death, they ultimately turned the viewer's gaze toward God.[58] Flayed bodies contributed to the presentation of anatomy as a reflexive practice: the vertical self-dissecting figure literally confronted the viewer "face to face" with death. The self-reflexive gesture, which allowed the *écorché* to delicately lift and peel back its own skin, lay in the ancient doctrine of *Nosce te ipsum* ("know thyself"). "What the device of self-demonstration guaranteed was a literal interpretation of the searching, inward gaze recommended by philosophical self-examination."[59]

In Vesalius's *Fabrica*, self-dissecting bodies are often set against rural landscape backgrounds akin to those of Hans Memling's *Crucifixion* panels (pl. 3). Urban features appear in the distance, setting the *écorché* in a liminal

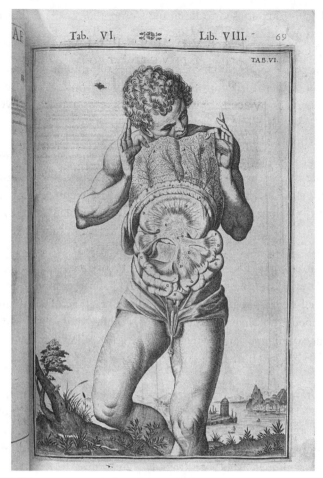

FIGURE 4.9 Self-demonstrating male figure from Giulio Casserio's *Tabulae anatomicae* (1627). The unveiled organs and their parts are mapped with letters of the alphabet.
Courtesy of US National Library of Medicine.

Arcadian space between nature and civilization, between life and death. Other times, these landscapes are peppered with classical ruins, speaking of the mutability of human affairs and the transiency of life. In other works, such as Estienne's *De dissectione partium corporis* (1545), landscape takes up further space, and its features seem to echo the mass of the body. Medical geographer Tom Koch suggested that the function of these landscapes was to remind the reader of the connection between microcosm and macrocosm; in other words, "to remind fellow physicians and students that humans—in health and disease—exist within the environment."[60] At the same time, these engravings mark the convergence of two acts of unveiling: the unveiling of the interior spaces of the human body through human anatomy, and the unveiling of nature through the use of linear perspective.

In the engravings in Giulio Casserio's *Tabulae anatomicae* (1627), such images achieve special sophistication. Dissected bodies reminiscent of ancient Greek sculptures immersed in verdant sceneries lift up their fleshy mantles, unveiling literal "maps" of inner organs (marked with letters, as was customary on maps and especially urban views of the time; see figs. 4.9 and 4.10). Sight, argued the Neapolitan scientist Giambattista della Porta (1535–1615), could reveal everything, provided it observed the world with "sharp eyes,"

FIGURE 4.10 Dissected female figure from Giulio Casserio's *Tabulae anatomicae* (1627). The woman's reproductive system is revealed to the viewer under the opening mantle of her skin and a second mantle of fabric.
Courtesy of US National Library of Medicine.

FIGURE 4.11 Frontispiece of Giulio Casserio's *Tabulae anatomicae* (1627). Two putti lift a drape over the dissection table, revealing the anatomist's tools under the direction of Anatomia (enthroned at the top between Ingenium and Diligentia).
Courtesy of US National Library of Medicine.

that is, "eyes sharp as the anatomist's scalpel, a cutting gaze that lays bare."[61] Both instruments are presented on the title page of Casserio's *Tabulae*: at the top of the composition, Ingenium, flanking Anatomia (in the center) and Diligentia (on the left), points at her own eyes; in the bottom section, scalpels, scissors, and the other tools of dissection are disposed on the operating table and uncovered by a drape lifted by two putti (fig. 4.11). The theatrical structure of the frontispiece bears a striking resemblance to that of Hobbes's *Leviathan* (see fig. 4.6)—only here the body of the king is replaced by Anatomia and his tools of governance by the surgeon's tools for dissection.

Confronted with the penetrative gaze of the scientist (and of the artist), nature herself spontaneously lifts up her veil, like Casserio's dissected bodies. The very moment landscape and human body are mastered and sketched through linear perspective, they become reified—paradoxically, they lose their veil.

Unveiling the Theater of the World

The spatial arrangement and iconography of the frontispieces to Hobbes's and Casserius's works reflect a distinctively Renaissance theatrical culture exacerbated by Baroque taste. Framed by velvety curtains and heavy drapes, the perspectival space of the Renaissance theater was, like that of the anatomical theater, a space charged with a deep moral meaning. By setting the spectator at a distance from the stage, the theater provided the rational detachment necessary to attain wisdom and the knowledge of truth. The *théatron* (literally "place for viewing") turned the viewer (*theatés*) into a sort of *theós*, or "divinity."

Extended to the global scale by way of the great Renaissance atlases, which often featured the word "theater" in their title, the metaphor turned the world into a stage for the lives, works, and salvation of its human inhabitants, as witnessed from an elevated point above its surface in flux.[62] Abraham Ortelius's pioneering use of the term in his *Theatrum orbis terrarum* (1570) explicitly connected his work to the ancient tradition of memory theaters, or devices for memorizing geographical information.[63] However, it also carried a more profound meaning: it coincided not only with the opening of geographical space through oceanic discoveries, but also with the "unveiling" of the self in the best neo-Stoic tradition.

Neo-Stoicism was a Renaissance philosophical current that owed to humanists' study of Cicero and other Latin authors and acknowledged that the wise man must be indifferent to mundane affairs, focusing instead on the universal unity and order that are the chief manifestations of Divine Providence. This could only be prompted by the realization of one's insignificance in face of eternity and of the vastness of the expanding world. For example, Ortelius inscribed his *Typus orbis terrarum* (the world map opening the *Theatrum*) with the Ciceronian citation "For what can seem of moment in human affairs for him who keeps all eternity before his eyes and knows the scale of the universal world?" The motto aptly appears below a vast, "still unknown" Antarctic continent (*Terra nondum cognita*).[64] If the anatomical performance was a *memento mori*, the geographical atlas was a memento of the limits of human knowledge.[65]

In Renaissance atlases and other geographical works, fabric took the liminal place that skin held in human anatomy. Or rather, maps became the skin of the world. Regardless of the medium on which they were painted or printed, late sixteenth-century world maps retained the original connotation of the word *mappa*, if only metaphorically. The Tuscan polygraph Tomaso Porcacchi (1530–85), for example, referred to the world map in his book of islands as *mappamondo*, "as though to mean a table or tablecloth [*tovaglia*] upon which all the places of the world are set."[66] Like the anatomist's dissection table, the geographic table and tablecloth both reduced the world to a flat surface upon which locales were arranged; they implied a superficial and synoptic vision. The tablecloth was analogous to the mantle in that it offered its surface to the eye, while concealing a solid yet invisible structure (the table, or the interior of the planet).[67]

On the maps of the great Renaissance atlases, precious silks and colorful velvets dressing exotic people form theatrical curtains opening onto the stage of the distant lands they inhabit.[68] Notable examples are found in the Mercator-Hondius *Atlas* (1616), in Blaeu's *Theatrum orbis terrarum sive Atlas novus* (1645), as well as in John Speed's *Prospect of the most Famous Parts of the World* (1631). As with other iconographic devices in Renaissance maps, these figurines and their costumes acted as mnemonic prompts. Opening a visual dialogue between map, frame, and viewer, they helped memorize the places to which they were connected by virtue of their peculiarities. At the same time, by framing the map, they contributed to create a space in which the viewer was raised above the earth's surface and encouraged toward distanced philosophical contemplation.

The most dramatic seventeenth-century symbiosis between curtains and cartography, however, is found not in maps themselves, but in seventeenth-century Dutch painting. For example, in Jan Vermeer's *The Art of Painting* (1666; fig. 4.12), the huge map of the Netherlands hanging behind Clio, the Muse of history, holds an astonishing textile quality and material presence. It is not a flat paper surface, but a giant drape wrinkling under the glowing light filtering from a window. In this map, art historians have seen "an image of human vanity, a literal rendering of worldly concerns." Its depiction of the northern and southern Netherlands has been interpreted as "an image of a lost past when all the provinces were one"—a melancholy echo of Ortelius's neo-Stoic dream.[69]

What has perhaps received less attention is the textile narrative underpinning the whole composition. The scene is revealed by a heavy piece of drapery with threads hanging loose from its back side. Like a stage curtain, the drapery opens up space. It guides the gaze into the painting via other pieces

FIGURE 4.12 Jan Vermeer, *The Art of Painting*, 1666. A heavy piece of drapery reveals the painter, Clio, and a map of the Netherlands notable for its tactile quality.
incamerastock / Alamy Foto Stock.

of fabric lying on the table and the silky dress of the model, to culminate and disperse it on the rippling surface of the map, itself "presented as a painting."[70] Jan van Staveren, an artist contemporary with Vermeer, likewise unveiled his *Astronomer* (1672) behind a dusty curtain lifted by a grotesque little creature (fig. 4.13). Here the opening curtain is mirrored in the astronomer's opening

FIGURE 4.13 Jan van Staveren, *The Astronomer*, 1672. A dusty curtain reveals the astronomer, as he leafs through the pages of his almanac.
© Sotheby's / akg-images.

gesture, as he leafs through the pages of his large almanac. The figure is set against an ink-black background into which our gaze this time is not allowed to penetrate.

*

The Renaissance has been described as "a time of fascination with interiority, of obsession with the idea of looking beneath the surface and seeing through," a time when "objects needed to be opened, veils torn, the visible incised to find the truth and its treasures."[71] The belief that knowledge means to break the shell to reach the kernel, that is, to pierce the veil, was a product of this culture and of the opening of space enacted at a variety of scales by explorers, artists, philosophers, and anatomists.

The shift in the perception of space explored in this chapter resulted in a dramatic spiritual revolution of which, in Alexandre Koyré's words, modern science was at the same time "the root and the fruit." Some historians, he notes, have seen its most characteristic feature in the "secularization of consciousness," its turning away from transcendent goals to immanent aims. Others have seen it in the discovery, by human consciousness, of its essential subjectivity and thus in "the substitution of the subjectivism of the moderns for the objectivism of medievals and ancients"; still others, in the change of relationship between *theōria* and *praxis*, whose seeds we can find in late Western medieval attitudes toward nature: "whereas medieval and ancient man aimed at the pure contemplation of nature and of being, the modern one wants domination and mastery."[72] Yet, the progressive domination and visual mastery of opening spaces implicitly turned the gaze inward; it led the modern subject to interrogate himself on his place in the world.

For Koyré, these are all expressions of an utterly spatial process—a progressive displacement caused by the Copernican revolution and the "*furor astronomicus*" that followed Galileo's invention of the telescope, when "man lost his place in the world, or, more correctly perhaps, lost the very world in which he was living and about which he was thinking."[73] As the skin of the firmament was pierced by the telescope and the reassuringly self-enclosed space of the ancient Aristotelian cosmos opened up, western Europeans were suddenly confronted with infinity. If the oceanic discoveries of the sixteenth century recast their place in a world more immense they had ever dreamed, by the end of the following century a far more dramatic transformation was taking place. Humans would eventually lose their central place in the universe, to discover themselves as insignificant dwellers of a planet floating in unbounded black spaces. Reduced to machine, creation was no longer confronted through prayer, but agnostically. As Pascal famously wrote, "The silence of these infinite spaces terrifies me."[74] The following chapter explores the unveiling of infinity and of nature's mysteries in the age of scientific revolution.

5

Drapes, Lights, and Shadows

In the seventeenth century the dialectics between invisibility and visibility activated by the act of unveiling underpinned not only geographical discovery and human anatomy, but also natural philosophy and the universe at large. The rise of experimental science and the improvement of scientific instruments gave rise to the belief that the human mind could penetrate the secrets of nature and thus, in Hadot's words, "raise the veil of Isis."[1] The verb "to unveil" entered the English dictionary as late as the 1590s (with the meaning of "to make clear"). By the 1650s it had acquired the sense of "to display" or "to reveal." Francis Bacon himself talked about "taking off the veil from natural objects, which are commonly concealed and obscured under the variety of shapes and external appearance."[2] As the English philosopher sketched his program of modern experimental science, the piercing Promethean gaze took over the contemplative Orphic look. Slowly, nature fell from her rank of goddess to become a machine governed by an external Creator.[3]

The Promethean gaze was not confined to the earth's mantle. It was also directed upward, to the dark depths of the universe. If in the sixteenth century the geographical discoveries had led to the opening of global space and the disruption of the closed cartographic order of the medieval *mappa mundi*, the scientific and philosophical revolution that took place over the following century brought forth the disintegration of the Aristotelian cosmos and the advance of a mechanistic world order. New things started to appear in the profundities of an enlarged universe. Through the lens of his telescope, Galileo discovered stars and, he believed, planets never seen before. Disturbingly, he also observed spots on the sun and the moon, which challenged ideas of incorruptibility traditionally ascribed to those heavenly bodies and to the rest of the extraterrestrial cosmos. In Marjorie Hope Nicholson's evocative words, "man looked up at night to skies both familiar and strange."[4] As the Aristotelian "circle of perfection" was shattered, the conception of the cosmos as a finite, closed, and hierarchically ordered whole was gradually replaced by one of "an indefinite and even infinite universe bound together by the identity of its fundamental components and laws, and in which all these components were placed on the same level of being."[5]

In 1620, the publication year of Bacon's *Novum organum*, Galileo's telescope had been in existence for less than a decade, and microscopes were circulating in western Europe. Empowered by technology, the Albertian winged eye was thus enabled to transcend human space and penetrate an infinitude of new worlds beyond direct visual reach. It became omnipotent. More than ever before, attentive observation was deemed the precondition for the acquisition of solid knowledge of the world and of the cosmos. In Peter Hulme and Ludmilla Jordanova's words, "Enlightenment was less a state than a process of simultaneous unveiling and observation. To look well and carefully, sufficient light was required, and looking in this way was deemed the only route to secure knowledge."[6] Upon this premise, an entire epistemology was built.

Unsurprisingly, over the seventeenth and eighteenth centuries images of theatrical unveilings became among the most popular subjects on the frontispieces and title pages of scientific treatises, atlases, and other compendia—from dramatic unveilings of Isis to heavy mantles and drapes solemnly raised over terrestrial and celestial globes, and over the open horizons of human knowledge itself.[7]

Such images have been generally interpreted as visual metaphors of a mechanistic worldview progressively revealed, or "brought to light," by the inquisitive eye of the scientist, in which Isis/Nature is identified with Truth. With their *chiaroscuri* and *trompe l'oeil* illusions, however, these drapes and veils can also be taken as metaphors for an age and an approach to knowledge caught between the metaphysics of the late Renaissance and the empiricism of Enlightenment inquiry. More specifically, the veil gives visual expression to the dialectics between light and darkness underpinning Baroque cosmographic discourse. It is but an illustration of an emblematic culture of "metaphysical enlightenment" in which knowledge was simultaneously "hidden and arcane and illuminating."[8] At once revealing and concealing, floating between beams of light and smoky clouds, between the obscurity of the night and the clear daylight, cosmographic veils and drapes acted as dynamic interfaces between the known and the unknown.

This chapter explores the visual and metaphorical interplay of light and shadows displayed through these textiles throughout the seventeenth and eighteenth centuries. In particular, it focuses on the role of vision and light in experimental science, cosmography, and the Enlightenment as spatial approaches to a still unfolding earth's mantle.

The Artificial Eye

We should also take into account that many things in nature have come to light and been discovered as a result of long voyages and travels (which have been more frequent in our time), and they are capable of shedding new light in philosophy. Indeed it would be a disgrace to mankind if wide areas of the physical globe, of land, sea and stars, have been opened up and explored in our time while the boundaries of the intellectual globe were confined to the discoveries and narrow limits of the ancients.[9]

In a famous passage from Bacon's *Novum organum*, quoted above, scientific discovery is likened to the discovery of unknown regions of the globe. The earthly globe traversed by increasing numbers of intrepid explorers and travelers is but the exterior projection of the modern subject's "intellectual globe." In the *Organum*, Bacon sought to establish a new method of natural philosophy that would avoid the scholastic drawback of interpreting everything in Aristotelian terms, as the old doctrine was being increasingly contradicted by new discoveries.[10] According to the English philosopher, "The experiment itself shall be the judge of the thing."[11]

Bacon's use of geographical imagery to describe the task is not surprising. If in the sixteenth century human anatomists compared themselves to oceanic explorers, in the seventeenth century the language of geographical discovery had ended up pervading science at large. As one of the very few subjects that linked scholarship with craft skill (and theory with direct ocular evidence), geography had itself become a sort of paradigm, or metaphor, for the new scientific method promoted by Bacon. In Livingstone's words, it stood as "a model for the type of investigation which allowed seventeenth-century practitioners to cast off the bonds of scholasticism and enter a new era."[12]

Fittingly, the emblematic frontispiece of the *Novum organum* features a ship passing between two columns (the Pillars of Hercules) and heading toward the open sea. The scene is accompanied by the motto "*Multi pertransibunt et augebitur scientia*" ("Many shall pass through and knowledge shall be increased"), referring to the explorers daring to trespass the limits of the familiar Mediterranean world set by the ancients. A similar iconography is found on the frontispieces of Bacon's other major works, but with some variants. On the frontispiece of his posthumous *Sylva sylvarum* (1627), the two pillars frame a globe struck by a beam of divine light illumining the Atlantic and northern Europe, while on the frontispiece of *The Advancement and Proficience of Learning or the Partitions of Science* (1640, originally published 1605) they support a drape inscribed with the book's title (fig. 5.1) depicted

FIGURE 5.1 Frontispiece to Francis Bacon's *The Advancement and Proficience of Learning or the Partitions of Science* (1640 [1605]): unveiling the expanding horizons of scientific knowledge.
Courtesy of Princeton University Library, Scheide Library, Department of Rare Books and Special Collections.

in the same fashion as those on the frontispieces to Hobbes's and Casserius's works (see figs. 4.6 and 4.11).

As the eye is invited to navigate beyond the pillars, the viewer is reminded that scientific research is an endless journey toward unknown and ever expanding horizons. In the works of contemporary commentators such as John Ray, gaze and intellect are pushed behind the veil of the already known and toward the infinite:

> Let us endeavour to promote and increase this knowledge, and make new discoveries. . . . Let us not think that the bounds of science are fixed like Hercules' Pillars, and inscrib'd with *Ne plus ultra*, [for] Treasures of Nature are inexhaustible.[13]

At once concealing and revealing what has yet to come, the drape hanging between the pillars reinforces the message. Looking at Bacon's frontispiece, however, the viewer is also reminded that learning is a journey for which one needs to be equipped with the proper instruments: at the top of the composition, the bright sphere of the visible world, complete with graticule and a clear outline of the continents, literally "shakes the hand" of the partly invisible intellectual world by means of "reason" and "experience."

For Bacon and his successors, clear vision was a prerequisite for rational knowledge, or, in his words, for "throwing light" upon philosophy.[14] This connection between reason, clear vision, and knowledge still permeates contemporary everyday speech. We can "see" (or not see) the value in certain arguments, "shed light" on a problem, be "suddenly enlightened," have a "bright" idea, or a "vision" for our future. Conversely, our memory can get "a bit cloudy," and we might end up seeing "as through a glass darkly." Why is this the case? Where does the link between sight and knowledge come from? And what do veils have to do with it?

Vision, as Dianne Harris and Dede Fairchild-Ruggles argue, is a powerful sense. "Humans have the ability to control vision and therefore feel empowered in ways that are less available to other senses. Sounds and smells can, for example, pervade spaces in uncontrollable manners, crossing architectural boundaries in ways that images cannot. To avoid a scent or noise in a room, one leaves the space, but to avoid a view, one can simply close one's eyes."[15] The eyelid is a sort of "veil" or "drape," which we can lower and lift at our command, allowing and obliterating views, literally, in the blink of an eye. This ease, Harris and Fairchild maintain, privileges the eye as an organ for analysis since we "own," control, and operate it in a more conscious manner than we can do with other senses.[16]

This simple fact was acknowledged by Aristotle, who elevated sight to "queen of senses" and placed it at the foundation of human knowledge.[17] As we have seen in the previous chapter, the equation between sight and knowledge was further reinforced in the fifteenth century, largely thanks to the mediation of surfaces, both physical and imaginary. The material surface of the Albertian veil enabled the mastery of three-dimensional space; the abstract texture of the Ptolemaic grid allowed the synoptic contemplation of the world. In the seventeenth century, the power of the sovereign eye was further amplified by the lenses of the telescope and microscope assisting the new experimental science. "Without some such Mechanical assistance," claimed the English physicist Henry Power (1623–68), "our best Philosophers will but prove empty Conjecturalists, and their profoundest Speculations herein, but gloss'd outside Fallacies; like our Stage-scenes, or Perspectives, that shew things inwards, when they are but superficial paintings."[18] For Power and his like, telescopes and microscopes enabled the eye to penetrate beyond the surface of reality and unveil orders hidden to the naked eye.

Bacon himself claimed that because sight held primacy over the other senses "as far as information is concerned," it was "the sense for which we must first find aids."[19] And it was indeed largely thanks to the English philosopher that one of these "aids," the microscope, was introduced and justified as an instrument of scientific reform. Of the newly invented device Bacon enthusiastically wrote: "By remarkably increasing the size of the specimens, [it] reveal[s] the hidden, invisible small parts of bodies, and their latent structures and motions. By [its] means the exact shape and features of the body in the flea, the fly and worms are viewed, as well as colours and motions not previously visible, to our great amazement."[20] The microscopic lens was thus envisaged as a prosthesis, or an "artificial extension" of the natural eye: "it did not distort vision, but rather corrected, improved and extended it. . . . [It conformed] to the same optical principles as the eye itself, essentially collapsing the distinction between body and instrument, natural and artificial."[21]

Bacon regarded microscopes and telescopes as "privileged instances" that "open doors or gates." Like the anatomist's scalpel, these optical devices were deemed akin to keys unlocking totally new worlds. As Dodd notes, "the metaphor of using these instruments to open doors or gates from the space of the visible everyday into the invisible and distant was a particularly useful one, since it functioned to present artificial observations as relatively unproblematic and direct forms of witnessing the world as it really was, countering enduring skepticism about the authenticity of artificial vision, and the nature and reliability of vision itself."[22] "Unveiling" metaphors, of course, bore similar connotations. Thus, when later in the century naturalists from Italy, the Netherlands, and England began to use microscopes to study biology, the theme of "Philosophy unveiling Isis" came to play a central role in the illustration of scientific books.[23]

FIGURE 5.2 Frontispiece to Antoni van Leeuwenhoek's *Anatomia seu interiora rerum* (1687). Isis (Nature) reveals herself under the lens of the microscope of Scientific Inquiry.
Courtesy of Linda Hall Library of Science, Engineering & Technology.

Almost all of the frontispieces of works by the famous Dutch microscopist (and discoverer of bacteria) Antoni van Leeuwenhoek featured unveiling themes. His *Anatomia seu interiora rerum* (1687), for example, is opened by an image of Isis unveiling herself as she holds her veil in her left hand, aided by an old man, perhaps Father Time (fig. 5.2).

Isis/Nature appears with her traditional attributes: the cornucopia, which she holds in her right hand, and her multiple breasts. On the viewer's left, under her flowery mantle, Philosophy (or the Science of Nature) holds a book emblazoned with the image of the Sphinx (according to Hadot, a symbol of perspicacity that unveils the secrets of nature). Philosophy points out with a wand to Scientific Inquiry the specimens overflowing from Nature's cornucopia: flowers, a snake, a frog, a butterfly. Inquiry eagerly looks at them through the lens of her key-shaped microscope, while drawing the shapes "unlocked" to her through the artificial eye.[24]

Iconographic elements such as the veil, Inquiry's keylike microscope, and Philosophy's wand are also found on the frontispiece of *Arcana naturae detecta* (1695; fig. 5.3). This image deserves special attention, not only because it introduces Van Leeuwenhoek's scientific *summa*, but also because, perhaps more than any other, it lays bare the aforementioned dialectics between visibility and invisibility, and the symbiotic alliance between sight, light, and scientific knowledge typical of the time.[25] The entire composition is articulated through a powerful play of *chiaroscuri* mediated through different sorts of fabrics and surfaces. A beam of light originating from the large disembodied Eye of Providence pierces the thick curtain of clouds of ignorance and illumines Inquiry, as she eagerly peeps through the lens of her microscope. Empowered by her optical device and instructed by Philosophy (which Van Leeuwenhoek calls "queen of sciences"), Inquiry is able to penetrate mysteries of a Nature half-hidden in the shadow by a heavy, velvety curtain. "Naked Truth" (who sits underneath) is thus revealed and made to shine "under the sun," as she tramples ugly Envy.

The act of scientific unlocking expounded by Bacon is here simulated through a dialogue between opening curtains of clouds and fabric. The veil half-concealing the mysteries of Nature and the tablecloth upon which specimens are set for scientific inspection conjure up a sort of miniature theater, a space for scientific and moral instruction.[26] In unveiling Nature, the scientist's inquisitive eye turns her into spectacle.

In the engraving, Inquiry features as a divine emanation. A diagonal visual axis departs from the Eye of Providence, hits the artificial eye of the microscope, and invisibly extends to the sun of Truth, indicating that, with the aid of the microscope, Inquiry brings reality to light, while blind Error (portrayed in the background) stumbles in the darkness. God and Inquiry are joined compositionally through light, and iconographically through sight. The giant Eye of Providence is refracted and multiplied on Inquiry's garment, as though to suggest that vision is a divine sense and interface between the human mind and the exterior world. While Van Leeuwenhoek's Eye of Providence floating

FIGURE 5.3 Frontispiece to Antoni van Leeuwenhoek's *Arcana naturae detecta* (1695). Guided by the light of Divine Providence and dressed in a garment covered with eyes, Inquiry penetrates the dark mysteries of Nature through the lens of the microscope.

Courtesy of Linda Hall Library of Science, Engineering & Technology.

in the sky is but one of many single, unblinking eyes looking down panopti-
cally over the contents of many early modern frontispieces, Inquiry's "eyed"
garment strikes us for its peculiarity.[27] In a way, it is reminiscent of the "fleshy
garment" of Argus, the giant of Greek mythology covered with one hundred
eyes, who appeared a few decades earlier on the frontispiece to Giovanni Battista Riccioli's *Almagestum novum* (1651; fig. 5.4).[28]

In this complex allegory of contemporary astronomical knowledge, we see
Argus as he observes the heavens with a telescope, and, on the opposite side of
the composition, Astraea, or *Dikē*, the goddess of Justice, dressed in a starry
garment. Astraea weighs Copernicus's heliocentric model in a balance against
Riccioli's modified version of Tycho Brahe's geo-heliocentric system, in which
the sun, moon, Jupiter, and Saturn orbit the earth, while Mercury, Venus, and
Mars orbit the sun.[29] Riccioli (1598–1671) was an Italian astronomer and Jesuit priest. His *Almagestum novum*, an ambitious encyclopedic work of more
than fifteen hundred folio pages covering every astronomy-related subject of
his time, provided the first explicit justification of Galileo's condemnation.
Among other things, the *Almagestum* included an extensive discussion on
competing cosmological systems. Here Riccioli scrutinized scores of astronomical, physical, observational, and philosophical arguments in favor of and
against the earth's motion. "By one count, no fewer than forty pro-Copernican
arguments were stated and criticized, and no fewer than 77 anti-Copernican
arguments were presented and developed. As a result of such critical examination, there was no doubt for Riccioli that the earth was motionless."[30]

On the frontispiece to the *Almagestum*, Justice's balance therefore tips in
favor of Riccioli's "Tychonian system."[31] Rendered obsolete by the telescope's
discoveries, the old Ptolemaic geocentric model lies discarded on the ground,
while the Alexandrian astronomer promises to rise again, if corrected
("*Erigor dum corrigor*"). As Riccioli himself clarifies in the dedicatory epistle,
"The ocular of the instrument is placed over Argus' knee, where an eye opens,
to mean that the scientist should always kneel down in the presence of God
and keep a reverent attitude."[32] Although subdued to biblical authority, Argus's fleshy mantle of eyes reflects the exhilarating powers of the telescope
in much the same way as Inquiry's garment of eyes celebrates those of the
microscope—as though the new ocular instruments "multiplied" the power
of vision to the infinite. As Joseph Addison wrote in his *Oration in Praise of
the New Philosophy* (1693):

> We not only have new heavens opened to us [thanks to telescopes], but we also
> look down on our earth, . . . where, by the help of the microscopes, our eyes are
> so far assisted, that we may discern the productions of the smallest creatures,

FIGURE 5.4 Frontispiece to Giovanni Battista Riccioli's *Almagestum novum* (1651). Argus, the mythical giant covered with one hundred eyes, pierces the veil of heaven through the telescope, while Astraea weighs Copernicus's heliocentric model against Riccioli's.

Courtesy of Linda Hall Library of Science, Engineering & Technology.

while we consider with a curious eye the animated particles of matter, and
behold with astonishment the reptile mountains of living atoms. Thus our
eyes become more penetrating by modern helps, and even that work which
Nature boasts for her master-piece is rendered more correct and finished. We
no longer pay a blind veneration to that barbarous Peripatetic jingle, those
obscure scholastic terms of art, once held as oracles, but consult the dictates of
our own senses, and by late invented engines force Nature herself to discover
plainly her most hidden recesses.[33]

Cosmography's Lights and Shadows

The frontispiece to Riccioli's *Almagestum* is articulated through a dramatic
tension between light and darkness, day and night, sun and moon, Argus's
eye-covered skin and Astraea's starry mantle. The former "belches forth the
word" (*eructat verbum*); the latter "reveals [divine] knowledge" (*indicat scientiam*) (Ps. 19:2). This theatrical play of *chiaroscuri* was one of the overriding
themes and tropes of Jesuit iconography. More broadly, it was a key feature
of seventeenth-century Roman Catholic cosmographic imagination. Mirrors
reflecting the light of redemption over celestial and terrestrial globes populate the ceilings of Baroque churches and the pages of Jesuit scientific works
alike. Split between the daylight and the dark night in patterns akin to the
Almagestum's frontispiece, these representations are traversed by shafts of divine light reflected by mirrors and lenses, while smoky clouds open celestial
space to speculative vision.[34] They betray the keen interest that many in the
order took in the issues raised by the new scales of nature that telescopic and
microscopic lenses opened to vision. Even before that, they convey Jesuits'
special fascination with illumination.

Such fascination is epitomized in Athanasius Kircher's *Ars magna lucis et
umbrae* (1671), "The Great Art of Light and Shadow."[35] Kircher was professor
of mathematics in Rome's Jesuit College and one of the most influential poly-
maths of his time. He designed numberless devices and machines, including
magic lanterns, "speaking statues," and magnetic toys. He studied Egyptian
hieroglyphs, wrote about music, and pioneered the study of germs and geology, among other things.[36] He also wrote about optics. His engagement with
light and lenses, however, transcended Baconian transparency. The wonderful works of light and shadow that Kircher expounds in the *Ars* disrupt normal visual expectations and certainties by means of optical illusions. According to Stuart Clark, Jesuits' attempts at unsettling the viewer had a religious
goal: stimulating questioning and inquiry, which would, in turn, dispel error
and superstition, "enlighten" the viewer, and thus reinforce true faith.[37]

Spreading the light of faith and dispelling superstition was indeed the self-declared mission of the Church in the Baroque period. As Cosgrove noted, such mission was global in scope: "to enlighten the spaces of European discovery, rendering the globe itself a symbol of belief."[38] The planetary outreach of "Catholic light" informed and structured the great cosmographic projects of the time. Encompassing the integrated study of earthly and celestial spheres, Renaissance cosmography offered a framework bridging the scales of locality, of the terrestrial globe, and of the universe intended as a perfect closed system. By the end of the seventeenth century, however, as the order of this universe was being shattered, cosmographers were struggling to cope with the increasing tension between their unifying framework and the continuously expanding horizons of human knowledge and geographical data. As in natural philosophy, in cosmography the omnipresent veil became, once again, a metaphor and an expression of this tension.

Flying veils, airborne banners, opening stage curtains, and heavy drapes hauled up by winged putti were ubiquitous presences in atlases, illustrated travel accounts, and other geographical or cosmographic works. In some cases, maps themselves were represented as drapes or stage curtains. The purpose of these veils was as manifold as their shapes. They could be used to build up anticipation, to signpost a journey, to convey the excitement of unfolding discovery, to trumpet the progress of a military campaign, to celebrate the advancement of geographical knowledge, or simply to displace the viewer. In all cases, these devices gave visual shape to the dialectic between light and darkness, between contraction and expansion at the heart of Baroque imagination. Their function was primarily rhetorical: like the map, the veil conceals and reveals; it is a metaphor for and a mediator between what can be seen and what is hidden from us. The veil is a thin layer that separates the present from the past and the future; it is a membrane that divides, and yet at the same time also dynamically connects, our world and other worlds.

In a way, the veil can be taken as a symbol for an early modern epistemology suspended between the "exoteric, empirical experimentation and discovery we associate with the Scientific Revolution" and "the still vital tradition of imaginative, esoteric speculation."[39] Among other things, the work of Kircher encompassed Ptolemaic celestial journeys, a geography of China, and a chorography of Latium. More singularly, it penetrated the surface of the earth, guiding the eye into the black spaces of unexplored subterranean worlds.[40] In other words, it invited the reader to follow the horizontal diffusion of Jesuit light throughout the globe, while at the same time venturing into the dark folds of the earth's mantle.

On the frontispiece of Kircher's *China illustrata* (1667), a drape featuring a map of China and the Far East at once screens and reveals the distant landscapes of that part of the world under the enlightened gazes of missionaries Matteo Ricci (on the right) and Johann Adam Schall (on the left) (fig. 5.5).[41] This "cartographic veil" is held by a flying putto, aided by the two Jesuit missionaries. Their figures are illumined by shafts of divine light radiating from a common source inscribed with the insignia of their order (IHS).[42] Above them, in an almost mirroring fashion, the ecstatic figures of Ignatius Loyola and Francis Xavier (the founders of the order) contemplate the divine light parting the clouds and choirs of cherubs. Staged on an ancient Greek proscenium, the scene speaks the rhetoric of Jesuit missionary enlightenment—"spreading the light of the Christian Empire across the surface of the earthly globe."[43] At the same time, it also speaks of Kircher's conception of global space and history.

In spite of his missionary aspirations, Kircher never traveled to China. Based in Rome, the heart of Catholicism, he nonetheless acted as a clearinghouse for information of all kinds. Thanks to an extensive network of missionaries, including Schall, Kircher gathered unprecedented numbers of texts and artifacts from the farthest reaches of geographic space and historic time, to the point that his collection of natural and ethnographic wonders became one of Baroque Rome's main attractions.[44] His life project was to form a syncretic understanding of creation's logic from the diverse materials that made up his collections and to trace a universal history of "global illumination"— a history of radiant truth and dark superstition, and their dissemination through space and time.[45]

China illustrata was part of this project. It was not a mere account of a country and its natural and ethnographic curiosities, but part of Kircher's spatialized universal history. The Jesuit believed that all polytheistic civilizations, including the Chinese, spread from ancient Egypt, the land of the descendants of Ham, the son of Noah who, according to Kircher, corrupted Adamic wisdom by combining it with the superstitious arts derived from Cain.[46] The initial parts of *China illustrata* thus offer a condensed account of "the introduction of Luminous Religion in China" triggered by the discovery of an eighth-century Nestorian monument in the northwest of the country. The rest of the book features a discussion of the conversion strategy of the Jesuit mission founded by Ricci in 1589, a long comparative survey of "dark" idolatrous religions across Asia, and an account of the marvels of Chinese nature and art.[47] Anticipating this wondrous journey to the East, the cartographic drape on the frontispiece mediates between different layers of time— and between light and shadow.

FIGURE 5.5 Frontispiece to Athanasius Kircher's *China illustrata* (1667), depicting the Jesuit missionaries Adam Schall and Matteo Ricci as they hold China's "cartographic veil."
Courtesy of Linda Hall Library of Science, Engineering & Technology.

In this image Ricci and Schall, the spreaders of light to China, occupy the standard place that allegories of Father Time and trumpeting Fame would take on later frontispieces. Their cartographic veil acts as a threshold of the Jesuit geographical imagination: it is an interface hiding and at the same time granting access to new exotic realms, bridging past, present, and future. The cartographic drape conflates object and metaphor in a way akin to the fleshy mantle peeled off the dissected human body and imaginatively flattened by the anatomist like a piece of fabric (see fig. 4.7). It is a reminder that while the "continuous disclosure of an ever vaster and more varied globe filled the cabinets of curiosities of seventeenth-century scholars and princes," much of the globe's geography remained "obscure, hidden, esoteric"—like the inner topographies of the human body and those gradually unveiled by the lenses of the microscope and telescope.[48]

Conflating and confusing map and drape, Kircher's *trompe l'oeil* expresses metaphysical truths by way of paradox; "complete deception in the service of utter veracity."[49] As with the illusions of light and shadows expounded in the *Ars magna lucis* (or, indeed, through Kircher's innumerable machines and illusionistic devices), the cartographic veil helped feed the diffused sentiment that "the uncertainty or duplicity of appearances was symbolic of the inconstancy of the world and the fecklessness of its inhabitants, when compared to the unchanging, immaterial deity and the eternal truths of religion."[50] As Kircher explained, "nature often paints images unto natural things."[51]

The cartographic drape, however, also fulfills a more mundane function—that of the frontispiece. As physical spaces, frontispieces and title pages, like stage curtains, build up expectation. They stir imagination. They anticipate the new worlds to be unveiled in the following pages. In a sense, frontispieces and title pages are themselves curtains—paper curtains, or mantles, lifted by the reader's hand as it impatiently turns the pages. Elsewhere, opening stage curtains, flying veils, drapes, and banners signpost the entire narrative of the book, as though to renew the sense of discovery and expectation at the turn of each page—like different acts of the same drama. In contemporary illustrated travel accounts, for example, we find title pages featuring heavy drapes lifted by trumpeting Fama over expanding horizons and echoed in the following pages by flying veils and blankets inscribed with page titles and maps.[52] The function of these veils is at once informative and rhetorical: appropriately, the "God's eye" view provided by the map comes from heaven by means of winged messengers—the same liminal figures marking the edges of the Ptolemaic maps of the previous century.

The most original usages of illusionistic "cartographic veils" and theatrical bird's-eye views were pioneered by the Venetian Franciscan friar Vincenzo

Coronelli (1650–1718), the official cosmographer of the Venetian Republic and surely one of the most prolific and original experimenters with maps, drapes, and veils of his time. Famous for the two monumental globes he built for the French king Louis XIV and for his vast production of geographical engravings (more than seven thousand in total), Coronelli was the founder of the Accademia degli Argonauti (1680), Europe's first "Geographical Society." It was under the society's patronage that Coronelli's enormous output of geographical texts, globes and maps was published.[53]

Characteristically, the emblem Coronelli designed for the Accademia features trumpeting Fama unfurling a banner inscribed with the motto *Plus ultra* ("Farther beyond") over the globe, while geographical and navigational instruments decorate the image, in turn represented within a giant unfolding drape. Within the image, "a large ship tops the globe's graticule, while Hercules' club and bearskin cast below suggest that the Moderns have superseded the great Classical heroes, passing beyond the Pillars named for him and reaching the invisible ends of the earth."[54]

This rhetoric of progressive global discovery permeates the pages of Coronelli's monumental *Atlante Veneto* (1690–97), a thirteen-volume work originally meant as the continuation of Joan Blaeu's *Atlas maior*. Bringing together ancient and contemporary geographical information, Coronelli conceived of his *magnum opus* as "a geographical, historical, sacred, profane, and political description of the empires, kingdoms, provinces and states of the universe, their divisions and borders, with the addition of all newly discovered lands, and expanded by many newly published geographical maps."[55]

The *Atlante* shares in the cosmographic dream of total knowledge and presents a spatialized approach to universal history; or rather, in Cosgrove's words, it reveals "cosmographic knowledge as a hierarchy of spatial representation proceeding from the divine Fiat, through the order of creation to the scale of individual regions and cities, palaces and rivers."[56] Different scales and perspectives (typically the bird's-eye view and the God's-eye view) are juxtaposed by means of cartographic veils and similar devices, while flying banners, unrolling scrolls, and undulating drapes simultaneously unfold old topographies, new discoveries, and geographical speculations—from the island of Jan Mayen in the North Atlantic to the mysterious sources of the Nile.

Veils and drapes also obscure the past. At the outset of his description of the city of Ravenna, for example, Coronelli observes how "the memories of the foundation of the city are buried under the darkness of a decrepit antiquity, to such an extent that instead of shedding light on the origin through the whiteness of their sheets, authors further adumbrated it with the darkness of

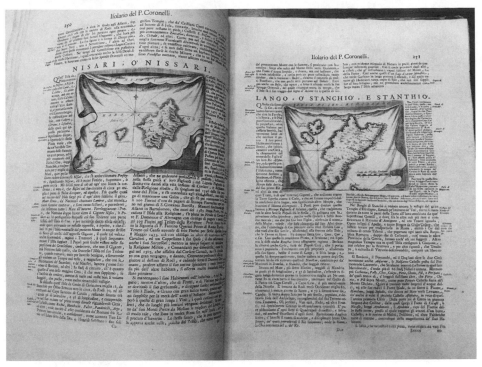

FIGURE 5.6 Vincenzo Coronelli, maps of Nisari and Lango portrayed on napkins, in *Isolario dell'Atlante Veneto* (1696).
By permission of Ministero dei Beni e delle Attività Culturali e del Turismo—Biblioteca Nazionale Marciana, Venice. Reproduction is forbidden.

their ink."[57] It is the cosmographer's task to penetrate this dark layer of misleading ink and bring truth to light. In the accompanying illustration a drape reveals a plan of the ancient city in utmost clarity, with each of its religious buildings carefully mapped out and numbered. Similar metaphors proliferate in the descriptions of the Greek islands, whose glorious memories are likewise "buried" under the veil of oblivion, due to the neglect of their current inhabitants oppressed by the Ottoman yoke.[58] Once home to the Knights Hospitaller but now reduced to ruin, the islands near Rhodes are characteristically portrayed on thin napkins, ready to fly away from the page, like ephemeral memories of themselves and their glorious past (fig. 5.6).

Coronelli's cartographic production and theatrical use of cartographic veils reached its peak in the context of the Ottoman-Venetian War of the Morea (1684–99). The conflict, which saw the creation of a multinational Holy League under the auspices of the Vatican, and the Venetian recapture of the Peloponnese and some of the adjacent islands, stimulated an unprecedented

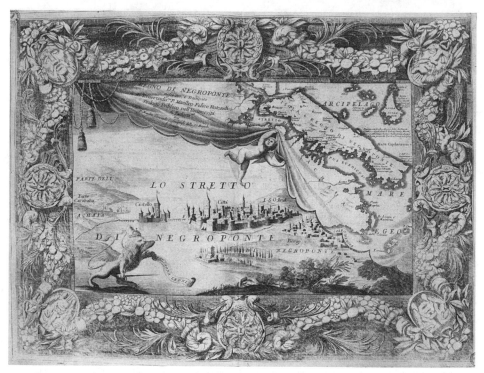

FIGURE 5.7 Vincenzo Coronelli, Regno di Negroponte, in *Isolario . . . dell'Atlante Veneto* (1696). A putto lifts a drape on one of the most strategic sites besieged by the Venetian Republic.
By permission of Ministero dei Beni e delle Attività Culturali e del Turismo—Biblioteca Nazionale Marciana, Venice. Reproduction is forbidden.

flourishing of propagandistic narratives inspired by classical mythology, of which Coronelli was at the forefront. During the years of the conflict, his workshop restlessly produced hundreds of maps and views of forts and territories seized by the Venetians.[59]

In his *Memorie istoriografiche* (1685) and *Teatro della guerra* (n.d.), of which many of the plates were recycled in the *Isolario dell'Atlante veneto*, the Morea appears as a cartographic stage always observed from a distance and mastered by allegories of Venice and by the viewer from a high-oblique angle. The entire conflict is narrated as a crusade, as an unfolding struggle between the expanding light of Christianity and the receding dark forces of Islam. Battles and newly conquered lands rapidly unfurl at the turn of each page, as a succession of thin temporal veils.[60]

Cartographic veils unfold territorial conquest and the progressive expansion of Venetian domains. At the same time, they also act as "curtains" revealing distant lands to the inhabitants of the motherland (fig. 5.7). Behind them, terrified Turks in vain seek refuge from the invincible winged lion of Saint Mark (fig. 5.8). Cartographic representation thus becomes a giant drape revealing both the bright and dark sides of the war, and, literally, turning territory into its own image.

In Coronelli's works, fabric signposts and at the same time adds further dynamism to the visual narrative. Time is constructed as fabric—as a constant process of unveiling triggered by the movement of the reader's hand turning the pages. Characteristically, on his terrestrial globe the Venetian friar chose to portray himself in the process of being unveiled by two little putti, while on the frontispiece to the *Atlante Veneto*, he uses his own body as a "veil." His benevolent gaze is turned to us, the readers, while his right hand, set on a globe, jealously covers the North Pacific, preventing us from seeing what he is pointing at with his compass. His protective gesture marks a liminal moment in Western intellectual history: the close of the great Age of Discovery and of cosmography. From now on, distant regions and their secrets will no longer be unveiled by the omniscient cosmographer, but by Father Time alone.

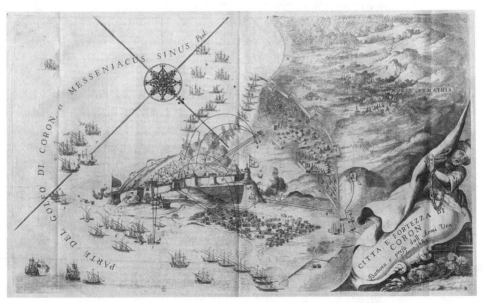

FIGURE 5.8 Vincenzo Coronelli, detail from map "Città e fortezza di Coron," in *Memorie istoriografiche delli Regni della Morea e Negroponte e luoghi adiacenti* (1686). Landscape becomes a giant veil covering the defeated Turk.
By permission of Ministero dei Beni e delle Attività Culturali e del Turismo—Biblioteca Nazionale Marciana, Venice. Reproduction is forbidden.

Unveiling Truth

If the sixteenth-century Ptolemaic *chlamys* opened spontaneously (like the self-dissecting figures in anatomical atlases), Baroque cartographic veils, such as Coronelli's, required external agency to unfold. The task was generally delegated to the putti. In the seventeenth and eighteenth centuries, these little winged creatures were found virtually everywhere: fluttering unto church altars, looping up curtains, looking down from some high pulpit, climbing upon fountains, mourning over funerary monuments.[61] In the pages of illustrated scientific works, we see them unveiling human skeletons, new lands (see fig. 5.7), and even new cosmic orders (see fig. 5.9). A Renaissance invention initially associated with love and playfulness, the Baroque putto came to represent the omnipresence of God. Yet, his seventeenth-century association with "cartographic veils" also signaled a subtle transition in the history of Western thought; it marked a shift toward an increasingly reified and mechanistic world, whose mantle was to be lifted by a detached, external observer.[62]

As the German art historian Wilhem Bode observed at the beginning of the last century, "the putto is at that enchanting age of childhood that is still innocent and without guile, when the consciousness of right and wrong still slumbers."[63] Within atlases and scientific works, the function of the pure, good-natured putto is simply to haul up the veil of truth. The innocent spontaneity of his gesture in a way balances the theatrical deictic gestures of his terrestrial counterparts, including allegories and exotic figurines pointing at places and continents, or scientists pointing at their objects of inquiry.

Take, for example, the frontispiece to Johann Gabriel Doppelmeyr's *Atlas coelestis* (1742; fig. 5.9). In the upper part of the composition, we see two putti unrolling the celestial drape featuring the Copernican model complete with the elliptical orbits of the planets and their satellites. In the lower part of the engraving, Copernicus points at his model in the sky; Kepler points out Copernicus to Tycho Brahe; Ptolemy is absorbed in his own world. The putti act as mediators between earth and heaven. Their presence marks an invisible threshold between the eternal celestial order they are disclosing and the terrestrial stage of human lives upon which the astronomers enact their performances. Between the enigmatic gazes of two sphinxes, an unveiled celestial globe bears witness to the gradual unfolding of human knowledge.

If putti were usually associated with "cartographic" (or cosmic) drapes and curtains, the unveiling of eighteenth-century globes was a duty generally reserved to Father Time, as if to stress progressive revelation, as opposed to sudden change. On Joannes Wolters's edition of *Strabonis Geographia cum notis Casauboni et aliorum* (1707; fig. 5.10), for example, we see Father Time with his

FIGURE 5.9 Frontispiece to Johann Gabriel Doppelmeyr's *Atlas coelestis* (1742). Two putti unroll the celestial drape with the Copernican model complete with the elliptical orbits of the planets and their satellites.
Courtesy of Linda Hall Library of Science, Engineering & Technology.

traditional attributes (the wings and the scythe) as he unveils Europe, Asia, and Africa. Female allegories point at their respective continents on a large globe, while History patiently records the unfolding of events in her book. At the same time, old Father Time conceals the western hemisphere with his mantle, a reminder that this part of the world was still unknown to Strabo.[64]

As with the mantle on Stradanus's frontispiece (see fig. 4.3), the mantle of Time here acts as a cosmographic curtain unveiling the theater of the world.[65]

FIGURE 5.10 Frontispiece to Joannes Wolters's edition of *Strabo's Geography* (1707). Father Time unveils
Europe, Asia, and Africa, but conceals the New World, as unknown to Strabo.
Library of the Holy Monastery of Docheiariou, Mount Athos.

Yet, while the former exalted the event (the discovery of the Americas as an
enterprise ascribed to Columbus and Amerigo Vespucci), the latter stresses
the slow unfolding of history. Time's mantle is not lifted all of a sudden, but is
gradually pulled aside by the measured gestures of the old man. Like nature,
knowledge of the globe is not given once and for all; it is rather "a process that
unfolds in time and is revealed to man only gradually and partially."[66] As with
Isis's opening veils on the frontispieces of Van Leeuwenhoek's books, here the
veil ultimately uncovers truth, in this case geographical truth. Or rather, it
reminds the reader of humankind's long march toward such truth.

The iconography of these engravings draws on a rich tradition of representations of the ancient motto *"veritas filia temporis"* ("Truth is the daughter of Time") and its corollary "Time unveils Truth." Erasmus of Rotterdam ascribed the origins of the adage to the Roman author and grammarian Aulus Gellius. As Fritz Saxl and Erwin Panofsky so richly illustrated, the motto was used by different classical authors and became especially popular during the Renaissance, when it also started to be represented in painting.[67] Over the seventeenth and eighteenth centuries, Time rescuing and unveiling Truth turned into a most fashionable theme of western European art, from Bernini's sculptures to Rubens's and Pompeo Batoni's paintings.[68] In these works, we see the old man continuing in his perpetual flight, as he hastily uncovers the naked body of Truth, now shining brightly "under the light of the sun." Occasionally, we also see him transfixing the masks of lie and hypocrisy with his terrifying scythe.

Francis Bacon, too, used the motto, though with a different meaning. While the ancients and most Renaissance authors ascribed a moral meaning to the formula (that is, that misdeeds will be unveiled in the long run),[69] for Bacon it denoted the progressive unveiling of the secrets of nature thanks to humanity's collective efforts:

> Through voyages and distant navigations . . . numerous things in nature have been revealed and discovered which may spread a new light for philosophy. . . . As far as authorities are concerned, it takes a great deal of cowardice to attribute infinite credit to them and deny its rights to the authorities of the authorities: Time. For it is rightly said, "Truth is the daughter of Time," and not authority.[70]

At the bottom of Walters's engraving, the oval portrait of the French philologist and commentator Isaac Casaubon thus shadows the portrait of Strabo, the ancient authority, amid burning incense, bones, shells, and other exotica brought to Europe from past voyages of discovery. Truth and the geographical space symbolized by the globe are made to coincide. The mantle of the earth is there to be lifted by a humanity ready to challenge and surpass its predecessors.

Elsewhere, Truth, Nature, and the globe are made to converge by way of optical devices—and mantles. On the frontispiece to Martin Frobene Ledermuller's *Amusement microscopique* (1764), for example, we see the goddess unveiled by Philosophy while sitting astride a globe partly veiled by her skirt (fig. 5.11). Above them are the busts of the pioneers of applied microscopy, including Van Leeuwenhoek. As on the frontispiece to his *Arcana* (see fig. 5.3),

FIGURE 5.11 Frontispiece to Martin Frobene Ledermuller's *Amusement microscopique* (1764): Isis unveiled by Philosophy sits astride a globe partly veiled by her skirt.
© The Royal Society.

here Philosophy's unveiling act is mirrored in the sky, as golden rays of glory part the clouds and reveal the omniscient Eye of Providence.

As Hadot observed, in the second part of the eighteenth century many engravings portrayed "the unveiling of Nature as the triumph of the philosophy of the Enlightenment over the forces of obscurantism," which was later to become one of the favorite themes of the French Revolution. In many of the later representations, Isis/Nature is denuded by a stout philosopher, "an impetuous champion who overturns Despotism" and crushes under his feet the masks of hypocrisy and lie, as Father Time did in the paintings of the previous century (fig. 5.12).[71]

The role of the hero-philosopher unveiling truth and liberating the human mind from the darkness of superstition and scholasticism is perhaps best outlined in Jean d'Alembert's "Preliminary Discourse" to Denis Diderot's *Encyclopédie* (1751), one of the Enlightenment's most important texts.[72] D'Alembert, a renowned mathematician, identified the model of enlightened philosopher in the very figure of Bacon. The English philosopher, d'Alembert tells us, was born "in the depths of the dark night," yet his mind was able to "see things on a large scale," to the point that "nothing concerning man was a matter of indifference to him: knowledge of economics, morality, nature, politics, in fact everything seemed to have fallen within the province of this luminous and profound mind."[73] Thanks to Bacon, continues d'Alembert, the natural sciences were systematically mapped in different branches for the first time; an immense catalogue of what remained to be discovered was compiled; and scientific experimentation was pioneered.

Bacon's initial efforts were followed by those of other enlightened philosophers.[74] "Without desiring to tear the blindfolds from the eyes of their contemporaries," d'Alembert writes, these individuals "worked silently in the remote background to prepare the light of reason which gradually and by imperceptible degrees was to illuminate the world."[75] To these heroic efforts, d'Alembert added his own and Diderot's: "to collect all the knowledge scattered over the face of the earth" and "spread Enlightenment throughout society."[76]

Drawing on the division of knowledge elaborated in Bacon's *Advancement of Learning*, Diderot devised a "genealogical or encyclopedic tree" that brought the branches of knowledge together and indicated their origins and relationships.[77] The frontispiece to the *Encyclopédie* visually summarizes these relationships through a hierarchy of allegories (fig. 5.13).

As he explained, at the top of the composition we see "Truth wrapped in a veil radiant with a light that parts the cloud and disperses them." On her right, we see Reason and Philosophy, "the one lifting the veil from Truth, the other pulling it away." On her left, Imagination "prepares to adorn her." At

FIGURE 5.12 Frontispiece to François Peyrard's *On Nature and Her Law* (1793). Isis is unveiled by a philosopher crushing under his feet the masks of lie and hypocrisy.
© Collections École Polytechnique, Palaiseau, France.

Truth's feet we see Theology, turning her back and waiting for light from on high, while "proud Metaphysics tries to divine her presence rather than see her." Underneath, different disciplines are arranged in rows: History (writing the annals and using Time as her support); below, Geometry, Astronomy, and Physics; a row down, Optics, Botany, Chemistry, and Agriculture; at the bottom, several Arts and Professions (portrayed as male allegories). On the other side of the image, below Imagination, we see the different genres of Poetry

FIGURE 5.13 Frontispiece to Denis Diderot's *Encyclopédie* (1751). Reason lifts the veil from Truth, and Philosophy pulls it away. The veil, however, is now transparent.
© Trustees of the British Museum.

and, in the row below, the Arts of Imitation: Music, Painting, Sculpture, and Architecture.[78]

Somewhat softened by the graceful bodies of the allegories, the act of tearing Truth's veil silently perpetrates the violence of the penetrative Promethean gaze expounded in Bacon's *Organum*. Unsurprisingly, the same violence recurs through the *Encyclopédie*'s entry on "Observation," with its strong emphasis on experimental observation as the key for attaining knowledge. Emphasis is placed on "opening interiors" ("the interior of geography," the "layers of the earth," the anatomist's "opening of cadavers") and on the instruments assisting the gaze to penetrate the secrets of Nature (the astronomer's telescope, the naturalist's magnifying glass, the physicist's microscope).[79] These instruments, the author of the entry tells us, serve not only "to render different objects of observation more concrete, but *to pierce the veil that hides them*."[80] On the frontispiece of the *Encyclopédie*, Truth, however, is no longer wrapped in a heavy drape or mantle, but draped in a diaphanous veil through which her light is made to shine for eternity.

*

The scientific revolution and technological advances that took place in western Europe over the seventeenth century and the intellectual developments of the following century ultimately led to the belief that the secrets of nature were no longer occult forces, but material entities made visible thanks to the microscope and the telescope. "The secrets of nature were finally uncovered," and, as Kepler had predicted, "man [was to become] the master of God's works."[81] Unfolding veils can be taken as visual metaphors for the progress of scientific knowledge and for the gradual "unveiling of truth." Yet, veils were not all the same, and this was far from a linear journey.

In the works of Catholic cosmographers, such as Kircher and Coronelli, unfolding drapes marked the global expansion of Christian light, which they articulated as a battle against the dark forces of Islam and superstition. Epitomizing the strongly tactile quality of Baroque visual experience, these veils and drapes also operated as illusionistic devices aimed at spiritual self-edification. They served as thresholds between the physical reality of creation and the metaphysical presence of the Creator.[82] In this sense, their role was akin to that of the putti with which they were often associated. On Dutch frontispieces to scientific and geographical works, the play between light and darkness was often staged by Father Time's unveilings—whether of Isis or of the globe. For French Enlighteners, the veil of Truth was still to be torn through collective action, yet it had also acquired transparency, as if mysteries had now been made visible to humankind. It is to this diaphanous veil that we now turn.

6

Romantic Veils

Of beams 'tis wove, and dews of morning sky—
From Truth's own hand the veil of poesy.[1]

On a bright morning a young man sets off from his alpine hut to climb the nearby peak. The fresh air blowing on his face shakes off the torpor of the night sleep. His pace is steady and fast, his spirits high. At every step he discovers a new little wonder. Blossoms full of dewdrops pave his way. The morning light wraps him in its warm embrace. As the wanderer progresses on his ascent, a streaky mist rises from the valley. It slowly spreads upward and surrounds him. The young man gradually loses sight of the landscape below; gone are the green pastures, the forest, the river, the village. All of a sudden, his little familiar world seems to be shrouded in a sinister impenetrable veil. But not for very long. Eventually, the sun comes out; it parts and dispels the fog. Although blinded by its dazzling glow, the young man dares to blink, and something extraordinary happens. Right before him, the most beautiful of women mysteriously appears carried by a cloud. Having seen her face, the young man decides that he will no longer speak of her as do the multitudes, who mistakenly believe they can name and possess her; he will rather keep this happiness for and by himself. The woman, however, rebukes him for his selfishness. Accepting her reproach, the young man eagerly embraces his duty to share his gifts with the rest of humanity. The woman thus presents him with the most precious gift of all: the splendid veil of poetry. The mist, the clouds, and the dazzling light have disappeared. The valley is open to view again. The trees, the river, and the houses of the village below all reacquire their crisp contours. The sky is high and bright.

The woman is none other than Truth, and the young man none other than Johann Wolfgang von Goethe (1749–1832). The story is the subject of the magnificent verses of his *Zueignung* ("Dedication"), a poem the German writer placed at the very opening of the first volume of his complete works

(1787). As literary critic Dorothea von Muecke noted, the *Zueignung* is to be understood as a framework for all his poetry and as an introduction to the revelatory power of poetry altogether: "the veil of poetry produces the illusion of a quasi-immediate access to the world. It produces a surface of reflection and refraction that offers an object to vision, rather than signs to be read and deciphered."[2] Goethe's verses can also be read as a statement on the socializing and didactic role of poetry: the poet is an individual gifted with a special sensibility who has the duty to "stretch the veil" over the world and thus enable humanity to look at it with new eyes.

While in the last part of the eighteenth century the understanding of nature and the "quest for truth" remained at the core of scholarly preoccupations, the Newtonian mechanistic worldview became a target of criticism. For Goethe and early Romantics, the secrets of nature were no longer to be "unveiled." Isis's veil was no longer to be pierced by the violent Promethean gaze of experimental science. The reason, Goethe believed, was that she had no veil. Everything was already manifest. "Respect the mystery," cried the poet. "Seek no secret initiation beneath the veil, leave alone what has been fixed," for "Isis shows herself without a veil, but mankind has cataracts."[3] For Goethe, it was not Isis that was veiled, but human eyes. In Hadot's words, nature revealed herself as she was; she consisted entirely of the splendor of her appearance. In order to see her, all one had to do was learn to look.[4]

If Bacon, Newton, and, after them, the French Encyclopédistes believed in the power of science and technology as the ultimate means to lift, or even tear apart, Isis's veil, for Goethe and his like, the veil of poetry and art alone had the power to disclose "absolute being, or unity between nature and man."[5] Unlike science, art and poetry did not unveil hidden laws or structures behind visible phenomena; on the contrary, they allowed one to see what was in broad daylight, yet people did not know how to see. They taught that "what is most mysterious and most secret is precisely that which is in broad daylight, or the visible."[6] While Galileo's telescope and Van Leeuwenhoek's microscope confronted humans with new worlds, Goethe's aesthetic lens confronted the subject with a new way of seeing the world—a world transfigured through the veil of poetic sensibility and poetic language.

As opposed to ordinary language and the dry language of the sciences, the perfected tongue of poetry was believed to "recover the experience of presence and communicate its contents immediately to intuition."[7] By the turn of the century, mantles and veils therefore had less to do with the rational process of scientific discovery and more to do with the bewilderment experienced in the face of mystery.[8] Whether standing on the brink of a dark abyss, on the top of a lofty mountain surrounded by whirls of clouds, on a shore contemplating the golden sun plunging into the immensity of the ocean, or simply lying on the moist grass with eyes fixed on the vault of heaven, the Romantic subject shuddered and rejoiced at the sheer sight of Isis. Her veil did not hide anything. It was not thick and opaque, but diaphanous and luminous, woven, in Goethe's words, "from the morning mist and the light of the sun."

While sharp shafts of light accentuated the folds and ripples of the heavy drapes and mantles on Baroque frontispieces, the veil of poetry emanated a suffused, rarefied glow that quietly illumined the textured surface of the earth. The insistent tactility of the former gave way to the ethereality of the latter—and yet, it was precisely thanks to this ethereal, transcendent glow that unseen details and textures were brought to view. In other words, the poetic veil did not hide, but revealed. Yet, what did it look like, and when did it start to stretch across the earth? Was it the preserve of poets and dreamers, or was it capacious enough to encompass scientists? How was the land transformed by its ineffable radiance? This chapter discusses the transfiguring powers of the veil of poetry and its ability to reenchant landscape as it unfolded across the globe—from the Alps and the English Lake District to the distant tropics. More specifically, it explores the veil as metaphor for a distinctively Romantic sensibility and mode of geographical knowledge.

Landscape and the Veil of Transparency

Goethe's encounter with Truth on that misty Alpine peak marks the emergence of a new poetic awareness and a new approach to nature, different from the approach of Baconian and Newtonian science. From the moment the poet was given the precious veil, the mantle of the earth began to shine with a new brilliance. It was precisely at this time that an entirely new model for imagining and describing the earth began to develop. Intellectual historian Chenxi Tang called it "a dynamic unity of man and earth."[9] As opposed to the mere description and classification of geographical facts and features, typical of eighteenth-century geography, the new model stressed interconnection and integration—of visible and invisible forces, of dark earthly materials and atmospheric phenomena, of human agencies and sensibilities.

The emergence of this new geography, Tang argues, is deeply indebted to the rise of Romantic landscape poetry and aesthetics, or, indeed, to the very concept of landscape. Unlike cognate concepts, such as "territory" or "environment," "landscape" is an inherently and insistently visual concept; it comes into being only at the moment of its apprehension by an external observer. Caught somewhere between the expanse of the land and the detached

gaze of the viewer, it mediates between subject and object, between proximity and distance, between place and space, between nature and culture. As
such, landscape holds a synthetic power; it is a *Zusammenhang*.[10] For Goethe
and early Romantics, landscape was "a symbol of nature constructed by the
subject, made out of the raw material of natural objects and phenomena, but
gaining shape only in subjective experience."[11] Romantic landscape poetry
thus transfigured the world into a space in which distinctions between the
human subject and nature collapsed in a single and absolute unity.

The search for this unity has been interpreted as a response to the increasing dominance of mechanistic science, which, by opposing mind to nature as
subject to object, undermined the basis on which the world's meaningfulness
had traditionally rested.[12] Inspired by landscape aesthetics, early nineteenth-
century geography gradually moved away from the traditional conception
of natural space as container of discrete objects and phenomena and started
to envisage it instead in terms of interconnections between the phenomena.
This approach was pioneered by the Prussian naturalist Alexander von Humboldt (1769–1859), himself a keen admirer and correspondent of Goethe. To
grasp the importance of his work and geographical science in shaping a new
spatial sensibility, it is worth pausing for a moment on what Tang called the
"deep archaeological layer" of modern geography, that is, eighteenth-century
landscape poetry.[13]

What did the Alpine landscape look like before Goethe stretched upon it
his veil "woven from the morning mist and the light of the sun"? Traditionally
avoided and despised as terrifying *loci horridi*, the Alps entered the Western
literary imagination toward the end of seventeenth century and by way of
different accounts. First were those of travelers on the Grand Tour, who were
shocked and overwhelmed by the encounter with what appeared to them an
otherworldly landscape. More significantly, there were the accounts of local
natural historians and natural theologians.[14] The modern appreciation of the
Alps owes its very origins mostly to these scholars, eager to study their phenomena and compile taxonomies of their productions—from mineral and
plant species to dragons.[15] Alpine landscapes had become the subject of poetry as early as the 1730s, when one of these scholars, the celebrated physician, anatomist, and natural historian Albrecht von Haller of Bern (1708–77),
embarked on a journey through Switzerland to study plants.[16] In addition
to two treatises on the Swiss flora and a rich travel account, the journey also
produced Haller's oft-cited poem "Die Alpen" (1732), a long encomium to an
idealized Alpine peasant community and its idyllic world.

In the poem, the virtue and innocence of the simple mountain folk (which
Haller opposes to the greed and lust of the rich inhabitants of the city) are

reflected in the purity of the landscape. In this sense, the Alpine landscape operates as a medium for moral edification and religious self-reflection. One of the most salient characteristics of the poem is its scientific legacy; poetry and the production of empirical knowledge are deeply intertwined.[17] In turn, empirical knowledge is granted by firsthand observation and transparency, which dominate the poem.

Textile metaphors are used throughout to stress the supremacy of sight, the rich textures of the landscape, and, not least, the creative yet inscrutable power of God. To Persian drapes decorating the walls of sumptuous palaces (an epitome of mundane futility) Haller opposes the natural garments that dress the mountains in the different seasons. In spring "the rose and the emerald spread all their beauties upon the meadows and the very rocks shine in a vestment of purple," while in summer, "the mower's scythe despoils the earth of her fair garment." When then "Autumn enwraps the sky in her vapoury robe," the thin floral carpet is replaced by the "more solid" fruits hanging from the trees. In winter a "mantle of frost" covers the land and a "crystal veil" conceals the currents of the flowing rivers.[18]

Veils of moisture and ice bring the minute detail into view. They also accentuate the crisp contours of larger features, in which the eye recognizes certain spatial structures, such as the vertical zoning of vegetation. At the same time, textile metaphors conjure up tactility. Sight and touch are stimulated simultaneously. For example, plant leaves are described not only in terms of color, but also as "satin." For Haller, the five senses were the basic tool for scientific observation. Sight, however, asserted itself as the chief sense, the only one capable of rational knowledge.[19] The stanzas of the poem are thus staged as a sequence of little scenes defined by Haller's field of vision. As a whole, the Alpine landscape is presented as a Baroque *theatrum mundi*. It is a stage commanded by the distanced eye of the scientist-poet. Only, here the stage is framed by natural curtains of clouds and vapors, rather than velvety stage curtains:

> Through the receding veil of the theatre, a whole world rises to the view! Rocks, valleys, lakes, mountains, and forests fill the immeasurable space and are lost in the wide horizon. We take at a single glance the confines of diverse states, nations of various character . . . till the eyes, overcome by such extent of vision, drop their weary lids.[20]

Haller's eyes, Tang argues, are Newtonian eyes. As they survey the landscape, like a camera obscura, they generate "objective, indeed mathematically measurable, representations of the world."[21] Their weakness is not so much a physiological limitation, as a symptom of the limits of human reason, which will never fully encompass the breadth of the horizon and comprehend the

mystery of creation. Haller's eyes are Newtonian also in their special fascination with light and colors. The space through which they move is not a flat,
monochrome Cartesian space with the primary qualities of size and shape,
but "a Newtonian space suffused with prismatic colors":[22]

> The checkered army of flowers seems to battle for primacy, a light heaven-blue
> puts a nearby gold to shame; a whole mountain appears, varnished by rain, to
> be a carpet embroidered with rainbows.[23]

As Haller's eye zooms in from the majestic mountain to the tiny detail of a
flower, the reader discovers similar patterns of colors and brilliancy wrapping
inner virtues. Distinguished by their "clothing of celestial blue," their golden
flowers forming "a gold crown," or their "satined leaves streaked with dark
green" and "liquid diamonds," gentians thus become emblems "of a beautiful
soul vested in amiable figure."[24]

Regardless of their scale, vegetal carpets and garments speak of the beauty
and rational order governing the theater of creation, a trope that was also
adopted by the Swedish botanist Carl Linnaeus two decades later.[25] Similar
textile metaphors are found again and again in contemporary idyllic and pastoral poetry across Europe. William Drummond (1746) and Johann Peter Uz
(1768), for example, greet Spring dressed in her "gleaming robe" and "mantle
bright with flowers"; Jean-Jacques Rousseau eulogizes the "earth in her wedding dress, the only scene in the world of which eyes and heart never weary."
James Thompson's sentimental and moralizing pen portrays heaven "in the
very act of giving earth a green gown," while in his *Irdisches Vergnuegen in
Gott* (1721–48), Barthold Heinrich Brockes likens the green roofing of the forest to "a living tapestry hanging there to the glory of God."[26]

These manifold mantles were unrolled by poets over the Alps and other
bucolic landscapes of Europe as proof of the harmony and abundance of creation and thus of God's perfection. In this sense, they were akin to the patristic mantle of the earth, at once wrapping God's unspeakable mystery and revealing his infinite Glory. For Haller, the Almighty "endows man with reason,
so that he can see and study nature, even though the human mind can never
penetrate and comprehend God's plan as a whole."[27] As he wrote elsewhere in
his collection of poems,

> Within nature no created mind can penetrate
> Happy is he to whom she shows
> Only her external envelope.[28]

Vegetable carpets, rain-varnished surfaces, veils of ice, curtains of fog are all
manifestations of this "external envelope." Haller pushes his Newtonian eye

from the nearby detail to the distant horizon and excavates the surface of mountains and rocks only to realize that he will never fully penetrate Isis's veil. The Orphic contemplation of the external surface of the envelope, with its bright colors and beautifully embroidered patterns, is enough to make him and his contemporaries cherish and strengthen their religious faith.

Between nature's external envelope and the retina of Haller's Newtonian eye, however, is another, almost invisible veil: the transparent veil of his descriptive poetry. As the veil of Truth on the frontispiece to Diderot's *Encyclopédie* (see fig. 5.13) effectively illustrates, eighteenth-century Enlighteners believed that virtue lay in transparency. And transparency, in turn, presupposed the neutrality of knowledge, as well as immediacy and objectivity of representation. For Diderot and his like, the signs of art and language were to be rendered as diaphanous as Truth's veil.[29] Words and allegory were not a shell or a mantle obscuring reality, but a thinnest diaphanous *peplos* akin to those of ancient Greek statues. As the English poet and translator John Hughes (1677–1720) wrote, "That the Allegory be clear and intelligible, the Fable being design'd only to clothe and adorn the Moral, but not to hide it, should methinks resemble the Draperies we admire in some of the ancient Statues; in which the Folds are not too many, nor too thick, but so judiciously order'd, that the Shape and Beauty of the Limbs may be seen thro' them."[30]

Veils of Mist and Light

Haller's veil of transparency differed from Goethe's poetic veil, as did their descriptions of the same Alpine landscapes. To start with, for Goethe there was no such differentiation between interior and exterior: "[Nature] has no pit or shell. She is all at once."[31] While he shared with his Newtonian contemporaries a keen interest in the morphology of plants and in the physics of color, the German writer criticized scientific experimentation for trying to unveil by violent and mechanical means something hidden behind the appearance of things.[32] Rather, he recommended attentive and protracted observation of forms in connection with other forms, as "series in which they take their place genetically, in order to see forms in their metamorphosis, see them born from one another . . . and to discover the simple and fundamental form, or *Urform*, from which the series of transformations develops."[33]

For Goethe, then, Nature was not an inert statue, nor was it an orderly array of phenomena, as in Haller's natural-historical vision. It was rather a vital breath animating the universe; it was a living entity in a state of perpetual transformation. Reality itself he challenged as something constantly in flux, as "a total process of interaction between mind and world."[34] This continuous morphing and interaction were reflected in the landscape—and in its veils. Haller's bucolic Alpine vignettes observed from the height of his "theater" were invested with an unfettered, dramatic emotional intensity. Pastoral idylls and the joys of spring dissolved together with literary conventions, including "green mantles" and "flowery robes." The landscape now surrounded a poet "buried in deep mist," "wrapped in silence," or in the spectral silvery sheen of a full-moon night.

> And night upon the mountains hung.
> With robes of mist around him set,
> The oak like some huge giant stood,
> While, with its hundred eyes of jet,
> Peer'd darkness from the tangled wood.
> Amid a bank of clouds the moon
> A sad and troubled glimmer shed;
> The wind its chilly wings unclosed,
> And whistled wildly round my head.[35]

Likewise, spring was no longer dressed in mantles of rococo flowery decorations, but in a veil of light shrouding the poet:

> Oh, what a glow
> Around me in morning's
> Blaze thou diffusest,
> Beautiful spring![36]

Unlike the green mantle of the earth, subject to the predictable cyclical changes of the seasons, in Goethe's writings veils are in a constant state of becoming. Changing clouds "continually pass over the pale disc of the sun, and spread over the whole scene a perpetually moving veil," while "webs of light" are formed by the sun, as it attracts the light mists evaporated from the glaciers, and by a gentle breeze "combing the fine vapours like a fleece of foam over the atmosphere."[37] Shrouded in these rarefied, almost impalpable atmospheric veils, the landscape becomes but a reflection of the inner state of the poet and his protagonists, to the point that inner and outer worlds blend into one another. The wilderness becomes a state of the land, as well as a state of mind. Indeed, Goethe believed, nature and landscape were not proper subjects for poetry when taken by themselves alone (as Haller and his contemporaries took them); they "must first be imbued with human feeling."[38]

This position had been pioneered by Goethe's tutor, Johann Gottfried Herder (1744–1803), who proclaimed: "Everything in Nature must be inspired by life, or it does not move me, I do not feel it. The cooling zephyr and the morning sunbeam, the wind blowing through the trees, and the fragrant carpet of flowers, must cool, warm, pervade us—then we feel Nature. The poet does not say he feels her, unless he feels her intensely, living, palpitating and pervading him."[39] In Goethe's verse and prose, inner life overflows into the outer world, in turn, bringing it to life. The landscape is filled with swirls of energy. It is not a passive backdrop, but a living texture always interwoven with the poet's (or his heroes') state of mind. From distanced observer, the poet becomes an organic part of its matter. The thin *peplos* of allegory and semantic explanation is stripped away. "Only landscape as such, the sensuous image of nature, remains."[40] The distance between the viewer and the land collapses. The Enlightenment aesthetics of transparency is superseded by self-referential signification, by an "aesthetics of autonomy" in which the primacy of eye is challenged by other senses.[41] Thus, the young Werther writes:

> I am alone, and feel the charm of existence in this spot, which was created for the bliss of souls like mine. I am so happy, my dear friend, so absorbed in the exquisite sense of mere tranquil existence, that I neglect my talents. I should be incapable of drawing a single stroke at the present moment; and yet I feel that I never was a greater artist than now. When, while the lovely valley teems with vapour around me, and the meridian sun strikes the upper surface of the impenetrable foliage of my trees, and but a few stray gleams steal into the inner sanctuary, I throw myself down among the tall grass by the trickling stream; and, as I lie close to the earth, a thousand unknown plants are noticed by me: when I hear the buzz of the little world among the stalks, and grow familiar with the countless indescribable forms of the insects and flies, then I feel the presence of the Almighty, who formed us in his own image, and the breath of that universal love which bears and sustains us, as it floats around us in an eternity of bliss; and then, my friend, when darkness overspreads my eyes, and heaven and earth seem to dwell in my soul and absorb its power, like the form of a beloved mistress, then I often think with longing, Oh, would I could describe these conceptions, could impress upon paper all that is living so full and warm within me, that it might be the mirror of my soul, as my soul is the mirror of the infinite God![42]

Nature is resolved into feeling and religion into pantheism, or, in Biese's evocative words, "Mind and Nature [are worked] into one tissue."[43] Lying on the grass in direct contact with the primordial earthly matter while gazing at the heavenly vault, the Romantic hero links the low to the high, the beautiful to the sublime. By solidly anchoring himself to the ground, he lifts his soul to the stars in a mystical celebration of communion with the totality of the cosmos.[44]

Goethe was not alone in turning landscape into "a mirror of the spirit."[45] By the early nineteenth century, Romantic poets traveling to the Alps were discovering secret correspondences between majestic peaks and lofty thoughts, or between deep gorges and the profundities of the human soul. Percy Bys-she Shelley (1792–1822) had a special fascination with those secret places of the earth. In "Mont Blanc," a poem he composed in Chamonix in 1816, the dark waters of roaring torrents rush through the human mind, and human thoughts, in turn, mingle with their restless flow, with the rocks, and with the trees. Eternity echoes in the vast caverns of the mountain; it is dispersed in the forest; it is reflected and refracted on its multiform surfaces and textures.

Wilderness, Shelley believed, spoke "a mysterious tongue."[46] The task of the poet was to bring this tongue alive without violating its mystery; or rather, to awaken the reader's sensibility to the mystery. But how to describe the mysterious? How to give voice to the inexpressible? Like Goethe, Shelley resorted to the veil. Veil metaphors stretch wide across his verse and prose. In his famous poem "Do Not Lift the Painted Veil" (1824), life itself becomes a fabric concealing a dark abyss, a vast cavern of nothingness, while his essay "In Defence of Poetry" (1821) resolves in an infinite act of unveiling—for poetry is itself infinite.[47] But veils are not only abstract metaphors. Shelley's physical landscapes too are shrouded in veils, especially atmospheric veils, for which he seemed to have a special predilection. Diaphanous veils of mist and light bestow a supernatural glow upon the landscape features of the wilderness; they add a touch of mystery to the already known; they materialize infinitude and bring eternity before the reader's eyes.

> Thine earthly rainbows stretch'd across the sweep
> Of the aethereal waterfall, whose veil
> Robes some unsculptur'd image; the strange sleep
> Which when the voices of the desert fail
> Wraps all in its own deep eternity.
> . . .
> Far, far above piercing the infinite sky,
> Mont Blanc appears—still, snowy, and serene.[48]

As the veil of poetry unrolled from the snow-clad Alps to the bucolic landscapes of Britain, cosmos and human soul interwove in new patterns. In the musical verses of Samuel Taylor Coleridge (1772–1834), veils of clouds and glare transfigure familiar cottages, rivers, and lakes into unfamiliar, magical landscapes.[49]

> Bright clouds of reverence, sufferably bright
> That intercept the dazzle not the Light
> That veil the finite form, the boundless power reveal
> Itself an earthly sun, of pure intensest White.[50]

FIGURE 6.1 Caspar David Friedrich, *Wanderer above the Sea of Fog*, 1818. The Romantic landscape is shrouded in veils of mist, fog, and light.
Kunsthalle Hamburg.

In William Wordsworth's (1770–1850) equally musical rhymes, poem and landscape seem to materialize out of the mist together:

> As silvered by the morning beam
> The white mist curls in Grasmere's stream
> Which like a veil of floating light,
> Hides half the landskip from the sight.
> Here I see the wandering rill,
> The white flocks sleeping on the hill,
> While Fancy paints, beneath the veil

> The pathway winding through the dale,
> The cot, the seat of Peace and Love,
> Peeping through the tufted grove.[51]

Wordsworth's verses transport Alpine drama to England's Lake District. The reader is walked to Snowdon under the moonlight, is offered views over misty cragginesses and dark abysses, and summoned up into the mountains where "he had been alone / Amid the heart of many thousand mists, / That came to him, and left him, on the heights."[52] Sometimes lonely characters emerge from the depths of misty, desolate rocky landscapes, like Caspar David Friedrich's *Wanderer above the Sea of Fog* (1818; fig. 6.1).

Other times they mysteriously vanish in the fog. Or, they simply merge with it. In "The Excursion," the Solitary, a disillusioned character who has chosen to withdraw from society, is given a glimpse of heaven through an impenetrable veil of mist. As peaks and clouds merge in a vision of the celestial city,

> My heart
> Swelled in my breast—I have been dead, I cried,
> "And now I live! Oh wherefore do I live?"
> And with that pang I prayed to be no more.[53]

Webs of Life

Haller's Newtonian lucidity of vision and the Romantic poetic gaze (notably, Goethe's gaze) converge in the work of Alexander von Humboldt. Their combination is perhaps best visually captured in his portrait by Friedrich Georg Weitsch (1806; pl. 5). Here Humboldt is immortalized in his early thirties as he collects plant specimens during his expedition to South America. The figure of the young Prussian naturalist is caught between a thermometer and the landscape in the distance. Giant banana leaves and craggy rocks frame a distant marine horizon shrouded in the humid haze of the tropics. Their dark shapes form a sort of small window, or natural stage curtain, through which the viewer is invited to look from an elevated viewpoint. The sharp contours of the plants collected by the naturalist explorer are juxtaposed to the soft light of the dusk, as it quietly filters through the pink veil of the atmosphere; the fine detail of empirical observation is set side by side the immensity of the universe unfolding beyond the misty horizon.

The birth of landscape as a geographical concept is located at the intersection between these two scales: the micro-scale of the flower, of the leaf, of the human body, and the macro-scale of the cosmos. Indeed, it was precisely through Humboldt's effort to harmonize these two scales that landscape was transformed, for the first time in Europe, from a purely aesthetic concept into a scientific one.[54] Between 1799 and 1804, the Prussian naturalist, together with the French botanist Aimé Bonpland, undertook an extensive journey of scientific exploration through South America, which resulted in an extraordinary accumulation of scientific and ethnographic observations. Although deeply committed to empirical observation, Humboldt, unlike eighteenth-century Linnean taxonomists, was interested not so much in the mere classification and description of phenomena, as in the principles underpinning their distribution. More broadly, he was interested in the functioning of the world as a whole, a great, integrated, "living whole, moved and animated by one sole impulse."[55] His goal was to uncover unity in diversity, the universal behind the particular, the invisible behind the visible.

Landscape enabled the marriage between detailed accuracy and Romantic grandeur. The poetic gaze over the land held a synthesizing power able to transform the multiplicity of natural objects and phenomena into a single, meaningful unity. In turn, landscape held the capacity to bring together incommensurate or even diametrically opposed scales and elements into a harmonious whole.[56] As in Weitsch's painting, landscape blended the focus of empirical observation with distant horizons forever shrouded in the haze; it combined sharpness of vision with invisibility. In Humboldt's words,

> When the human mind first attempts to subject to its control the world of physical phenomena, and strives by meditative contemplation to penetrate the rich luxuriance of living nature, and the *mingled web* of free and restricted natural forces, man feels himself raised to a height from whence, as he embraces the vast horizon, individual things blend together in varied groups, and appear as if *shrouded in a vapory veil.*[57]

The mantle of the earth was thus for Humboldt less a shiny surface than a substantial "web" made of different interrelated phenomena, both visible and invisible to the human eye. Nowhere did the web manifest its intricacy and biotic complexity more forcefully than in the Cordillera. Amid its deep ravines, the naturalist was offered commanding, bird's-eye views of the luxuriant tropical vegetation and its numberless species. In a single glance the eye surveyed majestic palms and humid forests of *Bambusa*, as well as more familiar oaks and medlars. Among the dark rocks and giant leaves of exotic plants, torrents and waterfalls exhaled their soft vapors, which tempered the effects of the light and smoothed the shapes of the surrounding landforms.[58]

In perpendicular rises of several hundred meters, the various climates succeeded one another, "layered one on top of the next like strata, stage by stage, like the vegetal zones, whose succession they limit; and there the observer may readily trace the laws that regulate the diminution of heat, as they stand indelibly inscribed on the rocky walls and abrupt declivities of the Cordilleras."[59] Witnessed from elevated spots, the variegated Andean landscape thus gave material shape to Humboldt's holistic vision. In it the Prussian naturalist saw vast observatories encompassing the entire variety of creation; spaces in which global order was made locally accessible.[60]

In Humboldt's descriptions of tropical landscapes, the earth's mantle expands from the surface to the entire biosphere. It acquires thickness, as well as complexity and dynamism. "In the midst of universal fluctuation of phenomena and vital forces—in that inextricable net-work of organisms by turns developed and destroyed," argues Humboldt, "each step that we make in the more intimate knowledge of nature leads us to the entrance of new labyrinths."[61] In *Cosmos* (1845–62) he envisages a "naked earth covered with an unevenly woven, flowery mantle, thicker where the sun rises high in a sky of deep azure, or only veiled by light and feathery clouds, and thinner towards the gloomy north."[62] His mantle, however, encompasses not only the geographical features one can observe upon the crust, but also hidden geological layers and the "elastic envelope of gaseous fluids" enshrouding the earth, with its fluctuating "vapors" and impalpable "light surrounding the horizon."[63] In other words, the earth's mantle is no longer a homogeneous surface, but an intricate three-dimensional texture—hence, Humboldt's preference for words such as "web" or "net-work."

Seeing the mantle as a web does not imply new discoveries, but rather a new way of seeing, an eye attentive to its fine texture and undulations. This way of seeing is best illustrated in Humboldt's pioneering profiles. Here the mantle's "web of life" is made visible in all its thickness. The *Tableau physique des Andes et pays voysins* (fig. 6.2) features Chimborazo, a 6,268-meter-high inactive volcano and the highest peak of the Cordillera, which was also believed to be the highest in the world.[64] Here the earth's entire climatic and vegetal spectrum is vertically mapped on the mountain.[65] The viewer is taken from its roots in the depths of the crust to the higher regions of the atmosphere by way of different altitudinal zones. The mountain appears as an ordered vertical microcosm made of interconnected elements: the snow-capped summit, the rocks, the different vegetal belts, the column of smoke quietly spewing from the chimney of Cotopaxi.[66] Phenomena are represented in relation to altitudinal change: the progressive diminishing of gravity and atmospheric pressure, the intensification of the blue color of the sky, the weakening of light.[67]

In the tableau the whole of the biosphere is brought under the commanding gaze of the viewer. Yet, as Sylvie Romanowski observed, what makes this

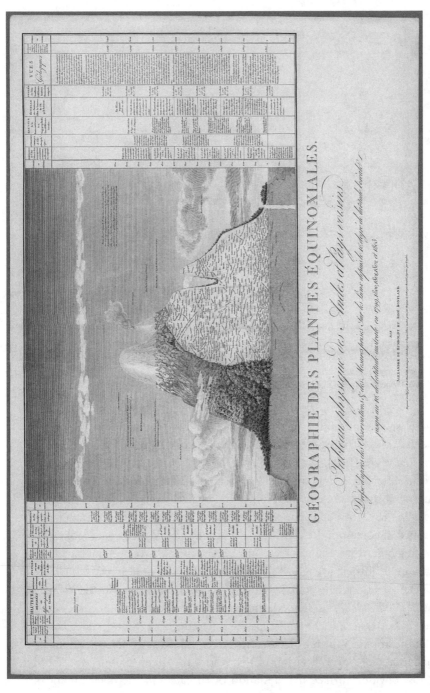

FIGURE 6.2 Alexander von Humboldt, *Physical Portrait of the Tropics*, 1807. Chimborazo, the highest peak of the Andean Cordillera, appears as a vertical microcosm encompassing the earth's entire climatic and vegetal spectrum.

Courtesy of David Rumsey Map Collection, David Rumsey Map Center, Stanford Libraries.

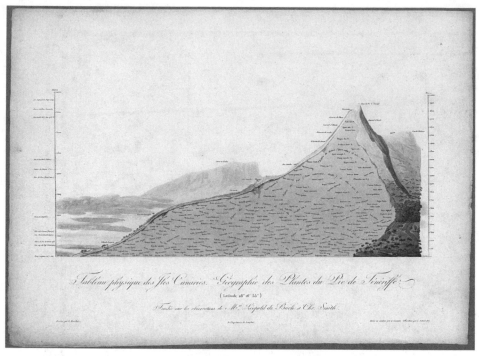

FIGURE 6.3 Alexander von Humboldt, *Tableau physique des Iles Canaries. Geographie des plantes du Pic du Ténériffe* (1827). The zoned green mantle of the Peak of Tenerife reveals clutters of plant names breaking through its vegetal belts.
Courtesy of David Rumsey Map Collection, David Rumsey Map Center, Stanford Libraries.

image of special interest is the organization of and the variations among its elements.[68] Most striking of all are the clutters of plant names breaking through the various vegetal belts. The latter are portrayed realistically, in different shades of green, almost in a three-dimensional fashion, while the former are scattered against an abstract white background. The ordered layering of the belts contrasts with the chaotic appearance of the names. In reality, however, each name is mapped in correspondence with a specific altitude. What we are looking at is nothing other than the reverse of the green tapestry. We are looking at the unglamorous, intricate tangle of threads that form the orderly pattern we see in vegetal zonation. The contrast is accentuated even further in the *Tableau physique des Iles Canaries* (fig. 6.3), where the tangle of plant names covers most of the Peak of Tenerife. The landscape fades into the background, veiled by the haze of the tropics. Why are plant names so central to these representations?

In Humboldt's overall vision of the world, the green mantle was "both the most obvious surface manifestation of climate and the determinant of many other natural and human features."[69] In his search for natural unity in diversity, Humboldt envisaged vegetative communities as both an expression of the transcendental coordinating principle of the cosmos and a field of rigorous geographical inquiry.[70] For Humboldt, the supremacy of plants also stemmed from their key position within the "thick web" of the biosphere. He envisaged the green mantle as a sort of interface between earthly matter and the atmospheric envelope wrapping the earth, as well as between the land and humans. Vegetation cloaked the earth's surface while governing the exchange of energy and materials between the atmosphere and the surface; it also conditioned the capacity of the land for human activities and shaped the aesthetic character of the landscape.[71]

Unsurprisingly, Humboldt dedicated the German version of his *Essay on the Geography of Plants* (1807) to Goethe. The frontispiece features an allegorical engraving of Isis unveiled by Apollo, at whose foot lies Goethe's book *The Metamorphosis of Plants* (fig. 6.4). As Humboldt later explained in the *Cosmos*, Goethe had treated the problem of metamorphosis "with more than

FIGURE 6.4 Frontispiece to Alexander von Humboldt's *Essay on the Geography of Plants* (Paris: Chez Levrault, Schoell & Co., 1805), vol. 1. Apollo, divinity of the sun, music, and poetry, unveils monstrous Isis. As Goethe commented, "Poetry might indeed raise Nature's veil." Huntington Library, San Marino, California.

common sagacity, and to the solution of which man is urged by his desire of reducing vital forms to the smallest number of fundamental types," that is, to Goethe's *Urformen*. As they contemplate the riches of nature and see the wealth of observations continually increasing before them, humans, Humboldt continues, "become impressed with the intimate conviction that the surface and the interior of the earth, the depths of the ocean and the regions of the air will still, when thousands and thousands of years have passed away, open to the scientific observer untrodden paths of discovery."[72] Unlike earlier, similar representations, however, on the frontispiece it is not the scientist-philosopher, nor Inquiry, to unveil Isis, but Apollo, the god of music and poetry.

Goethe himself commented on this engraving by saying that "poetry might indeed raise Nature's veil."[73] Like Goethe, Humboldt encouraged the poetic contemplation of nature. Poetic contemplation was an antidote to the cold mechanistic gaze of modern science that, Humboldt feared, was threatening the free pleasure of nature.[74] By stretching the veil of poetry over the Andes, Humboldt aimed at making those distant sceneries (and the knowledge of the earth) accessible to a generation of bourgeois educated in the Romantic taste of the sublime.[75]

> Enjoyment and emotions [awaken] in us, whether we float on the surface of the great deep, stand on some lonely mountain peak enveloped in the half-transparent veil of atmosphere or by the aid of powerful optical instruments scan regions of space, and see the remote nebulous mass resolve itself into worlds of stars.[76]

More significantly, poetic contemplation helped the scientist achieve an intimate, spiritual contact with the cosmos and therefore grasp its unity in diversity, the *Urform* behind the multiplicity of appearances. Whether made of vegetation or clouds, veils facilitated this contact. They unveiled nature's dazzling beauty while protecting her mystery:

> I would recall the deep valleys of the Cordilleras where the tall and slender palms pierce the leafy veil around them, and waving on high their feathery and arrow-like branches, form, as it were "a forest above a forest"; or I would describe the summit of the Peak of Teneriffe, when a horizontal layer of clouds, dazzling in whiteness, has separated the cone of cinders from the plain below, and suddenly the ascending current pierces the cloudy veil, so that the eye of the traveler may range from the brink of the crater, along the vine-clad slopes of Orotava, to the orange gardens and banana groves that skirt the shore. In scenes like this [the heart is moved by] the features of the landscape, the ever-varying outline of the clouds, and their blending with the horizon of the sea, whether it lies spread before us like a smooth and shining mirror,

or is dimly seen through the morning mist. . . . Impressions change with the varying movements of the mind, and we are led by a happy illusion to believe that we receive from the external world that with which we have ourselves invested it.[77]

Like Goethe, Humboldt was no detached observer; he was entangled in the web of life through a mystical "bond of union linking the visible and spiritual world."[78] His was a two-way relationship with the landscape. Yet, as Dettelbach pointed out, his poetic genius resided in his instruments; it was "through measurement and extrapolation that he strip[ped] nature bare in the altitudinal metamorphosis of plants on Chimborazo."[79] Or perhaps more precisely, his poetic genius resided in his ability to transcend binaries, in challenging d'Alembert's claim that "scientists study details and philosophers see the big picture."[80] Emblematically, Humboldt's views from elevated spots are always characterized by a mist or haze veiling the horizon, by a progressive loss of clearness and transparency as the distance increases. In stimulating the imagination, the misty veil mediates between the present and what has yet to come, between local scale and cosmos, as well as between scientific and poetic modes of inquiry. Landscape itself becomes a threshold, a space of possibility.

Characteristically, veils of haze and light also shroud the tropical landscapes painted by Weitsch and, more notably, those by the American artist Frederic Edwin Church (1826–1900).[81] Inspired by Humboldt's writings, between 1853 and 1857 Church traveled to South America seeking to capture the dramatic grandeur and the lively variety of the tropics. As with Humboldt's profiles, Church's paintings were not snapshots, but true "tableaux," that is, composites of the topographies observed during his journeys.

His huge canvases encompassed entire Andean landscapes, "from Arctic pinnacles, through temperate zones, to the palpable humidity of tropical valleys."[82] Sometimes unveiled behind velvety curtains (fig. 6.5), these large panoramas immersed viewers in the scene, while at the same time setting them at a distance.[83] They re-created the same paradoxical effect of immersion and detached mastery over the landscape found in Humboldt's accounts.

Adopting and magnifying the conventions of the picturesque, Church played with proximity and distance. For example, in *Tropical Scenery* (1859; fig. 6.6) the fine botanical detail in the foreground is juxtaposed to hazy slopes looming in the distance. Empirical observation is juxtaposed to imagination, as the light quietly filters through the layers of tropical humidity. As Humboldt had himself observed, the mild light imparted to the landscape through thinly veiled tropical skies gave "a mysterious power" to landscape painting.[84]

FIGURE 6.5 Frederic Edwin Church's *The Heart of the Andes* (1859), exhibited at the Metropolitan Fair in Aid of the Sanitary Commission, New York, in 1864. Curtains and handrails were added for its inclusion in the fine art gallery at the fair.
Collection of New-York Historical Society.

In *Cotopaxi* (1855), the viewer's eye is likewise guided from the dark silhouettes of the tall palms rising in the foreground to the cone of the snow-clad volcano calmly smoking in the distance. In between, the velvety mantle of the tropical forest clothes the land and reveals a waterfall shrouded in a thin veil of vapors; shapes of trees are reflected on the still surface of a small lake; tiny solitary clouds and flocks of birds traverse the hazy sky. In another view of Cotopaxi (1862), the volcano is dramatically captured during an eruption and the entire landscape transfigured by the red light filtering through the curtain of smoke.

Unmistakably naturalistic in intent and execution, Church's paintings give visual expression to the poetry enshrined in Humboldt's web of life. In Edward

Casey's words, "if nature for Humboldt could 'only be vividly delineated by thought clothed in exalted forms of speech,' the same nature was for Church clothed in equally exalted forms of image."[85] The intimate link between geography and painting was amply acknowledged by Humboldt himself. Notably, in the second volume of the *Cosmos*, the Prussian naturalist traced the history of mankind's poetic appreciation of nature through literary descriptions and the evolution of landscape painting in a breathless journey from classical antiquity to his Romantic present, by way of India, Persia, Palestine, and Arabia.[86]

As it traverses the different seasons of human history, landscape gradually expands, and with it human knowledge. The Hellenic scenery, Humboldt notes, "presents the peculiar charm of an intimate association of land and sea, of shores adorned with vegetation, or picturesquely girt round by rocks gleaming in the light of aerial tints, and of an ocean beautiful in the play of ever-changing brightness of its deep-toned moving waves."[87] And yet, in spite of its harmonious beauty, in Greek antiquity landscape was nothing but "the background of the picture of which human figures constitute the main subject."[88] In ancient Rome landscape morphed into a sequence of framed vignettes, whether the little "sketches of nature" traced by Cicero, or the numerous *topia*

FIGURE 6.6 Frederic Edwin Church, *Tropical Scenery*, 1873. The fine texture of the landscape is bathed by a golden light filtering through the veil of tropical humidity.
Brooklyn Museum, New York, Dick S. Ramsay Fund.

gracing the walls of the villas in Pompeii and Stabiae.[89] These representations, however, Humboldt dismisses as "nothing more than bird's-eye views
of the country similar to maps."[90] It was only in the Renaissance, he claims,
that the "geographical field of view became extended" and landscape painting
acquired autonomy. With the expansion of the world caused by the great geographical discoveries, landscape representation was broadened, encompassing "the individual character of the torrid zone, as impressed on the artist's
mind by actual observation."[91]

Landscape's progressive unfolding reflects the degree of human cultivation achieved at specific moments in history, which, inevitably, culminates
in the heightened poetic sensitivity of Humboldt's Romantic present: "the
animating influence of the descriptive element and the multiplication and
enlargement of views opened to us on the vast theater of natural forces may
all serve as means of encouraging the scientific study of nature, and enlarging
its domain."[92] Landscape emerges from the mists of the past and gradually
acquires ever broader and sharper contours. Using Haller's metaphor, nature
becomes a "splendid theater," and landscape a stage curtain opening on an
ever unfolding horizon—or perhaps that unreachable horizon itself:

> Long before the discovery of the New World, it was believed that the new
> lands in the Far West might be seen from the shores of the Canaries and the
> Azores. These illusive images were owing, not to any extraordinary refraction
> of the rays of light, but produced by an eager longing for the distant and the
> unattained. . . . At the limits of circumscribed knowledge, as from some lofty
> island shore, the eye delights to penetrate to distant lands.[93]

*

English historian Thomas Keith observed how eighteenth-century naturalists
constructed a detached natural scene to be viewed and studied from the outside "as if peering through a window [or a stage curtain] in secure knowledge
that the objects of contemplation inhabited a separate realm."[94] Romantic
poets, by contrast, not only came to inhabit the scene, but also became one
with it, as they considered landscape a mirror of their emotions. Measured
detachment gave way to pantheistic interfusion with the natural world and
its elements.

Lucid scientific observation and Romantic sentiment converge in Humboldt's work and in his use of textile metaphors.[95] On the one hand, web metaphors serve to convey his holistic vision of the cosmos; on the other, veils
of mist, haze, and light evocatively shroud the landscapes he contemplates,
blurring distances and scales, as well as the boundaries between subject and
object, inner and outer worlds. Unlike the heavy drapes lifted, or pierced, by

Renaissance explorers and anatomists, these atmospheric veils do not conceal, they rather refract and transfigure. They speak of a new poetic sensibility toward creation.

Set on the horizon, Romantic veils also had the power to trigger an "eager longing for the distant and the unattained."[96] Ultimately, Humboldt's lifelong quest was akin to the dream of omniscience haunting Goethe's Faust, the prophet of modernity.[97] Like Humboldt's readers witnessing the unfolding of history (and of landscape) from above, Faust is repeatedly raised to mountaintops and other elevated spots, and endowed with the diabolic privilege of the God's-eye view.[98] Humboldt's was an aesthetic gaze, but at the same time, like Faust's, it was also a universally valid scientific gaze that could be directed toward any region of the earth and toward any moment in history. And yet, potent as it was, this gaze could only attempt to chase a perpetually shifting horizon shrouded in the mist—what Farinelli called *bruma del mondo* ("the haze of the world").[99]

In the next chapter we will see what happened when, in the twentieth century, with the end of geographical discoveries and the advent of globalization, the Romantic mist dissipated, the earth's mantle ceased to unfold, and space closed once again.

PART III

The Surfaces of Modernity

The Surfaces of Geography

Geography is fundamentally the regional or chorological science of the surfaces of the Earth.[1]

Humboldt is often referred to as the forefather of modern geography. It was not, however, until the late 1880s that geography was institutionalized as an academic subject. As the discipline formally moved into the new century, the world was no longer the world Humboldt had known. It had changed, and it was continuing to change. With the exception of parts of Antarctica, there were no more *terrae incognitae* to penetrate, no stage curtains to open, no mystic veils to look through.[2] Improved transportation and new communication technologies were bringing places closer to one another. In an influential paper titled "The Scope of Geography and the Geographical Pivot of History," published in 1904, Sir Halford Mackinder (1861–1945), Oxford's first reader in geography, claimed that the Columbian epoch, or the opening of global space, had just ended. He prophetically warned his readers that "every explosion of social forces, instead of being dissipated in a surrounding circuit of unknown space and barbaric chaos, will be sharply re-echoed from the far sides of the globe."[3] The world had closed again, and in this closing gesture, it had made itself transparent, at least to the eyes of geographers. The horizon was no longer wrapped in misty veils.

The role of the geographer had changed, too. No longer was he a scrutinizer of internal forces and the mysteries of the cosmos, a seeker of unity in diversity. He was now an objective describer of the surface of the earth and its regions. Earlier in the nineteenth century, German geographical writing had been largely underpinned by the notion of a universal history providentially "unveiling truth," or divine design.[4] In 1845, for example, Ernst Kapp claimed that "it is only in history that geography reveals truth," while Karl Ritter (1779–1859) envisaged his monumental *Erdkunde* (1817–32) as nothing less than a global comparative study aimed at showing the providential connection between history and nature.[5] Mapped on global space, history, Ritter believed, progressively unfolded and unveiled God's plan; it was, in his words, a rich "fabric of events" stretching over the planet.[6] By the end of the century, by contrast, attention had moved away from pietistic narratives, veils, and ever-shifting frontiers. It focused instead on securing and justifying geography's status as a respectable scientific discipline.

According to Mackinder, with geographical exploration coming to an end, geography had to be "diverted to the purpose of intensive survey and philosophic synthesis."[7] However, while Humboldt and Ritter focused their work on the whole body of the earth, as the very titles of their *magna opera* (*Kosmos* and *Erdkunde*) suggest, the practitioners of the new institutionalized geography reduced the globe to a surface. In his 1889 opening address, Clements Markham, then president of the Royal Geographical Society, defined geography as "a description of the earth as it is, in relation to man, and a knowledge of the changes which have taken place on its surface during historical times. One aim of this science is to ascertain by what agencies and by what processes the earth has acquired its existing forms and characteristics."[8]

Legitimacy and authority were to be achieved in the field through the systematic study of landscapes. For Carl Troll (1899–1975), a leading geographer in the twentieth century, geographical synthesis meant that the emphasis was shifted from the examination of individual phenomena "to their accord in the spatial unity, in the *landscape*." With landscapes as natural regions, he argued, "geography . . . has finally found its own object that no other science can dispute."[9] Unlike the landscape in Humboldt's writings, however, the landscape studied by twentieth-century geographers was devoid of the subjective element. The viewer was made invisible. Flattened and reified, landscape was conflated with what were perceived to be more "objective" concepts, such as area or region. From a way of seeing and a hazy point of tension between scales and concepts, landscape was now transformed into an unambiguously bounded partition of the land (or the earth's surface).

At this point, the earth's mantle metaphor underwent a new transformation. Stage curtains, flying drapes, and veils morphed into surfaces, or coverings such as "tablecloths" and "carpets." Yet, over the following decades, even surfaces and their coverings morphed—or rather, the geographer's gaze morphed. With the development of new transport and communication technologies, especially powered flight, perceptions of landscape, and of the earth's surface in general, changed. Tablecloths rippled and flattened. Carpets vertically contracted and expanded. By the end of World War II, the terrestrial surface had acquired new thickness and consistency. This chapter charts these transformations in early twentieth-century geography.

The Surfaces of Modern Geography

According to Farinelli, modernity begins with the straight line. It begins as rectilinear inscriptions start to cover the surface of the earth; as territory is transformed into a map, that is, into a representation of itself.[10] According to other commentators, the hallmark of modernity is the planar surface, the precondition for cartographic representation. Inhabiting a world dominated by flat surfaces, whether a floor or a tennis court, a landing strip or the highway, is a fundamental condition of modernity. With modernity, the earth, argues B. W. Higman, "has been given a new skin, notably concrete and asphalt, designed both to obliterate what beauty had existed previously and to inhibit natural processes of change—to install flatness as a permanent condition."[11] The establishment and development of geography as an academic discipline took place as these transformative processes had been largely accomplished, to the point that in 1931 Mackinder envisaged the geographer of the future as a "world planner," a sort of Faustian world-scale extension of the urban and regional planner.[12]

Mackinder's vision stemmed from an emerging global consciousness. At the outset of World War I, the British geographer noted how the surface of the earthly globe was encircled by lines of all sorts: telegraphic lines, railway lines, steamboat lines, power lines, and, not least, lines of thought. It was the geographer's duty, he believed, to capture this new level of complexity.[13] The key task of the professional geographer was to bridge the gap between the human and physical realms, to visualize the morphology of the terrain and the colors of the landscape, as well as the traces of human action on those very landscapes. The modern geographer had ceased to be an explorer. His mission was no longer to unveil the unknown, but now to weave together different spheres of knowledge in a coherent disciplinary tapestry, interlacing the threads of hard science with those of art and world politics, constructing patterns of physical landscapes and human activities.[14] In Mackinder's words, geography was "the science whose main function is to trace the interaction of man in society and so much of its environment locally."[15] How was this achieved? How was this interaction expressed?

One of the main preoccupations of Mackinder and other early professional geographers was to define the scope of geography as a science and as an institutional discipline, distinct from, say, geology or the natural sciences. While geographers might have disagreed on the extent to which the environment shaped humans (or humans shaped the environment), they all agreed that, as a discipline, geography focused on the study of the earth's surface. In his presidential address to the Royal Geographical Society in 1920, Francis

Younghusband stated that "while Geology concerns itself with its anatomy, Geography, by long convention, restricts its concern to the earth's outward aspect. Accordingly, it is in the face and features of Mother Earth that we geographers are mainly interested." While geographers, he argued, must know something of the general principles of geology, as painters have to know something of human anatomy, their special business is "with the outward expression."[16]

As outward manifestations open to the human senses, surfaces are bound to aesthetics. For Younghusband, the characteristic of the earth's face most worth learning about and understanding was the beauty of its exterior traits and features. Indeed, the English geographer even developed a method for dealing with this aspect by comparing distinctive elements that made up the beauty of different regions.[17] Vaughan Cornish (1862–1948) took Younghusband's words seriously. Not only did he profusely write on the aesthetic appreciation of scenery, but he pioneered a whole new branch of geography concerned with the study of the surface's "ripples"—what he called "kumatology" (a neologism he coined from the Greek word *kyma*, or "wave"). Its objects of study ranged from the patterns of sea waves to the formation of sand dunes (fig. 7.1).[18] Cornish saw beauty in undulating surfaces, even artificially produced ones, such as Japanese roofs. Of rural England he wrote: "It is a country of gentle undulations where rivers flow quietly in winding curves, a land well timbered by deciduous trees of rounded form, of fields divided by a bushy fence, all in a climate of soft skies."[19]

The understanding of landscape as a malleable material surface was essentially a modern, secular incarnation of Gaia's mythical mantle and its biblical counterparts. As with their ancient predecessors, early twentieth-century geographers recurred to textile metaphors to illustrate the concept and the reciprocity between human action and the physical environment. "Imagine thrown over the land like a white tablecloth over a table, a great sheet of chalk," wrote Mackinder:

> Let the sheet be creased with a few simple folds, like a tablecloth laid by a careless hand. A line of furrow runs down the Kennet to Reading, and then follows the Thames out to the sea. A line of ridge passes eastwards through Salisbury Plain and then down the centre of the Weald. . . . The powers of air and sea tear our cloth to tatters. But as though the cloth has been stiffened with starch as it lays creased on the table, the furrows and ridges we have described have not fallen in. Their ruined edges and ends project stiffly as hill ranges and capes.[20]

The "tablecloth" metaphor expresses the three-dimensionality of the terrestrial surface and its dynamic nature. The form of the lithosphere, explains

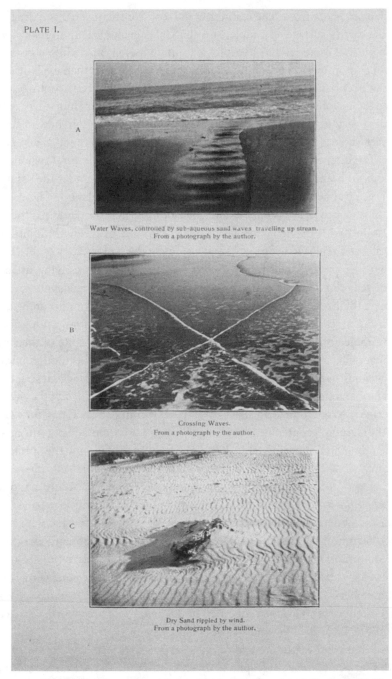

PLATE I.

A

Water Waves, controlled by sub-aqueous sand waves travelling up stream.
From a photograph by the author.

B

Crossing Waves.
From a photograph by the author.

C

Dry Sand rippled by wind.
From a photograph by the author.

FIGURE 7.1 The rippled surfaces of kumatology: crossing waves and sand rippled by the wind in Vaughan Cornish's essay "On Kumatology" (*Geographical Journal*, 1899).
Courtesy of Royal Geographical Society.

Mackinder, is not fixed. The features of the land constantly change.[21] Imagining the earth's surface as a tablecloth illustrates geological and erosive processes; it makes aerial and hydrological influences visible, while constantly drawing upon their interlinking relationship.[22] Mackinder then explains how human occupancy has subsequently adjusted to the physical contingencies of the "tablecloth." The scientific value of geography, he concludes, lies not in the discovery of new facts and data, but in pointing out certain new relations between those data: geography must be "a continuous argument" that reveals the causal interactions between human communities and natural regions.[23]

The old mantle of the earth thus becomes a tablecloth under the scrutiny of the geographer's omniscient gaze. Humboldt's contemplation of landscape as a theater made of visibilities and invisibilities, the Romantic pantheistic interfusion with its elements, the mystical curtains of haze and light have now all been replaced by the systematic study of the morphology of the tablecloth, of its ripples, folds, and tattered edges, of the features lying upon it. While veils can float in the air, can reveal, refract, and even transfigure reality through their diaphanous textures, tablecloths leave no room for ambiguity. Their function is simply to cover a table. There is nothing numinous about them. Tablecloths conjure up physical consistency, but, in spite of their ripples, they also evoke the flatness of the surface they cover—as though underlying flatness were the precondition for building up geographical description.

The metaphor was used by other geographers in the first half of the twentieth century to emphasize precisely this aspect. Writing on savannas, for example, the German geographer Alfred Hettner (1859–1941) noted how "the Spanish word *sabana*, meaning tablecloth, originally connoted only the surface of a plain." Leo Waibel (1888–51) pointed out how savanna attains its characteristic appearance only if trees are scarce and observed how "according to the *Encyclopedia Britannica*, for instance, 'savanna or savannah (Span. *savana*, a sheet, . . . a linen cloth) [is] a term applied either to a plain covered with snow or ice, or, more generally, to a treeless plain.' "[24] Mackinder similarly described uniform uninhabited areas as "mantles of deserts and wildernesses"—a sort of planetary extension of the medieval *loci non vestiti*. These encompassed eastern Siberia, Alaska, the Canadian Shield, and the arid and semiarid western United States. Grouped together, he argued, these areas formed a giant "mantle of vacancies" beyond whose edges humanity dwelt.[25]

Not only did textile metaphors help express the mutable nature of the earth's surface and its geographical diversity, they also gave expression to its complex "texture," of which humans were part. Friedrich Ratzel (1844–1904), the pioneer of the German *Anthropogeographie* and father of environmental determinism (the study of how the physical environment, and climate

in particular, predisposes societies toward particular development trajectories), described "regions of civilization" as "a great web" made of "an infinite
number of threads."[26] His American pupil Ellen Churchill Semple (1863–1932)
went as far as claiming that "man is a product of the earth's surface" and that
the earth "has entered into his bone and tissue."[27] The texture of the land and
the fabric of culture she understood as being deeply intertwined into one
another in the most visceral sense. An evocative example is her discussion of
the production of rugs by the pastoral tribes of Persia and Central Asia. The
manufacturing of these crafts, she argues, is interlaced with the land and climate of those regions. Indeed, a fundamental factor is "proximity to the fin-
est raw materials whose quality is equaled nowhere else, because it depends
upon the character of the pasturage, and probably also upon the climatic conditions affecting directly the flocks and herds."[28]

While allowing humans further agency in shaping the land, the French
geographer Paul Vidal de la Blache (1845–1918) resorted to similar narratives
and metaphors. "When one considers all that is implied in this word 'environment,' and all the unsuspected threads of which the fabric that enfolds us is
woven," he asked, "how could any living organism possibly escape its influence?"[29] For Vidal de la Blache, the environment was a complex fabric, "holding together heterogeneous beings in mutual vital interrelationships." Every
region of the planet was a domain where "many dissimilar beings, artificially
brought together, have subsequently adapted themselves to a common existence."[30] The geography he promoted thus rested on the comparison between
different human regions. These were characterized by distinctive *genres de vie*,
or "functionally patterned [modes] of living—nomadic, agricultural, and so
on—that constitute an integrated web of physical, social, and psychological
threads."[31]

Genres de vie left their distinctive imprint on the land, making regions
akin to medals "struck in the likeness of a people."[32] The resulting *paysages*
were complex fabrics whose human threads had the power to transform, thus
increasing regional differentiation. The appearance of the landscape, argued
Vidal de la Blache, owed in the first place to "the nature of the soil" and second
to the typology of vegetation. More significantly, it was a surface inscribed
with the marks of history, whether traces of ancient glaciers or the furrows
produced by "the plough mov[ing] easily over the table lands."[33] For Vidal de
la Blache and his followers, the terrestrial surface as a whole was therefore a
colorful patchwork, whose patterns and embroideries were left to the geographer to identify and chart.

The American geographer Carl Ortwin Sauer (1889–1975) called the ability to detect such patterns "the morphological eye," for, he argued, "geography

is always a reading of the face of the earth" and its forms.[34] The geographical bent, he believed, "rests on seeing and thinking about what is in the landscape, what has been technically called the content of the earth's surface."[35] For Sauer, landscape was "the material of geography" in the most visceral sense. More specifically, it was a physical area, a container of material objects, which gave it a distinctive appearance, or "personality." Like Vidal de la Blache's *paysage*, Sauer's concept of cultural landscape emphasized human agency, as opposed to mere adaptation to the environment. His main concern was "man's record upon the landscape," or rather, the "carving" of cultural landscapes out of the land by different culture groups.[36] The scientific validity of landscape, argued the American geographer, lay not in its individuality, but in patterns of relations, that is, in its texture.[37] Thus understood, landscape provided geography with the substantial content needed to legitimize itself as a respected scientific discipline; it ensured both empirical focus and material substance.

Sauer was one of the most vocal advocates of fieldwork of his time. In order to fully understand the landscape, the professional geographer, he believed, had to immerse himself in its textures. He had to traverse it afoot, sleep outdoors, sit about camp in the evening, see the lay of the land in all its seasons.[38] The result was an appreciation of the earth's surface that was at once scientific and aesthetic:

> Why can't a geographer working in the Great Plains convey to the reader the feel of horizon, sky, air and land that William Johnson did? . . . Aesthetic appreciation leads to philosophic speculation, and why not? Are not the compositions of nature, the lines and colors of terrain and of mantling vegetation proper things to consider? . . . There is an aesthetic morphology of the assemblage of forms, an aesthetic morphology of landscape, latterly often violated by industrial civilization.[39]

In this, Sauer echoed British geographers. "Every irregularity on the surface of the earth, every hummock and valley, stream and pond has a cause, perhaps discoverable," an Oxford geography student wrote in 1914. "Once a week we leave the School of Geography and proceed . . . on excursion . . . for the study of geomorphology—of landforms and their causes." Direct exposure to natural surfaces, however, turned the excursion from mere scientific feats into a deeply aesthetic experience. "Week by week," the student continued, "beauty has deepened—blue of hyacinths—delicate tracery of young beech leaves against the sky, the burning bush of the gorse, sky larks and cuckoos, wind on the open downs, mingled rain and sunshine on old stone."[40]

Being a professional geographer meant appreciating the textures of land-scape, as well as seeing the bigger picture, that is, elevating oneself over the surface. "When I was in Africa I remember seeing before me a great billowing slope, clothed with dense forest, dark green and burnished in the sunshine," recalled Mackinder.

> I entered and traversed that forest for a long day. When I emerged and looked back there was the same forest, and yet to my vision it was not the same, for I could now appreciate its texture; I had not merely sight of it, but insight. So it is with the trained geographer: he starts on the shoulders of the scientific specialists, he traverses his natural regions, and emerges with a new grasp and insight of the world as a whole.[41]

The concept of landscape as a richly textured and diverse surface was further discussed by the American geographer Richard Hartshorne (1899–1992). After Albrecht Penck, Hartshorne compared the earth's surface to "a great, irregular carpet." The metaphor was akin to Mackinder's "tablecloth," but placed more emphasis on the fine detail (for example, vegetation and human features) than on the physical topography of the terrain—in other words, on texture and patterns, rather than on folds and ripples:

> The form of the surface is determined primarily by the relief of the land, but it is also affected in minor degree by the height of forests and in urban areas particularly, of man's buildings. The material character of a landscape is expressed by color and texture and may be observed by sight and feeling. To designate the material character of the landscape apart from a surface configuration, we might use the term "landscape cover." Over most of the world, this consists of the uppermost surface of vegetation—whether natural, wild, or cultivated—and of surface water. In lands lacking in vegetation—whether permanently or seasonally—the "landscape cover" consists of bare ground, snow, ice or of the surface of the works of man.[42]

Flatness and Verticality in an Air Age

One of the many paradoxes of modernity, observes Higman, is that in the effort to reshape the surface of the earth, "human beings have frequently sought to create an artificial flatness or smoothness—in spite of the purported aesthetic attraction of diversified, elevated landscapes."[43] Whether as a tablecloth or a carpet, early twentieth-century geographers understood and described landscape topographically, that is, in terms of its complex physical and human processes. When it came to applied geography and planning, however, in most cases landscape lost its texture and three-dimensional aesthetic appeal: it became an abstract isotropic surface—that flat surface hidden beneath the tablecloth or the carpet. It became pure geometrical space. In this sense, the difference between the topographic and geometric understanding of the terrestrial surface was akin to Kandinsky's postulations on realistic and abstract art: "in the former the sound of the element in itself is veiled, thrust back; [by contrast,] when abstract, it attains its full, unveiled sound."[44] The transformation of the earth's surface into a flat geometrical plane—into "unveiled sound"—was the key, if not the precondition, to location theory and regional planning.

In 1909 the German economist, sociologist, and geographer Alfred Weber (1868–1958) published his theory of industrial location, assuming all transport was by rail on "a mathematically flat plain where the mountains are razed, the valleys filled, and the swamps covered."[45] Twenty-four years thereafter, Walter Christaller (1893–1969) developed a "central place theory," which sought to explain patterns of spatial distribution of human settlement and their hierarchical order. He argued that the settlements functioned as "central locations" providing services to surrounding areas in patterns of concentric circles marking the cost of movement to services. "Since these circles would necessarily overlap, he found the hexagon to be the most efficient straight-sided geometric shape to divide the space equally."[46]

As with Weber's theory, Christaller's model presupposed an isotropic surface, as well as an evenly distributed population and resources.[47] The flat landscape of southern Germany, where the theory was conceived, offered a good approximation of these conditions. As Christaller joined the Nazi Party in 1940 and went to work on the proposed transformation of the settlement landscape in Poland, the formal geometry of his theory was "co-opted to create a spatial vacuum, directed at removing people (mostly Jews and Slavs) from conquered lands in order to create an 'empty space' to be re-occupied by Germans." This, Higman notes, was represented "not only by depopulation, but also in the shape of the landscape, which was literally bulldozed."[48] The land was to become the model, and the model the land.

The irony of modern planning is that horizontal flatness could only be "seen," or indeed achieved, by way of verticality. While Weber's and Christaller's models were constructed at ground level, they presupposed an all-seeing eye set perpendicularly above their isotropic surfaces. More dramatically, it was the airplane that for the first time revealed to the architect Le Corbusier "the flattened and jumbled city, terrifying in its confusion."[49] The aerial view made it possible to grasp urban and rural areas in their entirety and to visualize the grid—the most distinctive imprint of Western order on newly colonized (and flattened) landscapes.[50] A pivotal weapon in planning the destruction of cities in enemy countries during the two world conflicts, aerial photography played just as crucial a role in their reconstruction. As the poet Archibald MacLeish claimed, "The airman gained a unique perspective of the earth to see things 'as they truly are.' Removed from the petty squabbles and concerns of daily life on the surface, the airman could dream of utopian futures."[51]

Inevitably, the development of flight and other modern transportation technologies was deeply interrelated with modern geographical imagination and practice. Different technologies produced different experiences of verticality and horizontality, and shaped different perceptions of the "face of the earth." In preparation for his *Tableau de la géographie de la France* (1903), for example, Vidal de la Blache explored the landscapes of France by train. During these long journeys, he observed the landscape unfold through his carriage window, like a film. Solitary mountain hikes and coastal walks complemented the view from the railway and enabled him to engage more closely with the land and its matter. The resulting descriptions were insistently visual and tactile. As he journeyed down the Durance valley in southeastern France, for example, the geographer noted the silvery olive trees, the barren mountains, the small villages perched on craggy peaks and their gray houses "almost blending with the rock."[52] Although it moved along a straight line and required a smooth surface to operate, the railway actually heightened the geographer's topographical perception; it produced an understanding of the earth's surface akin to Mackinder's rippled tablecloth. It helped the geographer turn the map into a terrain relief model.

In 1936 the British economic geographer Stanley Beaver devoted an entire journal article to the study of landscape from a railway train.[53] His goal was to "relate British railway lines to their geological environment" and "describe from the geographical point of view the scenery and cultural landscape which are to be observed from trains traveling over these lines."[54] If on the map the railway appeared as an abstract straight line cutting across the country, at ground level, from inside the train, it allowed the geographer to appreciate the lie of the land in its manifold variety. The geographer was taken through broad cultivated fields carpeting stony soils spread over the chalk; he was carried across undulating fields bearing witness to former cultivations; he admired "mantles of forests" clothing the land; he spotted yellow and brown soils, and egg farms, engineering works, and silk mills. Beaver's descriptions were complemented by cross-sectional geological diagrams of his routes, as though the railway line enabled the geographer to take better notice of the undulations of the earth's surface and its different consistencies.[55]

If the railway contributed to a three-dimensional understanding of the terrestrial surface, this inevitably remained an earthbound understanding. The true revolution in spatial perception was to come from below and especially from above the surface. Less than a decade after the publication of Beaver's article, Harvard geography professor Derwent Whittlesey (1890–1956) advocated an expanded "three-dimensional sense of space." Air power and new communication technologies, he argued, demanded a new geography—a geography able to challenge, or rather to vertically extend, traditional understandings of the earth's surface.

While geographers, including Beaver, used aerial photography to complement what could be seen in the field, Whittlesey's claims were far more radical.[56] Humans, Whittlesey explained, only occupy the land surface of the earth, that is, a two-dimensional plane. At the same time, however, advanced societies use not only the land surface, but also other parts of the biosphere: "its surface waters, the depths of the sea and land, and the heights of the air."[57] The more "civilized" and technologically advanced a society, the more "vertical" its engagement with the terrestrial surface. In the past, two dimensions described both the narrow worlds of small societies and the great regions of Eurasia: "human beings could not descend beneath the surface of oceans, lakes, or streams for more than a few minutes, and they had no means of reaching levels of the atmosphere above their citadel towers. . . . In the vertical plane, society had not progressed significantly beyond the stage of primitive man."[58]

Recent scientific discoveries and mechanical inventions, by contrast, opened the way to utilization of what Whittlesey called the "third or vertical dimension of the earth's surface":

> One of the first uses of the coal-burning steam engine was to pump water out of a coal mine in order to dig more coal. Nowadays petroleum-powered pumps draw additional petroleum from far greater depths. Forced ventilation and compressed air make possible deep mining and tunneling. Many cities now have several storeys of service beneath their principal streets and buildings. Gas engines have made it possible to explore the ocean in the bathysphere, and to carry on warfare by deep submarine. Exploration of the atmosphere began with balloons, and flying (made possible by the gas engine) has both stimulated and aided study of the air. The convectionless stratosphere has been discovered and distinguished from the troposphere. Radio communication, by making novel use of the third dimension, aids flying and navigation.[59]

Penetration of the vertical dimension therefore not only produced a new "geographic sense of space," but also materially transformed the texture of the earth's surface by way of mechanical hardware and infrastructure. Central to the development of this infrastructure, especially communication and transport technologies, was warfare. World War II involved for the first time effective use of radio communications, submarines and undersea weapons, and even more dramatically, of aviation. Powered flight and firebombing enabled the leveling-out of cities and the "flattening" of entire regions ("carpet bombing" was a phrase coined in 1943 to describe precisely this process).[60] No less dramatically, powered flight reduced continents to "arenas," and oceans to "bridges."[61] As President Franklin Delano Roosevelt famously announced in his first fireside chat in 1942, this was "a new kind of war," a war "different from all other wars of the past, not only in its methods and weapons but also in its geography. It is warfare in terms of every continent, every island, every sea, every air lane in the world."[62] In other words, it was a truly global war fought and imagined across the whole surface of the planet, as well as across the whole vertical spectrum of the biosphere. It was "the airman's war."

Ultimately, powered flight and the ability to move freely over the surface of the earth produced a new global consciousness and geographical imagination: the terrestrial surface was no longer a flat plane covered by a wrinkled tablecloth, as early twentieth-century geographers imagined it, but a curved surface. The orthographic views in popular magazines and newspapers, pioneered by American artists such as Richard Edes Harrison and Charles Owens, effectively convey this new perception of the terrestrial surface from cockpit level (fig. 7.2). Capturing continents and regions from unusual viewpoints, and emphasizing relief over the smooth surface of silky seas and oceans, Harrison's maps explicitly stressed "the geographical basis" of world strategy. They combined global consciousness with three-dimensionality. In Susan Schulten's words, they "instantly reminded the user that the world was round and that aviation had created new realities of travel and movement."[63]

After the war, as jet engines entered commercial use, the opportunities opened up by aviation continued to capture the imagination of many geographers, including Ellsworth Huntington (1876–1947). In the last article he wrote before his death, this American scholar speculated on the future consequences of intensified powered flight. For example, he believed that aviation would promote the creation of "one world," but also accentuate contrasts between regions "of different geographic types." Perhaps more intriguingly, Huntington envisaged new mobility patterns, including "vertical commuting" between the plantations of "unpleasant or unhealthy tropical lowlands" and new and healthier "mountain centers" built for the workers' families several hundred feet up. Aviation, he concluded, "is writing the final chapter in making the world smaller," and yet it does "not annul the principles of geography, but merely gives them a new application."[64]

Europe from the Southwest

FIGURE 7.2 The surface of the globe unveiled by the airman's view: Richard Edes Harrison, "Europe from the Southwest." In *Look at the World! The Fortune Atlas for World Strategy* (1944). A. Knopf / Rumsey Map Collection.

From Surface to Palimpsest

Not only did the airman's view expand space upward, or extend time into a utopian future. It also extended space and time in the opposite directions— downward, or below the surface, and backward, into the past. The airman's view served planners busy reconstructing postwar landscapes, but it also served the students of their past. The methodical application of aerial photography to archaeology began in the early twentieth century and was boosted by military aviation in its initial decades.[65] Osbert Guy Crawford, an English wartime pilot and former geography student at Oxford, for example, began a systematic aerial survey of Wessex in the 1920s, after he noticed from the plane patterns of ancient Celtic settlements invisible at ground level. At about the same time, the French Geographical Society commissioned Antoine Poidebard to conduct an aerial survey of Syria, in the course of which, from his biplane, he traced ancient Roman roads and irrigation systems.[66] In revealing ancient vestiges and otherwise invisible patterns, aerial photography opened up (or simply amplified) a fourth dimension of the earth's surface and its landscapes: time.

As the stage upon which humans enact their lives, landscape seems to crystallize the flow of time, and yet be inherently dynamic. In their sheer di-

versity, landscapes are the product of continuously evolving visions and actions. They are "not records, but recordings."[67] Whether modeled by the wind
and the water, by the weight of the plow, or by the brutality of the concrete
and asphalt, landscapes are constantly shaped and reshaped. In this material
sense, landscapes are always temporal. Like human memory, they are also
multidimensional. They are constantly built and rebuilt upon the vestiges of
what they used to be. Detected by the lens of an airman's camera beneath
the brown blanket of the English moorland, or underneath a carpet of fresh
grass undulating away to the horizon, traces of preexisting villages and ancient enclosures added further texture to the land. They also compelled a
new group of geographers and local historians to reconstruct the past "face"
of their country.

Landscape history and historical geography emerged in Britain between
the 1930s and the 1950s as somewhat overlapping subdisciplinary branches
preoccupied with the geographies of the past and, in particular, with transformations of the landscape caused by human action. These included changes in
settlement forms, in cultivation and field systems, and in boundaries of fields
and larger units, such as parishes or counties. The respective pioneers of the
subdisciplines, W. G. Hoskins (1908–92) and H. C. Darby (1909–92), studied
these transformations with a focus on England from the Middle Ages to the
Industrial Revolution.

Darby, a professor of geography at Cambridge, London, and Liverpool universities, elaborated a method that involved the detailed study of past geographies of different regions using "cross-sections," or "snapshots" of a specific
region at different moments in time. These "horizontal" cross-sectional reconstructions could be connected into "vertical" historical sequences by way of a
narrative outlining the changes between the periods studied.[68] For example,
in his seminal paper "The Changing English Landscape" (1951), Darby took
readers on a journey from the arrival of the Anglo-Saxons to Britain's industrial present through different stages, including wood clearing, marsh drain-
age, land reclamation, landscape gardening, and the construction of industrial
settlements. These, Darby believed, were the processes that had forever altered
"the face of Britain"—for better or worse.[69]

As with Vidal de la Blache and Sauer, Darby's understanding of the landscape was deeply material. His cross-sections were akin to geological strata,
or layers of a palimpsest, which the historical geographer had the duty to
uncover from the depths of oblivion: "Not all the ancient writing is legible
through what has been written since, but much of it is, and still more of it
is for those who have eyes to see."[70] Thus understood, landscape was a thick
surface, or rather, a stack of surfaces. Under the scrutiny of well-trained eyes

and with the aid of aerial photography, the land acquired depth, it extended downward in a seamless vertical movement.

The palimpsest metaphor was famously used by nineteenth-century legal historian and medievalist F. W. Maitland to describe the rich texture of the Ordnance Survey map. It became a tradition within British history to envisage features in the landscape as survivals, or "fossils," from previous eras. In his *Tudor Cornwall* (1941), historian A. L. Rowse, for example, stated that "there is no research more fascinating than attempting to decipher an earlier, vanished age beneath the forms of the present and successive layers that time has imposed."[71] The most evocative and vivid uses of the metaphor, however, are found in the pages of Hoskins's celebrated work *The Making of the English Landscape* (1955).

Britain's first professor of English local history (at Leicester University), Hoskins, like Darby, made reconstructing, or rather "unearthing," past landscape layers his mission. His guiding research question was a seemingly simple one: "Why does landscape look the way it does?" The answer, he argued, was to be dug out of the land; it was to be found beneath the surface.[72] The task of the landscape historian, however, he pointed out, was different from that of the geologist. "The geologist explains to us the bones of the landscape, the fundamental structure that gives form and color to the scene and produces a certain kind of topography and natural vegetation." By contrast, the historian, argued Hoskins, occupies himself with "the flesh that covers the bones." His task is to "show how man has clothed the geological skeleton during the comparatively recent past."[73] Hoskins thus guided his readers through strange layers of familiar landscapes: "Many villages have changed their sites for different reasons, but most probably have not. They have been built over and over again and their true age lies buried several feet down in the humus and debris of village gardens and the floors of houses."[74] The result was a transformative journey. "Veils were stripped away as I read the book," the Scottish novelist William Boyd commented in the preface to a late edition of *The Making of the English Landscape*:

> I started looking around me at whatever landscape I happened to be in with new intensity, new insight. The familiar English countryside, in whatever regional variation, became a form of historical palimpsest. . . . It was as if the landscape were, all of a sudden, an archaeological dig . . . and the book in which that history was written was the very land itself.[75]

Hoskins's landscape was articulated between visibilities and invisibilities; it was a landscape made of traces, fossils, and vestiges. Distant periods, Hoskins believed, "are embedded in the landscape around us"—and in the

most physical sense.[76] Beneath the typical English landscape made of enclosures and hedgerows (largely the result of an eighteenth-century parliamen-
tary act), lay the "fossilised" furrows and hedgebanks of ancient open fields
and the footprints of bygone medieval villages, while the dead of the Bronze
Age silently rested "under the heathery blanket of a burial mound."[77] What we
see is merely a sheet covering deeper strata buried in the soil. For Hoskins,
reconstructing these past landscapes meant imaginatively "digging them out"
and clothing them in their original garments. But this was no simple sport.
"Unless facts are right," warned the historian, "there is no pleasure in this imaginative game, if we clothe the landscape with the wrong kind of trees."[78]

Textile metaphors here are used to convey the relative thinness of the historical layer (as compared with the underlying geological skeleton), as well as
the softness of the landforms visible to the eye. This aspect is intensified by
the numerous full- or half-page black-and-white aerial photographs illustrating the book (fig. 7.3). Unseen at ground level except to the eye of the attentive
viewer, fossilized landscapes are fully revealed only from above. Yet, the airman's view is far from being merely documentary; it is also deeply aesthetic.
The photographs used by Hoskins invite readers to "touch" the landscape with
their own eyes, to caress their velvety surfaces and the undulating patterns
of furrows and ridges concealed under the soft grass, as ashes and elms cast
their long shadows over the land.

The softness of a grassy mantle coating ancient arable strips, the smoothness of the turf covering the chalk country, the delicacy of old patterns of light
and shadows projecting on the land at dusk, all speak of a slow time. They
speak of a "slow landscape," of ancient symbioses between the land and the
people. Soft textures conjure up the work of unhurried agencies and natural
processes; they invite melancholy meditation on what is no more.

> Landlords come and go, but farming goes on. Every harvest was a gamble
> whatever work went into the preparation: villas may have been destroyed or
> even decayed like many a modern house, but the tenant farmers go on from
> year to year and generation to generation. Across in Suffolk a massive em-
> bankment, deeply ditched on both sides, divides the hundreds of Bosmere
> and Hartismere and it has its own name—Hundred Lane—followed by parish
> boundaries of the tenth century. It is slowly being destroyed by modern farm-
> ing machinery.[79]

Hoskins's English landscape was not simply a green carpet gently covering
rolling hills and fossils of the past. It was also a wounded landscape; it was a
landscape threatened by the flattening action of modernity; it was a landscape
marked by the scars inflicted by postwar industrial society and human greed.

FIGURE 7.3 Landscape as a palimpsest: the site of the deserted village of Lower Ditchford (Warwickshire) revealed by aerial photography under the green mantle of grass. The village probably disappeared sometime in the mid-fifteenth century. The ridge-and-furrow pattern, marking the former open-field arable ground, is also revealed in this view. From W. G. Hoskins's *The Making of the English Landscape* (1955). © Cambridge University Collection Air Photographs.

Hoskins's vivid ruminations on the wounded English countryside are, in a sense, akin to Alain de Lille's medieval descriptions of Nature's torn garment. Human degradation is laid bare with the same intensity of pathos:

> What else has happened in the immemorial landscape of the English coun-
> tryside? Airfields have flayed it bare. . . . Poor devastated Lincolnshire and

Suffolk! And those long gentle lines of the dip-slope of the Cotswolds, those misty uplands of the sheep-grey oolite, how they have lent themselves to the villainous requirements of the new age! . . . England of the Nissen-hut, the "pre-fab," and the electric fence, of the high barbed wire around some unmentionable devilment; England of the arterial by-pass, treeless and stinking of diesel oil, murderous with lorries; England of the bombing-range wherever there was once silence. . . . Barbaric England of the scientists, the military men, and the politicians; let us turn away and contemplate the past before all is lost to the vandals.[80]

As David Matless observed, here Hoskins presents a postwar "alliance of planning, science, industry and the military tearing up the country."[81] Far from being the God's-eye hero of reconstruction promising citizens a better future, the planner emerges from the dusts of the war as a dark character flattening and erasing what is old and precious.[82] But the scars modernity inflicted on the land were not as fresh. For at least two generations, lamented Hoskins, "every single change in the English landscape has either uglified it or destroyed its meaning, or both":

Only the great reservoirs of water for the industrial cities of the North and Midlands have added anything to the scene that one can contemplate without pain. . . . The country houses decay and fall: hardly a week passes when one does not see the auctioneer's notice of the impending sale and dissolution of some big estate. The house is seized by the demolition contractors, its park invaded and churned up by the tractors and trailers of the timber merchant. Down come the houses; down come all the trees, naked and gashed lies the once beautiful park.[83]

Darby too denounced the disfiguring of the English landscape caused by industrialization.[84] In "The Changing Landscape of England," aerial views of the countryside are dramatically juxtaposed to high-oblique shots of grim industrial landscapes, while long excerpts from nineteenth-century accounts are used to convey the dreary character of the latter. The effect is powerful and disturbing. As various commentators have noted, however, there is a distinctive and poignant melancholy to Hoskins's writings that sets them apart from other works.[85] There is a sense of nostalgia for a world that is quickly vanishing. But there is also an uncanny aesthetics and a poetic halo shrouding the industrial ruins of a not-too-distant past, as they become quietly assimilated to the texture of the surrounding landscape:

Just across the Devonshire border is the old mining landscape of Blanchdown, west of Tavistock, where, in the middle decades of the nineteenth century, the Devon Great Consols was the richest copper mine in the world: now its miles of spoil-heaps have created a silent and desolate beauty of their own, and foxes and snakes haunt the broken buildings and the glades between. There is a point, as Arthur Young saw, when industrial ugliness becomes sublime.[86]

✻

Surfaces are the skins of tangible objects; they are what our bodies come up against in touch; they are the outward appearance of things—in other words, what we see. As visual psychologist James Gibson noted in the 1950s, surfaces lie between medium and substance. They are where matter and energy interact; where light is absorbed and reflected; where textured patterns are revealed to the gaze.[87] In nature, textures, that is, the structures of surfaces, are always heterogeneous. A perfectly homogeneous and perfectly smooth surface, argues Gibson, is an abstraction. Even the surfaces with which humans carpet the earth—the pavements of concrete, asphalt and other aggregates—are "differently textured."[88] Gibson uses a series of six photographs depicting different kinds of familiar natural surfaces to illustrate the point. One shows the transverse surface of sawn wood, another shows clouds in the sky, another a field of mown grass, another a woven textile, another the rippled surface of a pond, and another a patch of gravel. In each case, comments Tim Ingold, "we can recognize the texture visually because of the characteristic scatter pattern in the light reflected from the surface."[89]

The practitioners of modern geography studied the textures and patterns of the terrestrial surface. Yet, the surface of the earth reflected back not only patterns of light, but also different ways of seeing and thinking about the environment. Early professional geographers, such as Semple and Mackinder, envisaged the terrestrial surface as a relief map (or tablecloth) defined by craggy peaks and narrow valleys that trapped populations and by the open plains and broad waterways that impelled them to move. Others placed further emphasis on the action of humans and their ability to imprint the surface in their own image, or weave their own image into its texture. Others reduced its topography to a flat geometrical surface—to an impossible cartographic abstraction that nonetheless had the power to "flatten" the actual landscape. Prompted by, or responding to, new technological developments, others extended their gaze upward, into the atmosphere; still others downward, into the vestiges of landscapes buried by distant pasts. Some cherished human action, others mourned it. What happened when technology allowed humans to pierce the terrestrial crust and the mantle of the sky? The following chapter explores the vertical expansion of these visions and the return of unveiling metaphors.

Pierced Surfaces and Parted Veils

On October 4, 1957, Sputnik, the first artificial satellite, was successfully put in orbit by the Soviet Union. The news came first from Radio Moscow. The following day, a headline in the *New York Times* reported confirming details: "Soviet Fires Satellite into Space: It Is Circling the Globe at 18,000 M.P.H.; Sphere Tracked in 4 Crossings over U.S." According to the Soviet Union, the article stated, "its sphere circling the earth had opened the way to interplanetary travel."[1] In piercing the veil of the atmosphere, Sputnik inaugurated a new age and a new perception of space. As rockets turned their trajectories 90 degrees upward, the horizontal perspective of the European landscape tradition and the high, oblique airman's view of the early twentieth century both shifted to an altogether different axis, what Fraser MacDonald aptly termed "the perpendicular sublime."[2] From surface, tablecloth, or carpet, the earth's mantle became a breached shell, or a parted veil.

From the Age of Discovery through the mid-nineteenth century, geography and exploration were largely a surface matter; veils and curtains opened on a predominantly horizontal plane. Even earlier in the twentieth century, when the "mantle" acquired thickness, it nonetheless remained an involucre wrapping the planet—a sort of vertical extension of its surface.[3] In an influential article published in 1931, for example, Mackinder expanded his "tablecloth" metaphor to "a bubble of varying thickness at the surface of the globe, roughly corresponding to the combined space which mankind uses and occupies." This bubble, the hydrosphere, argued the British geographer, was "the chief instrument in weathering, transporting, and shaping the surface mantle of the lands and shallow seas." Encompassing the watery element in all its manifestations, from oceans and rivers to glaciers, clouds, and rain, it operated "constantly to amend the soil, the natural resource from which mankind draws most of its sustenance and much of its shelter."[4]

By 1950, Carl Troll defined geography as "the science that studied the phenomena of the earth's envelope—lithosphere, hydrosphere, and atmosphere—in their local differences and functional interconnections."[5] Shortly thereafter, Gibson turned his attention to the space between the earth's surface and its outer envelope—the atmosphere. Influenced by the recent impacts of powered flight, he interrogated airmen's spatial perceptions. "The abstract question of how one can see a third dimension based only on a pair of retinal images extended in two dimensions," he believed, had become "very concrete and important to the man who was required to *get about* in the third dimension."[6] These three-dimensional geographies nonetheless remained confined to the interior of the envelope and deeply anchored to the ground. "A pilot who cannot see the ground or sea," Gibson concluded, "is apt to lose touch with reality in his flying."[7] Even the "vertical geographies" advocated by Whittlesey were limited to the interior of the earth's envelope. It was not until the late 1950s that exploration turned to the vertical axis, piercing the envelope and "cutting adrift the solidity of the earth."[8]

By the end of World War II, Whittlesey had acknowledged the inadequacy of Mackinder's "hydrosphere" as the framework through which to explain modern human activities. The watery bubble, he believed, failed to capture their range and scope. The American geographer deemed Vladimir Vernadsky's "noosphere" a better concept for capturing the space of human activities, since it allowed indefinite further expansion, so as to encompass other planetary spheres, as far as the human mind (Greek, *nous*) could stretch.[9] Although Whittlesey did not live enough to witness the launch of Sputnik, the little Soviet satellite became the starting point for the space race and the icon of the outward expansion imagined by Vernadsky. The noosphere's frontier, however, also moved in the opposite direction. Four years later, as Yuri Gagarin was orbiting the planet, a team of American scientists attempted to pierce the terrestrial crust down into the geological mantle. This mirroring downward movement was to unveil a realm as mysterious as outer space.

Paradoxically, the continued vertical expansion of human frontiers into these, and other, realms of the unknown was both triggered and enabled by another veil—the Iron Curtain. Dating to the eighteenth century, the phrase originally referred to fireproof curtains in theaters. It was used already in the early twentieth century to denote closed geopolitical borders, but its popularity as a Cold War symbol is attributed to a speech Winston Churchill gave in 1946 (fig. 8.1).[10] By the early 1950s, the phrase was deemed "a rather old metaphor which, because of its apt description of the relations, or rather lack

FIGURE 8.1 Leslie Gilbert Illingworth, "Churchill Having a Peep under the Iron Curtain," *Daily Mail*, 6 March 1946.

of relations, between East and West Europe, has caught the public fancy."[11] In Patrick Wright's words, to penetrate the Iron Curtain was "to make a journey that could be charted on a conventional map, yet it was also like falling off the edge of the known world."[12]

Besides secrecy, the curtain metaphor also carried its original theatrical connotation. Massive shows of technological might were enacted on both sides of the curtain.[13] From time to time, breaches or programmed "unveilings" would reveal monstrous weapons meant to strike terror in the opponent.[14] The introduction of the hydrogen bomb made it increasingly impossible for the rival blocs across the curtain to face each other militarily. Doing so would have led to nuclear annihilation.[15] As a result, the ideological struggle had to be played out in other ways, including a symbolic competition theatrically pushed to new, extreme domains: the polar regions, outer space, and the earth's interior. In closing up, the globe opened up new profundities. This chapter traces the resurgence of the Age of Discovery's "veil" and "curtain" metaphors and their new incarnations in these three domains of Cold War scientific exploration.

Lifting the Veil of Ice

The 1950s saw unprecedented scientific investigation into the earth's fabric and outer space alike. Much of the impetus came from the International Geophysical Year (IGY) of 1957–58, a "scientific Olympics of sorts," or, as a contemporary observer called it, "a symphony" making "a concerted, globe-girdling study of the so-called 'earth science.'"[16] The feat involved 60,000 scientists and amateurs from sixty-seven different countries in a worldwide enterprise of data collection, analysis, and exchange.[17] The idea of "experiments in concert" as the most effective way in which to understand the world was attributed to Francis Bacon. For its commentators, the IGY marked the fulfillment of his vision and "the growing awareness that only this sort of approach could bring substantial advantages in a number of sciences."[18] Advertised as "man's most ambitious study of his environment," "the world studying the world," "a new age of exploration," or as an "assault on the unknown," the IGY was largely presented to the public as an international cooperative venture transcending geopolitical divides and ideological barriers.[19] "Tired of war and dissention," a dedicated issue of the *National Geographic* proclaimed, "men of all nations have turned to mother Earth for a common effort on which all find it easy to agree."[20]

The IGY ushered into existence a new phase of "vertical" exploration, from the oceanic depths to the higher regions of the atmosphere, from the deeper layers of the terrestrial crust to the infinite blackness of outer space. Developments in space technologies and sounding rockets provided a detailed picture of the atmosphere and of the earth from space, while geological strata were breached vertically through radio waves.[21] In this way, a new global imagination, made of integrated "shells" and "layers" waiting to be pierced and penetrated, was crafted and promoted in the popular media: "Scientists of the world are going to take a long and special look at our earth—at its wrinkled crust, its hot hearth, its deep seas, its envelope of air, its mighty magnetism, its relationship to outer space."[22] The ultimate goal was to turn the earth into a "great laboratory" and contribute to "man's unceasing search for clearer understanding of his environment and to the fuller appreciation of its practical and spiritual values by all peoples."[23]

If science was deemed a powerful instrument of peace able to "break," or at least "soften" the Iron Curtain, the IGY was conceived of as the largest and most dramatic manifestation of this vision.[24] Of course, the IGY was the exception, rather than the rule, and even some of the projects falling under it heightened tensions and suspicions across the curtain, boosting competition between the two blocs. This led to a new alignment between "Big Science" and the military, as well as to huge investments in certain areas of research.[25]

While inevitably underpinned by the Cold War "veiling" culture—what Michael Gordin called "knowledge about knowledge"[26]—the IGY triggered a vocabulary of unveiling akin to that of the Renaissance Age of Discovery. As the popular image of the scientist as "the explorer of the unknown" took precedence, the old motif of "unveiling" the secrets of nature gained new currency and appeal.[27] Coupled with militaristic tropes, the unveiling act came to enshroud a new epistemic violence, as well as a new promissory potential. In the 1954 IGY planning meeting in Rome, for example, it was noted that remote regions, otherwise inaccessible, would "be open" during the IGY.[28] Among these regions were the polar regions, and especially the vast and still largely uncharted Antarctic continent.

Unveiling metaphors underpin the history of polar exploration from its inception in heroic Victorian Britain to the IGY scientific expeditions. For example, in the 1890s John Murray claimed that "the key to many puzzles, the end of many controversies affecting the theory of the phenomena of the whole world, lies behind the vast Antarctic veil" and that it was "the duty of the human race to lift that veil, wherever there be much or little behind it."[29] In the Victorian geographical imagination, the Arctic was likewise shrouded in an icy veil waiting to be raised by intrepid explorers. Rhymed poems and cartoons in the satirical magazine *Punch* portrayed the region as a veiled "Polar Queen to be won," the "White Ladye of the Pole," or as a "Bride of Snow and Death" drawing "manly hearts with strange desire to lift her icy veil."[30]

In an 1876 illustration, for example, the mighty Ice Queen is depicted atop an iceberg holding her scepter and looking out over her unwelcoming dominion. At her foot, an expedition party struggles its way through the treacherous landscape of ice (fig. 8.2).[31] An icy veil falls from the queen's crown over her shoulders; it blends her body into the iceberg, so to make the woman appear a natural continuation of the landscape she silently commands. The scene is lost in the desolation of a black Arctic night. As Heidi Hansson comments, "the Arctic is represented as a virgin land, available to male exertion and British colonization. . . . The Ice Maiden is colorless, immobile, captured by the snow, silent and mysterious. Her power lies in the implication that she is a *femme fatale* who lures men to their destruction by keeping her secrets."[32] The queen's veil reinforces this sense of mystery. It reminds readers that the secrets of the Arctic lie forever locked in the ice.

This and similar cartoons were produced shortly after the unveiling of the memorial to Franklin and his crew, who had lost their lives thirty years earlier during an ill-fated expedition in search of the Northwest Passage (1845–46). As such, they were part of a broader narrative seeking to justify lack of success through descriptions of unbearable Arctic conditions.[33] The icy veil

A COLD RECEPTION
(ARCTIC REGIONS, 1875.)

FIGURE 8.2 "A cold reception (Arctic Regions 1875)," *Punch*, 11 November 1876, 203. The Ice Queen waits to be unveiled by the British explorers.
Author's private collection.

nonetheless remained a popular trope in twentieth-century polar exploration. Roald Amundsen (1872–1928), for example, spoke, in his memoirs of the
South Pole expedition, of "unveiling" Antarctica.[34] His rival, Captain Robert
Falcon Scott (1868–1912), likewise recorded in the diary of his tragic expedition, "Dear Victoria lift[ing] her veil and display[ing] her charming contours,"
after a morning of fog and mist.[35] By then, "lifting the veil" as a metaphor for
discovery was firmly established in Anglo-Norwegian polar discourse.

In the 1950s, with polar science and exploration assuming increasingly militaristic overtones, the metaphor adjusted accordingly. For example, a special color map (pl. 6) featuring in a glossy issue of *National Geographic* dedicated to the IGY was pompously subtitled "Expeditions Led by Adm. Richard E. Byrd Tore the Veil from Half of Antarctica."[36] The youngest admiral in the history of the US Navy, Byrd (1888–1957) led a number of Antarctic expeditions between 1928 and 1957, including the first flight over the South Pole and the exploration of the uncharted western portion of Antarctica east of the Ross ice shelf. As part of the multinational collaboration for the IGY, in 1955–56 Byrd commanded the US Navy Operation Deep Freeze I, which established three permanent Antarctic bases, including the Scott-Amundsen base at the South Pole. Author of several books, articles, films, lectures, and even a radio program shortwaved from his camp base station, Byrd was one of the key popularizers of Antarctic science and probably the main individual responsible for bringing the southern continent into the homes of thousands of Americans.[37]

The *National Geographic* map belongs to this popularization campaign. It features the southern continent as a blank surface penetrated by the colored "fingerprints" of Byrd's expeditions. Dark green, fiery red, and other vivid colors mark the progress of the admiral's explorations and produce a stark contrast with the pure whiteness of the Antarctic continent. "Putting the fingers" on the white continent resonates with the violent act of "tearing the veil apart" in order to penetrate nature's mysteries. It also resonates with the militaristic lexicon surrounding Cold War Antarctic exploration and with the many photographs illustrating Byrd's dramatic accounts: from icebreakers cutting through the veil of pack ice (fig. 8.3) to members of the expedition dynamiting crevasses in order to allow their tractors into the continent's interior.[38]

Severed by technology, the Antarctic "white mantle" is described as an alluring yet treacherous surface. This ambiguity is once again conveyed through the personification of Antarctica and the use of textile metaphors, such as the soft snow "blanketing" the pole and precluding aircraft from landing, the white mantle "gently shrouding" the body of Scott buried at the South Pole, or the "ghostly veil of snow" blowing from the "jagged face" of the Ross shelf.[39] As in Victorian polar accounts, in its various guises, here the icy veil acts as a barrier concealing polar secrets, and yet this time being torn by means of grand-style "assaults."[40]

For Byrd, however, the southern continent was more than an arena for physical struggle and heroic conquest. Its white mantle also held a spiritual value. "I am hopeful that Antarctica, in its symbolic robe of white, will shine

FIGURE 8.3 "Smashing through the ice six feet thick, *Glacier* batters down Antarctica's guard." In Richard Byrd, "All-out Assault on Antarctica: Operation Deep Freeze Carves Out United States Bases for a Concerted International Attack on Secrets of the Frozen Continent," *National Geographic*, August 1956, 170.

forth as a continent of peace," the American explorer wrote.[41] Earth's only continent without a native human population, by the late 1950s Antarctica came to enshrine the values of international cooperation promoted by the IGY and the subsequent Antarctic Treaty.[42] Before the space race turned public attention upward, the southern continent was the key focus, and in a sense, the icon, of the IGY. Over the duration of the program, twelve countries, including the United States and the Soviet Union, and no fewer than five thousand scientists, established more than fifty research stations on the continent, in order to investigate its geophysical qualities and the electromagnetic properties of the planet.[43] Antarctica suddenly turned from an unexplored geographical area into a vast laboratory for science and peace.

This discursive shift translated into a shift from a horizontal to a vertical axis. One of the major outcomes of IGY scientific investigations in Antarctica was the measurement of the thickness of the ice sheet and the subsequent "unveiling" of the subterranean landscapes it concealed. The technique that was initially employed was seismic shooting, or explosion seismology. This technique entailed the physical piercing of the icy veil (or rather, of its upper part). A series of holes were drilled at different points along a transect. Explosives were placed in the holes and detonated. Ice thickness at the various locations was established by measuring the time it took the resulting shockwaves to travel down through the ice to the underlying rock strata and return up to be detected by geophones.[44]

Seismic shooting was used by the French in the late 1940s to sound the Greenland ice cap. The resulting image was that of an "ice-filled bowl rimmed by coastal ranges" (fig. 8.4). Scientists compared the ice cap to an "alabaster shroud" overlaying four-fifths of the island and burying "a lost world under thousands of feet of ice."[45] Seismic sounding in Antarctica was pioneered during the 1933–35 Byrd expedition, but it was vastly expanded during the IGY "assaults," as US naval planes and overland tractors crossed the continent taking soundings along new and longer transects. The results were revelatory: the ice sheet turned out to be much thicker than had been estimated, averaging more than 2,000 meters.[46] Further unveilings, however, were to follow with the development of another technique, radio-echo sounding (RES).

The origins of RES are shrouded in tragedy. In the 1950s, radar altimetry turned out to work unpredictably in the Antarctic, as the ice sheet is almost transparent to radio waves. The instruments thus frequently recorded the bottom surface rather than altitude above the ice, causing aircrafts to crash on the ground. The erroneous working of the device was nonetheless exploited to take vertical measurements through the ice sheet during (and after) the IGY. Radio engineer Amory Waite (1902–85) produced a modified radar altimeter to investigate the bottom surface by way of radio waves.[47] As a result, "beneath the white and pristine Antarctic surface an entire new world was uncovered in greater resolution than ever previously possible, a world made of valleys, mountains, lakes and plateaux in which the ice goes as deep as 4,776 meters."[48]

Seismic shooting and radio-echo sounding led to the emergence of new three-dimensional subglacial maps and cross-sectional representations of

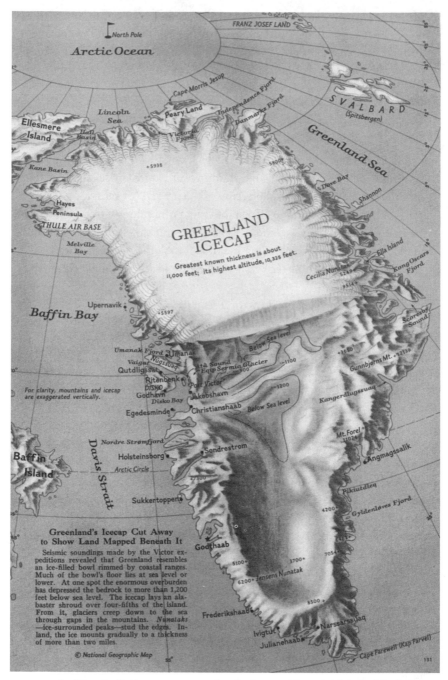

FIGURE 8.4 "Greenland's icecap cut away to show land mapped beneath it," in Paul-Emile Victor, "Wringing the Secrets from Greenland Icecap," *National Geographic*, January 1956, 121.

the Antarctic continent fundamentally different from earlier ones, which had imagined it as an all-encompassing white blanket. The very idea of geographical exploration was extended vertically, to the new, uncharted territory under the ice.[49] Probing the face of Antarctica meant lifting its white veil in the literal sense, as invisible radio waves penetrated into the icy mantle, bringing invisible landscapes back to the surface. As Antarctica was discursively transformed from an unexplored geographical area into a scientific lab, the horizontal icy veil morphed into a vertical translucent archive enshrining the secrets of the planet and of its past ages. As a *Time* correspondent wrote in 1960,

> Only a handful of hardy explorers had ever probed the cruelly impassive continent of Antarctica and they scarcely guessed what mysteries the immense ice mass concealed. But the IGY brought a concert of the world's scientists and armed forces, mobilized in an unprecedented example of nations' cooperation, to unveil the land that had lain in forgotten, frozen slumber.[50]

Piercing the Earth's Envelope

Radio waves not only penetrated the polar ice; they also pierced the upper layers of the atmosphere. Although radio signals were initially thought to travel in a straight line, in 1901 Guglielmo Marconi had been able to receive a signal sent from across the Atlantic, beyond the earth's curvature. The existence of a reflecting layer "high in the sky" was therefore postulated, and scientific interest moved to the unexplored upper regions of the atmospheric "bubble." In 1923, amateur radio communications led to the discovery of short-wave propagation via the ionosphere, a region including four electrically conductive layers located between 60 km and 1,000 km above ground level and strongly influenced by solar activity (see fig. 8.5).[51] This newly discovered region nevertheless remained wrapped in mystery. Experiments, including sounding rockets and high-altitude explosions, were proposed in the 1930s, but eventually abandoned, because of the Great Depression and the war. One of the main objectives of the IGY was thus "piercing" the upper veil of heaven and studying the "vertical geography" of the atmosphere and the electromagnetic radiation on it, as well as solar storms and auroras, among other phenomena. It was therefore proposed that the IGY be scheduled in 1957–58, in order to make it coincide with an expected peak of solar activity, which would have offered optimal conditions for study.

Intriguingly, the same set of textile metaphors employed in Antarctic exploration was also used to describe this still largely uncharted vertical domain.

Besides conveying the sense of mystery and enabling readers to visualize otherwise invisible phenomena, mantles, veils, and curtains embedded that deep aesthetic quality that had made them so appealing throughout Western history—from the Church Fathers to Humboldt. An article in *Life* magazine, for example, poetically likened the atmosphere to a "canopy of air," or an "airy mantle," while identifying it as the ultimate factor for much of the beauty of the environment:

> Every mood and spectacle of nature that most deeply touches [humans'] es-
> thetic spirit springs from the union of light and air. The blue sky and the blue
> sea, white clouds, green twilight and pearly November mists, rainbows, the
> flash of lightning—all these flow from the palette of the encircling air. For the
> atmosphere intercepts the sun's radiation which contains all the wavelengths
> of the visible spectrum. It happens that air molecules are the proper size to
> intercept the short blue wavelengths, deflect them on their flight to earth and
> scatter them across the sky. The azure "firmament" is thus no glassy dome or
> vault, but a gauzy, glowing fabric spun of blue light and air.[52]

Auroras were likewise described as beautiful textiles. The same author, for example, noted how "a drapery aurora develops when parallel streamers of light, spreading upward from horizon to zenith, fall into a graceful pattern that appears to shimmer like the folds of a great curtain hanging from the sky and stirred by a silent wind."[53] In his commentary on the achievements of the IGY, *New York Times* correspondent Walter Sullivan similarly referred to auroras as "curtains of light that hung almost nightly over northern Scandinavia." The effects of magnetic storms he described as "gigantic sheets of electricity" enveloping the night side of the earth, or as "heavy blankets of light" sinisterly advancing from the poles to the tropics.[54] Byrd himself saw in auroras a manifestation of unspeakable beauty—what he called the aesthetic side of magnetic storms—while also alerting his readers that "when the aurora appears, an 'electric curtain' will shut off our radio signals for days from the world."[55]

Like the icy veils shrouding the poles, in the twentieth-century American popular imagination, the invisible atmospheric layers and electric veils wrapping the planet remained surfaces waiting to be pierced. Space, it has been noted, was always a metaphorical extension of the American western frontier. Frederick Jackson Turner's claim that the frontier was central to American national identity could thus be equally applied to twentieth-century space exploration.[56] According to the French science historian Jacques Arnould, however, "piercing" the upper veil of the atmosphere was a more universal task. As the ultimate frontier of human exploration, the celestial veil has always held an almost sacred quality. If sacred is that which is set apart, "how then could the sky not be declared sacred, perhaps even first, before anything else? Its commensurable elevation and its terrifying infinity, its cold immutability and its formidable power imposed themselves on the newly-budding human consciousness, in a sort of primordial separation."[57] As such, the sky naturally occupied, and continues to occupy, the dreams, hopes, and longings of humans. Yet, Arnould observes, the sacred cannot exist without transgression. "One needs only dare and have the means to cross thresholds . . . to tear open veils."[58]

In the 1950s, technological advances set the conditions to "tear open" the celestial veil. Long-range missiles had emerged as standard weapons during World War II. Aimed straight up, they could virtually penetrate the entire spectrum of the atmosphere. V-2, a missile "kidnapped" from the Germans at the end of the war and put to the service of the American military and scientists, was the first two-stage liquid-propellant rocket and the first object to penetrate into space. In 1950, another rocket, Bumper WAC, broke extant records, reaching an altitude of 224 miles. Equipped with special measuring instruments, sounding rockets could provide direct evidence of what happens above the atmosphere.[59] In this way they instilled a new, vertical imagination. They gave unprecedented uplift to all sorts of modernist exploration fantasies, "from walking on the moon to interplanetary travel."[60] Rather than turning the cosmos into "space," they seemed to accentuate its sacredness. As a *Life* commentator wrote in 1953,

> So, in dreaming of a passage beyond his natural medium of existence, man will indeed be casting himself into an environment for which he was never designed. Ancient cosmologists were not altogether wrong when they regarded the blue vault of heaven as the palpable roof or ceiling of the world. For the blue sky does indeed mark the upper boundary of the useful, life-giving atmosphere—the surface of the airy ocean on whose floor man dwells and above which he may not rise unarmored into the hostile and limitless domain of interstellar space. In its combination of beauty and mystery, the ever-changing canopy of air that encompasses and shelters the earth, that gives it light and warmth, color and rain, and provides the breath of all living things, has always stirred the emotions of perceiving man. "Sometimes gentle, sometimes capricious, sometimes awful," wrote John Ruskin of the sky, "never the same for two moments together: almost human in its passions, almost spiritual in its tenderness, almost divine in its infinity."[61]

The IGY dramatically accelerated the vertical ascent to the unknown. Popular magazines routinely announced "scientists firing missiles equipped with electronic eyes and ears" and "probing mysteries on the borders of outer space."[62] At the outset of the IGY, thirty-four American Aerobee rockets were said to belch into the sky "to measure the ionosphere electromagnetic fields," while two years earlier the White House announced the launch of an artificial satellite to take place during the IGY.[63] In the meanwhile, Sputnik caught the world by surprise.

In breaching the veil of the sky and entering outer space, Sputnik highlighted the thickness of the Iron Curtain and the tensions and suspicions on both sides of it. After its unexpected launch, the little Soviet satellite traced a path of anxiety across the night sky of the western hemisphere.[64] In spite of the IGY cooperation imperatives, Russians were unwillingly to share the satellite's orbital data with the rest of the world scientific community, as they feared the data could be used to identify the launch site and therefore disclose secret military information. The following year, Americans successfully launched their own satellite. By the end of 1959, they had fired almost 300 research rockets. The Soviets had launched no fewer than 175.[65] Dramatic space records followed one another at breathtaking speeds: the Russians soon sent the first animal into space (1957), followed by the first human in space (1961), and they achieved the first spacewalk (1965), while President John F. Kennedy pledged that the United States would land a man on the moon by the end of the decade.[66]

In addition to "giant rockets," the war had produced many other weapons and instruments that could be put to the service of space exploration. As Sullivan evocatively put it, "there were many other new eyes with which to peer into the mysteries of science."[67] During the IGY, manned balloons and unmanned "rockoons" (hybrids of balloons and rockets), for example, were sent up into the troposphere, breaking previous records and allowing prolonged observations over a specific area from "the fringe of the great void."[68]

As balloonists were for the first time given to see the different shadows of the sky at unprecedented altitudes, cross-sectional charts of the atmosphere and its different layers (fig. 8.5) populated the pages of scientific and leisure magazines, along with visual renderings of revived Baconian *topoi*, such as "unveiling" or "unlocking" the secrets of the universe (fig. 8.6; see also figs. 5.2 and 5.3). Through a compelling exchange of metaphors, new space technologies were said to have broken "the ice of ignorance where it happens to be thinnest," while parts of the Antarctic continent were claimed to be "no more familiar than the moon."[69]

Scientists, however, were not content to penetrate the "upper veil" through rockets, stratospheric balloons, or the powerful lenses of their new telescopes.[70] They endeavored to part it, to tear it asunder—and artificially craft new ones. One of the most important discoveries that took place within the

IGY framework was the discovery of the Van Allen radiation belt, a zone of energy-charged particles originating from the solar wind and held around the earth by its magnetic field at an altitude of about 500 km to 58,000 km. Soon after the discovery, an experiment involving a series of very high-altitude atomic tests over the South Atlantic was proposed and secretly approved by the Pentagon. Code-named Argus, after the hundred-eyed giant of classical mythology (see fig. 5.4), the operation was aimed at seeing what the effects of the explosion would be on the Van Allen belt. "It might be amusing to end the IGY by destroying some of the radiation field first discovered during the IGY," physicists Edward Ney and Paul Kellog sarcastically commented.[71]

Nicholas Christophilos, who conceived of the experiment, believed that high-altitude nuclear detonations would produce radiation belts similar in effect to the Van Allen belt. Given forthcoming bans on high-altitude atomic tests, the operations were conducted in secrecy within a mere half-year of conception. The IGY was used to "cover" the experiment's military purpose. Artificially produced radiation belts were indeed viewed as having possible tactical use in war, including "degradation of radio and radar transmissions, damage or destruction of the arming and fuzing mechanisms of ICBM warheads, and endangering the crews of orbiting space vehicles that might enter the belt."[72] The experiment indeed caused radio signals across the Atlantic to fade, and three artificial electron shells were said to have been "secretly wrapped around the earth."[73] An invisible mantle of planetary dimensions had for the first time been woven by human hands.

Once made public, the experiment was greeted by the American press as "an intellectual triumph . . . an experiment that enveloped almost the entire planet."[74] It also opened the way to even bolder Faustian visions and more ambitious projects. In May 1962 the US government announced its intention to conduct a series of very high-altitude atomic tests in the Pacific. Detonated some 250 miles above Johnston atoll, the first bomb produced a giant blood-red artificial aurora and initiated a magnetic storm in the central Pacific. The experiment aroused controversy. Bernard Lovell, the British scientist who had famously tracked Sputnik, feared the impact of the explosions on the makeup of the atmosphere. He also criticized the unilateral way in which the experiment had been announced. "The much-vaunted dedication of the United States to peaceful use of extra-terrestrial space," he complained, "will now be seen as a veil that can be torn asunder at the convenience of the American militarists and their attendant scientists."[75]

For their part, American commentators welcomed such experiments as "a start in the exciting business of modifying the upper atmosphere."[76] In their shameless boldness, these experiments are to be placed within the broader

FIGURE 8.5 Detail of map of North America featuring the layers of the atmosphere and ionosphere traversed by radio waves and solar radiations. *Life*, November 1960.

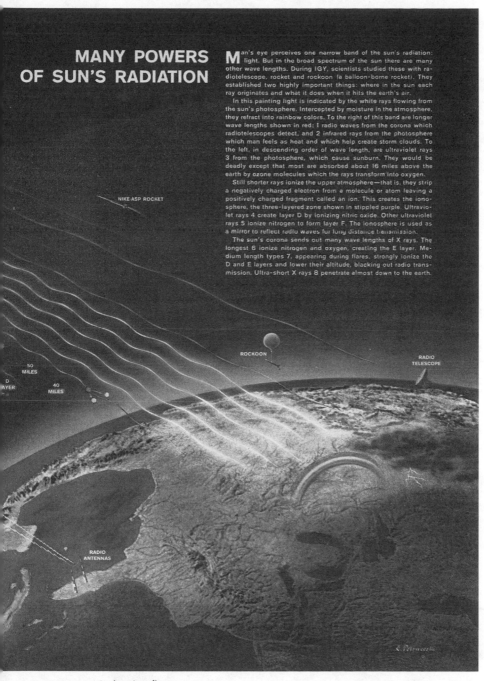

MANY POWERS
OF SUN'S RADIATION

Man's eye perceives one narrow band of the sun's radiation: light. But in the broad spectrum of the sun there are many other wave lengths. During IGY, scientists studied these with radiotelescope, rocket and rockoon (a balloon-borne rocket). They established two highly important things: where in the sun each ray originates and what it does when it hits the earth's air.

In this painting light is indicated by the white rays flowing from the sun's photosphere. Intercepted by moisture in the atmosphere, they refract into rainbow colors. To the right of this band are longer wave lengths shown in red: 1 radio waves from the corona which radiotelescopes detect, and 2 infrared rays from the photosphere which man feels as heat and which help create storm clouds. To the left, in descending order of wave length, are ultraviolet rays 3 from the photosphere, which cause sunburn. They would be deadly except that most are absorbed about 16 miles above the earth by ozone molecules which the rays transform into oxygen.

Still shorter rays ionize the upper atmosphere—that is, they strip a negatively charged electron from a molecule or atom leaving a positively charged fragment called an ion. This creates the ionosphere, the three-layered zone shown in stippled purple. Ultraviolet rays 4 create layer D by ionizing nitric oxide. Other ultraviolet rays 5 ionize nitrogen to form layer F. The ionosphere is used as a mirror to reflect radio waves for long distance transmission.

The sun's corona sends out many wave lengths of X rays. The longest 6 ionize nitrogen and oxygen, creating the E layer. Medium length types 7, appearing during flares, strongly ionize the D and E layers and lower their altitude, blacking out radio transmission. Ultra-short X rays 8 penetrate almost down to the earth.

NIKE-ASP ROCKET

50 MILES

D LAYER

40 MILES

ROCKOON

RADIO TELESCOPE

RADIO ANTENNAS

FIGURE 8.5 (*continued*)

FIGURE 8.6 "Unlocking the Secrets of the Universe," advertisement for Union Carbide & Carbon Corp., New York. *National Geographic*, August 1956.

framework in which the IGY took place—what Klaus Dodds and Lisa Funnell called "elemental geopolitics," the control over resources coupled with the idea that humans could radically transform, or indeed engineer, the earth's mantle by way of atomic power.[77] At a time when the earth was commonly associated with resource exploitation, and environmental plunder was decisively accelerated, there were enthusiastic proposals on both sides of the curtain, for example, to use atomic explosions to dig a new Panama canal, to

PLATE 1 The Second Coming with the heaven "departed as a scroll when it is rolled" (Rev. 6:14). Chōra monastery (Kariye Djami), Istanbul, fourteenth century.
Photo: Fusion of Horizons.

PLATE 2 Giotto, *Last Judgment*. Two angels on the top roll up the skin of the firmament, which continues behind the window. Cappella degli Scrovegni, Padua, 1305.
Photo by the author.

PLATE 3 Hans Memling, *Crucifixion*, 1470. Panel featuring Saint Veronica holding her holy veil. Here the verniclе is no longer a planetary blanket imprinted with the Creator's vestiges (as in the Ebstorf *mappa mundi*), but an object in the landscape.

Samuel H. Kress Collection, National Gallery of Art, Washington, DC.

PLATE 4 Alejo Fernández, *The Virgin of the Navigators*, 1531–36. Featuring Christopher Columbus, Amerigo Vespucci, the Pinchon brothers, and their patrons, as well as the indigenous people of America, the painting explicitly connects the Marian mantle with oceanic exploration.
Source: Wikimedia.

PLATE 5 Friedrich Georg Weitsch, *Portrait of Alexander von Humboldt*, 1806. The young naturalist is captured between the micro-scale of his botanical specimens and the macro-scale of the cosmos beyond the hazy horizon.

Alte Nationalgalerie, Berlin.

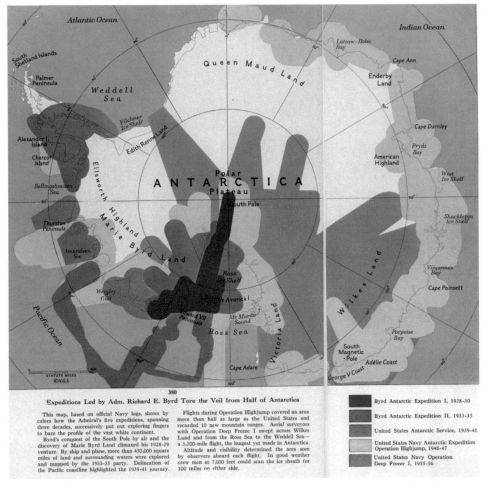

380

Expeditions Led by Adm. Richard E. Byrd Tore the Veil from Half of Antarctica

This map, based on official Navy logs, shows by colors how the Admiral's five expeditions, spanning three decades, successively put out exploring fingers to bare the profile of the vast white continent.

Byrd's conquest of the South Pole by air and the discovery of Marie Byrd Land climaxed his 1928-29 venture. By ship and plane, more than 450,000 square miles of land and surrounding waters were explored and mapped by the 1933-35 party. Delineation of the Pacific coastline highlighted the 1939-41 journey.

Flights during Operation Highjump covered an area more than half as large as the United States and recorded 10 new mountain ranges. Aerial surveyors with Operation Deep Freeze 1 swept across Wilkes Land and from the Ross Sea to the Weddell Sea—a 3,200-mile flight, the longest yet made in Antarctica. Altitude and visibility determined the area seen by observers aboard each flight. In good weather crew men at 7,000 feet could scan the ice sheath for 100 miles on either side.

■ Byrd Antarctic Expedition I, 1928-30

■ Byrd Antarctic Expedition II, 1933-35

■ United States Antarctic Service, 1939-41

□ United States Navy Antarctic Expedition Operation Highjump, 1946-47

■ United States Navy Operation Deep Freeze I, 1955-56

PLATE 6 "Expeditions led by Adm. Byrd tore the veil from half Antarctica." In David Boyer, "Year of Discovery Opens in Antarctica," *National Geographic*, September 1957, 380.

PLATE 7 The Atlantic from more than fifty miles above the earth. The photograph was snapped during Alan Shepard's Freedom 7 mission in 1961, the first US manned space flight, by the porthole camera, two and a half minutes after lift-off. *National Geographic*, September 1961, 433.

PLATE 8 Sidney Meteyard, *I Am Half-Sick of Shadows*, 1905. The Lady of Shalott is immortalized as she gently falls asleep before her web and the nocturnal view of two lovers reflected in her mirror. Painters / Alamy Foto Stock.

PLATE 9 Alighiero Boetti, detail from *Mappa*, 1989–94. The irregular texture of Boetti's cartographic textile works betray their manufacturing by different individual Afghani embroiderers.
Private collection, Florence; photo by the author.

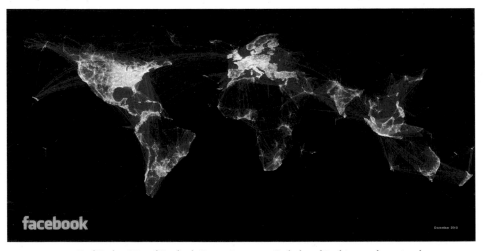

PLATE 10 Paul Butler, *Map of Facebook Connections*, 2010. Each thread in this complex network represents a human relationship.
Source: https://www.facebook.com/notes/facebook-engineering/visualizing-friendships/469716398919/.

melt the Antarctic ice sheet in order to access potential uranium sources, or to create artificial gulfs and water basins, among other nuclear engineering projects.[78]

Dreams of control over and large-scale modifications of the ancient earth's mantle were nonetheless balanced, or rather enabled, by the dream of piercing beyond it into the unknown—of lifting Isis's veil. In this respect, the

FIGURE 8.7 "Seeing beyond the Stars": parting the veil of heaven as an enduring trope for space exploration and scientific discovery in the 1980s. Front cover of *Life*, December 1985.

International Geophysical Year proved pivotal. As Sullivan noted, "the IGY had broken the bonds that had tied man to the earth. . . . It had given the world a peek beyond the enveloping atmosphere, a glimpse of the universe in its naked glory. . . . Having had one look, we cannot close the door again" (fig. 8.7).[79]

Drilling into the Geological Mantle

Nuclear experiments physically pierced both upper and lower veils. Until the 1950s, geologists had to rely on natural earthquakes in order to study the interior of the planet. As the Soviet geologist Anatolii Malakhov (1907–83) evocatively put it, "each tremor of the earth pierces the globe through and through."[80] Traversing different underground layers, the seismic waves carried back to the surface important information waiting to be decoded. From the speed with which the waves were reflected and refracted, scientists could infer the depth, structure, and composition of the layers. For example, in 1909, during a major earthquake in the Pokuplje region, about 40 kilometers southeast of Zagreb, the Croatian seismologist Andrija Mohorovich (1857–1936) observed how seismic waves traveling about 30 kilometers underground moved faster than those above that depth, indicating a change in the constituency of the rocks. He had discovered the upper boundary of the geological mantle, the Moho discontinuity, as it was christened after him.[81]

In the 1950s, however, as enthusiasm for nuclear engineering was flying high on both sides of the Iron Curtain, underground explosions came to be envisaged as the "future" of earth science research—a sort of global-scale analogue of the seismic-shooting technique employed by glaciologists in the polar regions. The seismic waves generated by such explosions, the American seismologist Bruce Bolt (1930–2005) believed, "dramatically opened up new experimental methods for studying the deepest parts of the earth," as they freed scientists from their "ancient dependence on the vagaries of natural earthquakes."[82] In 1955, a group of seismologists put forward the idea that nations might cooperate to carry out four atomic explosions at "seismically useful locations." Two years later, the seismic waves generated by a detonation in the remote region of Maralinga in southwestern Australia made it possible to determine the thickness of the earth's crust in that continent.[83] Likewise, an explosion at Pokrovsk-Uralsky in the Soviet Union made it possible to elucidate the thickness of the crust in the Ural region. The explosion, Malakhov noted, was "a signal sent into the unknown."[84]

These results notwithstanding, knowledge of the interior of the planet was only partial and indirect; it inevitably rested on inference and speculation. The dark matter of the earth remained wrapped in mystery. While geologists

FIGURE 8.8 Henry Hess, a founding father of plate tectonics theory, explains the Mohole project. Fritz Goro / The LIFE Picture Collection via Getty Images.

were in general agreement that the mantle, which enshrouds the core and makes for 84 percent of the planet's volume, was made of solid rock, nobody knew how that material looked or what it felt like. Even the lower strata of the terrestrial crust remained unknown. The only way to gain direct contact with those hidden regions would be to retrieve rock samples from the geological mantle. This could only be achieved by piercing through the crust, that is, 20–40 kilometers down into the land, or 5–10 kilometers into the ocean floor, where the crust is thinner. Such an operation would entail unprecedented efforts. Crustal drilling in those days was primarily undertaken for oil and gas exploration, and the maximum depth achieved was no more than 2 kilometers into the ocean crust, or one-third of the way to the mantle.[85]

The idea of drilling a hole through the crust, in order to "touch the hem" of the geological mantle, was launched in the United States in 1957, during the IGY. It came primarily from the physical oceanographer Walter Munk (1917–2019) and from Harry Hammond Hess (1906–69), one of the founding fathers of plate tectonics theory (fig. 8.8). Frustrated by what they saw as a stream of worthy yet ordinary research proposals in the earth sciences, they suggested what they thought would be a project able to meet the criticism: a "Mohole," literally, "a hole through the Moho."[86] Experimental drilling was to

be conducted from a specially designed unanchored barge (CUSS1) off Gua-
dalupe Island in the eastern Pacific, where the crust was only 5 kilometers
thick. This initial phase, which took place in 1961, would help establish the
feasibility of the project. But why drill a hole through the crust? And why at
this moment in time?

While the earth sciences were to become closely allied with the Cold War
(given, for example, the necessity to distinguish natural earthquakes from secret underground nuclear tests), US geologists lacked access to the growing
federal support made available to space researchers and nuclear physicists. To
tap those resources, earth scientists felt they needed a glamorous project—
something able to arouse the imagination of the American public much in
the same way rockets, artificial satellites, and other "technological spectacu-
lars" were doing.[87] As geophysicist Gordon Lill claimed, "geology needed a
single, major project that would open new avenues of thought and research
in many fields." A hole through the crust, he believed, "would do that—and
more."[88]

Framed as "the perfect anti-analogue of a space probe," Mohole fit, and
in turn fed, the then pervasive imaginary of Cold War competition.[89] The
launching of the project largely depended on this imaginary and on the play
of visibilities and invisibilities articulated by the Iron Curtain. At a meeting
in Toronto, while Americans were still debating the feasibility of the project,
a Soviet geophysicist claimed that his country already had the equipment to
drill such a hole and they were "just looking for the place." "Ten years later,"
American journalist Daniel Greenberg sarcastically remarked, "the Russians were possibly still looking for the place, but East-West competition was
money in the bank."[90] Indeed, Mohole engineer Willard Bascom (1916–2000)
recalled how the Soviet remark pricked the pride of the American oil industry. " 'Anything they can do we can do better' was the instant reply, and before long a group of Texas oilmen held a meeting to ask themselves critically,
'What are the limits of deep drilling?' " Bascom concluded that "perhaps there
will be a race to the mantle."[91]

The hole in the crust was nothing but the natural downward extension
of the vertical sublime. It was a continuation of the narrative of unveiling
and penetration that had enthralled the American public during the IGY and
the then nascent space race. Whether consciously or unconsciously, scientists, policymakers, and journalists all constantly framed Mohole in terms of
analogies between inner and outer space. "I took a quick look at the project
and decided, Why not? It's only going to run about the cost of one space shot,"
a White House science adviser later confessed.[92] The correlation echoed
through the media. *Frontiers of Science*, a popular Australian comic strip syn-

dicated to hundreds of newspapers around the world, for example, reported that earth scientists were embarking on an adventure "as exciting to them as a space probe to the moon" (fig. 8.9). Recuperating an analogy used in the IGY Antarctic context, John Steinbeck claimed that "we know less about [the earth's interior] than we do about the moon."[93] In his glossy account of CUSS1 for *Life* magazine, the American journalist noted that

> this expedition will cost less than one single glittering missile blasting from the launching pads at Canaveral, and yet this project is surely an adventure towards the discovery of a new world as were the three little lumbering ships of Columbus. And this world is here—not a million miles away. . . . If we can seriously plan and design stations in space and on the moon, we are surely capable of mining a few thousand feet under water. . . . We spend treasures daily on fantastical sky rockets aimed feebly towards space. . . . We spend and devise and dream towards the nearest star unreachable in a lifetime of travel, and meanwhile we know practically nothing of far the greatest part of our home planet covered by the sea.[94]

Piercing "the earth's skin" activated vertical narratives perhaps less spectacular and yet no less compelling than severing the celestial veil. It also activated similar vocabularies and Baconian tropes, including "unveiling" and "unlocking" mysteries, "interrogating" and even "torturing stones," or "probing the unknown."[95] The sheer elemental nature of the project provided these metaphors with an insistently material constituency. The mantle was not simply a rhetorical cover, but the thickest and most solid of all involucres—and scientists were after its matter.

Not only did the geological mantle conceal mystery, it was a mystery in itself. It was the unseen generative matrix from which the crust was formed. The planet's history was therefore shrouded in its deep folds. Journeying to its upper fringe meant journeying into darkness:

> Down to a depth of 560 feet, the drilling went rapidly through the soft clay— the green clay of the ocean floor. At that point it slowed abruptly as the bit chewed into the hard second layer of the earth's crust. The scientists wanted eagerly to see the core. A sample of the earth's second layer would provide evidence to crack a great geophysical mystery.[96]

Public responses to the project varied. Some suspected the team of mining for diamonds; others, of looking for hidden treasures; still others, of surveying the sea bottom for permanent guided missiles.[97] Indeed, Bascom amusingly recalled how an advertising campaign for a company producing potentiometers announced that a "MOLE project" was going to put "an end to the threat of war" thanks to a "radically new" system of missiles launched downward

"FRONTIERS of SCIENCE"

FOURTEENTH WEEK'S RELEASE: "THE MOHOLE PROJECT"

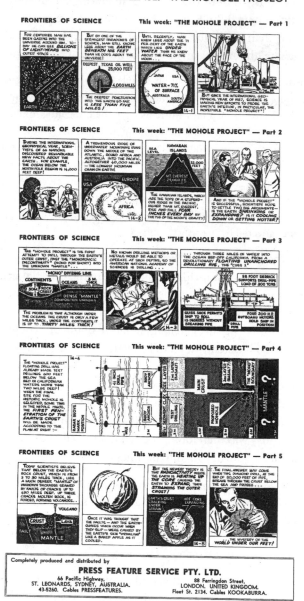

FIGURE 8.9 "Scientists are embarking on an adventure as exciting to them as a space probe to the moon." Vignette from the popular science cartoon "The Mohole Project," *Frontiers of Science*, 11 December 1961. University of Sidney Library, Rare Books and Special Collections.

underneath the crust.[98] Some feared that the ocean would be sucked into the hole; others, that the drilling into the mantle would set off a full-scale volcano; and still others, that it would cause the earth to burst like a toy balloon.

Apocalyptic forecasts aside, the experimental phase of the project was a widely acclaimed success, which won the team President Kennedy's congratulations. Cores were drilled into unprecedented water depths (950 meters and 3,560 meters), and precious specimens of basalt were brought to the surface from the abysses of space and time.[99] Unlike its outer space cognates, the project, nonetheless, never came to completion. Over the years costs skyrocketed, and the project shipwrecked amid disputes, bad administration, and political scandals of sorts—to the point that in 1966 the US Congress took the unprecedented step of decreeing its termination. The Moho frontier was never reached. The Mohole nonetheless provided the basis for other more modest and yet highly successful American deep sea programs.[100] It also succeeded in breaching the curtain and stirring the Soviet imagination in much the same way as it did in the West.

"The whole world first began to talk about the remarkable Moho project in 1957," Malakhov wrote in his book *The Mystery of the Earth's Mantle* (1966). "A super-deep drill hole was to pierce the earth's crust right to the mysterious mantle. . . . The Americans wrote that they would penetrate into the substance from which all the rocks of the earth originated."[101] In both the American and Soviet imagination, the mysterious mantle was nothing but Chthonia herself, the shapeless primordial matter hidden beneath the surface of the earth—a monstrous Isis without veils. Asked to prepare the script for a film play illustrating different theories about what the mantle was made of, Malakhov imagined different possible scenarios following the successful drilling of a super-deep hole in one of the Kuril islands in eastern Siberia.

In his first scenario, Malakhov tells the story of the town of drillers that had sprung up on the island. In the center of an immense oval basin stood the derrick for the super-hole, around which the life of the entire village revolved. The village was wrapped in a disquieting silence. This silence was periodically broken by the creepy sound of a loudspeaker announcing how much had yet to be drilled to reach the mantle. At 20 meters from the Moho, villagers were evacuated, as a precautionary measure for the catastrophes that might occur with the opening of the Moho frontier.

> At last the long awaited moment came. The loudspeaker had just announced: "Five meters left to the opening of the Moho frontier . . . four meters . . . three meters . . . two. . . ." Everything was set in motion. It was impossible to see any details. Only the cine-film, which recorded these unique moments at a

speed of about five hundred million frames per second, showed afterwards how the lattice-work of the derrick slowly (or so it seemed on the screen) dissolved and vanished into thin air and how a torrid whirlwind demolished all the structures surrounding the drill-hole. Simultaneously, a dazzling bright fiery column of magma flew upward. . . . The din and rumbling grew in intensity. . . . The air shook with thunderclaps accompanied by sparkling streaks of lightning. . . . The explosions continued, and the earth trembled.[102]

Such was the cry of the wounded earth. Piercing her skin meant awakening Chthonia's destructive underground powers. It meant facing the abyss, seeing the naked earth in her primordial state, without her geographical *peplos*:

A red hot stream began to fill the basin in which the drill-hole, the shops, and the houses had been situated. Before the eyes of the onlookers, everything that an immense body of men had worked so hard on was demolished by the raging chaos. . . . The earth shuddered again. Then came silence, complete silence. . . . Before us lay a lake of primordial substance. Mankind saw for the first time the mysterious substance that lies under the earth's crust everywhere.[103]

Other scenarios imagined by Malakhov included the retrieval of a colossal core made of an extraordinary substance encompassing all the elements of the periodic table and impossible to break, and the discovery of unimagined natural resources: petroleum, nicknamed "the Queen of Depths," vast quantities of diamonds, and a massive electric field deep "in the bowels of the earth."[104] Malakhov nonetheless concluded that he could not really produce a movie, as more experiments were needed: "We must drill! . . . We must drill deep holes as soon as possible to open the substance of the earth's mantle."[105]

The plea seemed to materialize few years later. In the prospect of finding untapped resources, in May 1970 super-deep bore holes began to be drilled in the Kola peninsula, in the far northwest of Russia. In 1989 the deepest hole reached 12,262 meters, only half the distance, or less, to the mantle in that region. Today it is still the deepest artificial point on earth. Like its American counterpart, however, the Soviet project never came to completion. It was stopped in 1992 when drillers encountered higher temperatures than expected. In 1995 the project was formally terminated, due to the dissolution of the Soviet Union. The site has since been abandoned and is now considered an environmental hazard.[106] The derelict gray concrete structure silently towers over the desolate wasteland, as a lonely monument to a bygone past and to human hubris.

★

Empowered by technology and pressed by geopolitical competition, Cold War scientific practice and imagination were pushed to extreme frontiers. In

an epic vertical flight from the highest regions of the atmosphere and the unbounded spaces of the universe down to the mysterious pits and bottomless
abysses of the earth's interior, rockets and drilling machineries pierced tangible and intangible veils, surfaces, and mantles. A new technological sublime
blended with Humboldtian drama. Old Baconian metaphors were revived
with a new intensity of pathos. Unsurprisingly, "unveiling of the mysteries of
the earth" was made the *raison d'être* of the International Geophysical Year.
It was, in a sense, the thread joining the different elemental realms in which
IGY scientists labored.

The first in a series of themed articles that appeared in *Life* magazine in
1960 explicitly defined the IGY as "a New Age of Exploration moving with
fantastic speed into realms undreamed of by navigators of old." The new explorers, the author claimed, "were pursuing the secrets of the stony globe that
whirls man with comparative safety and comfort through the bleak, inhos-
pitable cosmos." Piercing veils meant domesticating nature; it meant making
the earth a vast laboratory, or rather, as the *Life* commentator suggested, placing the world "under man's multi-instrumented eye, stud[ying] it as though it
were a beetle under a microscope."[107] During the IGY the planet was "girdled,"
"wrapped," but first of all, it was "probed." The mantle had opened once again.
Space had extended into new dimensions. But, as we shall see in the next chapter, the cry of the wounded earth did not go unheard.

9

The Green Mantle

As the American public's attention was split between the slow downward progress of the Mohole drill and the supersonic upward trajectory of Russian astronaut Yuri Gagarin, somewhere, on a quiet shore of Maine, a woman was fighting her last battle. In her lonely summer cottage, amid the rhythmic sound of the sea waves and the cries of the seagulls dispersed by the wind, marine biologist Rachel Carson (1907–64) was secretly struggling against terminal breast cancer. Emaciated by the illness and by the many pressures of life, she was also struggling toward the completion of what would be her last book.

Carson had gained fame for awakening in her readers a sense of wonder toward the marvels of the sea and the mystery of life. Her lonely retreat on the edge of the ocean, atop a rocky bluff overlooking tidal pools and lobster boats, had certainly offered inspiration. So many times had she walked in the moonlight along the shoreline at low tide and observed its small creatures as they clung to the rocks or swayed in the pools. Watching them closely restored to Carson an image of harmony, balance, and community, in which everything was mutually interconnected—a lively "web of life," as she liked to call it. In it she found solace and comfort. The magic whisper of that little world wafted through the pages of her books, bringing the fresh breath of the ocean and its awesome beauty into the homes of thousands of Americans.

Her last book, however, was different. *Silent Spring* (1962) was not a celebration but a warning. Or rather, a harrowing cry. The setting was no longer a deserted beach, or the vast horizon of a blue marine expanse, but a checkerboard of prosperous farms deep in the continent. In the opening, Carson asked the reader to imagine an idyllic small town in its midst. This town, she explained, lived in perfect harmony with the rhythms of nature, until one day it was suddenly wrapped by the shadow of death. Road vegetation withered; animals, and subsequently humans, were struck by mysterious diseases. Some of them died. "There was a strange stillness. The birds, for example—where had they gone? Many people spoke of them, puzzled and disturbed. The feeding stations in the backyards were deserted. . . . It was a spring without voices. . . . Only silence lay over the fields and woods and marsh."[1]

The sinister silence of that spring without birdsong brings to mind Malakhov's apocalyptic tale. This, however, was no science fiction, Carson warned. It was a collage of various disasters that had actually happened in different places and at different times as a result of the fallout produced by nuclear tests and of chemical contamination—and, she feared, a stark reality every American citizen might come to know, sooner or later. During World War II, the US military had used DDT to kill fleas and mosquitoes, and to protect soldiers and civilians against outbreaks of malaria and other epidemics. After the war, the pesticide had entered commercial use, with more than two hundred types for use by farmers, foresters, and suburban lawn owners. Despite their heavy usage across the country, a decade later few Americans were aware of the side effects and dangers of pesticides. And yet, Carson was at pains to show, toxic substances entered the food chain and quickly spread through the intricate web of life—traces of DDT were found in human breast milk, for example. Clearly, the uncontrolled use of pesticides posed a serious threat to animals, to the environment, as well as to humans.[2] For this reason, the American biologist had chosen to devote her last energies to the cause.

Carson's mission, however, went beyond sensitizing public opinion to the dangers of the uncontrolled use of chemicals. Hers was far a more titanic effort. She demanded from her audience nothing less than a shift of paradigm, or rather, of perspective. She invited Americans to redirect their gaze from the bottomless abysses of outer space to the very surface of the earth they inhabited. More precisely, she urged them to think about the earth's mantle less as a veil to be torn, or a surface to be pierced, than as a marvelous, tightly woven green fabric pulsating with life, and yet finding itself under unprecedented threat. In asking her readers to replace arrogance with humility, she compelled them to focus on the fine textures of the mantle, respecting the mystery of what lay beneath it.

It is easy to unpick the gendered aspect of Carson's tale—the female "gentle subversive" quietly weaving her web of life, as opposed to noisy, male, mainstream scientists violently tearing veils apart.[3] Carson was nevertheless not alone in her task. If science and exploration in the 1950s witnessed the resurgence of Baconian "unveiling" metaphors stressing conquest, penetration, and mastery over nature and its secrets, in the following decades a new set of metaphors borrowed from Romanticism came to define a new "popular ecology" and environmental consciousness.[4] These metaphors highlighted interconnection and interdependence. They helped make visible the close interrelations between humans, other creatures, and the biosphere. Ironically, awareness of this holistic relationship between bodies, the environment, and global space was a direct consequence of the Cold War and Atomic Age. The growing realization that seemingly disconnected nuclear events could (and did) have serious repercussions on the rest of the planet, and potentially on the lives of everyone, led to the disruption of modernist ideas about the discreteness and impermeability of bodies and spaces. It led to the rise of a new "ecological thinking" in the United States.[5]

Rather than concealing mysteries to be unveiled, textile metaphors were used by supporters of the new environmental movement in their ancient sense, as tightly woven crafts. The mantle of the earth turned green, and it became an intricate living fabric: the *hyphasma* celebrated by the Greek Church Fathers, or Humboldt's magnificent web. But there were two fundamental differences: first, its integrity was now being irremediably compromised by human action; and second, the green mantle wrapped a planet seen, for the first time in human history, from space. Paradoxically, as it has been observed, "the surveillance imperative that thrived in the context of superpower competition helped create an image of the earth as a fragile, complex entity and to highlight the power of human agency to harm the planet."[6] In other words, piercing the veil of heaven became a precondition for seeing the new green mantle. This chapter charts the evolution of the metaphor in modern ecology and environmental writing.

Green Mantles

The popularization of the phrase "earth's green mantle" in twentieth-century public environmental speech owes much to Carson, who chose it as the title for a chapter of *Silent Spring* on the effects of chemicals on plant life. The metaphor cast the greening of the planet as a divine marvel; thus, it heightened the contrast with human destructive action, which the American biologist was keen to bring to the eyes of her public. For the same reason, "the green mantle" also attracted twentieth-century ecologists and environmental writers concerned about its preservation. The metaphor nonetheless boasts a much longer history, stretching, as we have seen, as far back as Scripture and patristic writing. How did it make its way into modern-day environmentalism?

The "green mantle" is found as a poetic figure for terrestrial vegetation in various Western literary traditions. Shakespeare, for example, employed it in

King Lear (1606) to describe the surface vegetation covering the still waters of a pond.[7] In most instances, however, the metaphor carries its original patristic meaning of beautiful divine handicraft, or outer manifestation of God's creative powers. The metaphor therefore generally appears in religious literature, especially in seventeenth- and eighteenth-century Italian sermons and biblical commentaries, as well as in morally edifying poems, from Haller's floral carpets gracing the Alpine slopes, to John Gilbert Cooper's elegiac verses celebrating the rebirth of life.[8] Varied as it is, this literature features the green mantle in all its splendor: now as an antidote to tribulation "embroidered with drops of dew akin to precious gems," now as the shiny vegetal garment with which God robed the earth at Creation, now as a synecdoche for the regenerative beauty of spring, or "the couch where life with joy reposes," now as a *topos* in descriptions of promised lands, cultivated fields, and other *loci amoeni* tamed by the human hand and gaze.[9]

A liminal surface forever suspended between life and death, by the mid-nineteenth century the green mantle was, unsurprisingly, appropriated by American nature writers, including the two great pioneers of modern environmentalism: the transcendentalist Henry David Thoreau (1817–62) and, especially, the conservationist John Muir (1838–1914). As with their European Romantic counterparts, Thoreau and Muir consistently employed metaphors and personifications in order to connect humans to the wilderness.[10] Metaphors and personifications allowed these authors to apply human feelings to nature and thus "transform a divorced and alien world, in which the human being would feel himself to be outcast, into a congenial world, in which he can feel thoroughly at home."[11] Downsizing the mantle metaphor from the scale of the earth to that of a pine tree, Thoreau thus describes the plant as it is being felled:

> It lies down to its bed in the valley, from which it is never to rise, as softly as a feather, folding its green mantle round it like a warrior, as if, tired of standing, it embraced the earth with silent joy, returning its elements to the dust again. But hark! There you only saw, but did not hear. There now comes up a deafening crash to these rocks, advertising you that even trees do not die without a groan. It rushes to embrace the earth, and mingle its elements with the dust. And now all is still once more and forever, both to the eye and the ear.[12]

By likening the pine tree to a dying warrior, Thoreau creates an immediate emotional bond between the plant and the reader. The dignified demeanor of the tree reminds both author and reader that every being (themselves included) is part of the same earthly matter to which all things shall return. The earth's mantle of humus and vegetation is more than a silent backdrop; it comes to life as a fluid primordial entity, always in the making—a quality that is also stressed by Muir in his unpublished diary:

> February 19. Clouds none, or only a few scarcely discernible touches. Earth's green mantle is rapidly deepening. Three flowers today—Dodecatheon, Ranunculus, and a crucifer. The construction of a very fine quality spider web is going on rapidly at present. A species of spider making an elegant circular web upon the dead branches of Eriogonum has the power of invisibility by rapid swinging in the center of its web.[13]

In late antique *hexaemera*, from Basil the Great to George of Pisidia, both green mantles and spiders' webs were privileged *topoi* for God's creative power.[14] In Muir's description, the image of the green mantle fades into the spider's web in a seamless scalar continuum, as if to reinforce the dynamic interconnectedness of the works of creation. Elsewhere in his diary, the writer compares the iconic Yosemite's Half Dome to an ancient deity, clad "not in vasty, inseparable masses of dim sublimity, but in ten thousand things of equal, separate, outspoken loveliness—feathery pines, . . . masses of the Caenothus and Manzanita bushes." Never, he concludes, "have I beheld so great and so gentle and so divine a piece of ornamental work as this grand gray dome in its first winter mantle woven and jeweled in a night."[15]

As in patristic writings, the green mantle wraps and gives visibility to divine invisibility. In Thoreau's and Muir's writings, however, it is not the body of Christ to secure the unity of creation, but a transcendent pantheism permeating, or rather forming the core of, all things. The more isolated and deprived of human presence a place may be, the more intense the spiritual experience it promises. Mountains, in particular, were deemed privileged sites for epiphany by both authors. On Mount Katahdin, the highest summit in Maine, for example, Thoreau said to have encountered "that earth we heard, made out of Chaos and Old Night. Here was no man's garden, but the unhandselled globe. . . . It was matter, vast, terrific."[16] In other words, what the American philosopher saw on the mountain was primordial Chthonia still uncovered by Zeus's polychrome mantle. From that otherworldly landscape he learned that "any efforts to tame and control the world, to civilize it are doomed to failure."[17]

Muir similarly regarded mountains as sacred sites in which Zeus's mantle progressively thinned with altitude:

> In cold mountain altitudes, Spirit is but thinly and plainly clothed. As we descend down their many sides to the valleys, the clothing of all plants and beasts and of the forms of rock becomes more abundant and complicated. When a portion of Spirit clothes itself with a sheet of lichen tissue, colored

simply red or yellow, or grey or black, we say that is a low form of life. Yet is it more or less radically Divine than another portion of Spirit that has gathered garments of leaf and fairy flower and adorned them with all the colors of Light? . . . All of these varied forms, high and low, are simply portions of God, radiated from him as a sun, and made terrestrial by the clothes they wear, and by the modifications of a corresponding kind in the God essence itself.[18]

For Muir, life was but a manifestation of spirit. Like spirit, life was indivisible, while at the same time growing in richness through the evolutionary development of more intricate forms and sophisticated garments.[19] Such variation and variety could only be observed in the fine textures of pristine wilderness—and especially in the Humboldtian vegetal belts of mountains. But the green mantle was also a fragile artifact. A pioneer of environmental conservation and the individual responsible for the establishment of Yosemite Park and many other natural reserves, Muir was concerned about the integrity of the green mantle of wilderness and the preservation of its splendid diversity. Thus, questioning the impact of humans on the mountains, during his early campaigning in 1875 he provocatively wrote:

I often wonder what man will do with the mountains—that is, with their utilizable, destructible garments. Will he cut down all the trees to make ships and houses? If so, what will be the final and far upshot? Will human destructions like those of Nature—fire and flood and avalanche—work out a higher good, a finer beauty? Will a better civilization come in accord with obvious nature, and all this wild beauty be set to human poetry and song?[20]

Both Thoreau and Muir used the earth's mantle as a poetical figure evoking the beauty and complexity of vegetation and of creation in general. Whether as a compost of humus, a verdant surface, or a fragile garment, the green mantle was ascribed a moral connotation; it was a precious and rich heritage to be preserved. As such, the metaphor passed on to early twentieth-century environmental conservationists to stress the aesthetic value of certain types of areas. It reflected an attitude toward the environment as "wilderness" to be preserved—if only within the bounded space of national parks and natural reserves, the sanctuaries of the young American nation.[21]

By the 1930s wilderness was understood primarily as an uncultivated region uninhabited by humans in which the "primitive environment" was preserved "as nearly as possible."[22] Wilderness preservation was promoted mostly on moral and aesthetic, rather than scientific, grounds. For example, Robert Marshall, one of its main proponents, argued for the immense psychological and physical benefits that periodical retreats into "the undisputed beauty of the primeval" offered to an increasingly urbanized population. Wilderness,

he concluded, provided humans with the highest form of aesthetic enjoyment: "One looks from outside at works of art and architecture, listens from
outside to music or poetry. But when one looks at and listens to the wilderness, he is encompassed by his experience of beauty, lives in the midst of his
esthetic universe."[23] Wilderness, especially "green" wilderness, was therefore
an important resource, and both agricultural exploitation and urbanization
needed to be balanced by its preservation through thoughtfully planned spatial enclosures.

While lacking the transcendence and nationalistic rhetoric of early American environmentalism, similar views were held on the other side of the Atlantic. In his book *Earth's Green Mantle* (1939), Sydney Mangham, a pupil of
the famed English botanist and ecology pioneer Arthur Tansley (1871–1955),
explicitly identified uplifting wilderness with green landscapes. "Year after
year," he observed, "plants have been exploited in the service of mankind for
the increase of his pleasure." Every summer, growing numbers of people venture to the countryside in pursuit of health and relaxation "in places made
lovely by plants," while "beauty spots" are "set aside" by governments as national parks or reserves, "so that they are preserved for public enjoyment and
study forever."[24]

The book, an introduction to plant science for the general reader, outlines
the evolutionary story of the earth's vegetal mantle, as well as the story of
"how man has attempted to discover and exploit [it]."[25] Like Muir, Mangham
uses the green mantle metaphor to initiate his audience into the splendid
diversity and complexity of the world of plants, though this time in a thoroughly scientific way. "Spirit" and the transcendent here are replaced by a
wholly mechanistic conception of nature in which plants are likened to "machines," and diseases to "engine troubles."

Whether as a source of food, fuel, shelter, and clothing, or of aesthetic
pleasure, Mangham understands the green mantle first of all as a vital and
"almost illimitable" reserve available to humankind, or indeed, as a precondition for its very existence.[26] The green mantle is nevertheless far from being
a static and homogeneous entity (fig. 9.1). "Of all the earth's varied raiment,"
observed the British botanist, "grasslands are perhaps the most widely worn.
They cover enormous areas of the globe, and vary much in color and appearance according to soil and climate and the uses to which they are put."[27]
And grasslands are just a part of the green mantle. The reader is conducted
from forests and jungles to velvety tapestries of flowers and lichens "weaving
coats of many colors"; from floating fabrics of algae, "the first threads of the
mantle," all the way down to the kaleidoscopic beauty of the tissues revealed
by the powerful lens of the microscope.[28]

PLATE 5.—IN THE LIBYAN DESERT.
Above : Spring flowers in profusion, at end of damp season. *Mesembryanthemum* dominant in central photograph.
Left : A small wadi (dried river-bed) with bushes of *Ephedra* (a gymnosperm) and tussocks of *Pennisetum* (a grass). *Right :* Innumerable pieces
of petrified wood, the remains of an ancient forest vegetation. *(Phot. F. W. Oliver.)*

FIGURE 9.1 Different types of "green mantles" in the Libyan Desert. In Sidney Mangham, *The Earth's Green Mantle* (1939), pl. 5.

The reader is likewise familiarized with "useful" plants and techniques for improving their productivity, as well as with noxious ones threatening crops. Central to Mangham's vision is ecology, that is, the mutable and complex relationship between plants and their environments.[29] The earth's green mantle, Mangham shows, continuously spreads and adapts its colorful patterns to the changing environment of which it is part. Indeed, he talks about "successive green mantles" clothing the planet:

> In its youth the earth was hot and unclad, and only in later life did it gather raiment. Fashion followed fashion, and the mantle today is but the latest in a wardrobe unfailingly replenished throughout the ages. This mantle is an intricate patchwork of small and large pieces, and it is intermittently altered and repatched. Sometimes a large patch replaces smaller ones, as some exceptionally vigorous and successful type of plant comes into existence and in the due course invades and conquers new territories.[30]

The most spectacular manifestation of the power of natural vegetation, Mangham argues, occurs in the tropics, where entire cities and cultivated lands enfeebled by disease and wars are rapidly invaded by the encroaching jungle. Entire civilizations were swallowed by the green mantle, as "the lusty jungle surged forward, overwhelming in its path, and hid for centuries the handiwork of man, till some traveler, more inquisitive than most, lifted a corner of the mantle and peered into the past."[31]

Throughout the decades following the publication of Mangham's book, the green mantle metaphor continued to be used by professional botanists and ecologists, both as a device helping them communicate scientific knowledge to the general public and as an evocative figure of speech retaining its original meaning as "natural heritage."[32] "Green mantles" featured in this latter sense on the title of a number of publications covering different countries and regions of the earth, from the Mediterranean to New Zealand, from Scotland to Hertfordshire, England.[33] Tansley himself titled one of his many books Britain's Green Mantle (1960). As he acknowledged in the introduction, he had borrowed the phrase from Mangham. Unlike his old pupil, however, Tansley was concerned solely with the "natural green mantle" of native vegetation, as opposed to artificial crops and farmland. In other words, he was interested in "the precious heritage of wild nature," which had survived modernity and the postwar environmental damage condemned by scholars like Hoskins and Darby. Rather than a purely economic resource, for Tansley, Britain's green mantle was first of all a relic that needed to be preserved—if not in its entirety, at least in its "historical character."[34]

Webs of Life and the Wounded Green Mantle

The proliferation of "green mantle" metaphors in scientific and other publications between the 1960s and 1980s coincided with the progressive awakening of a new popular environmental consciousness, which paradoxically emerged and flourished at the height of the Cold War and Atomic Age. Much credit for this awakening has been ascribed to Carson's Silent Spring. Translated into twelve languages in its first year of publication and posthumously elevated to canon status, the book has been defined by environmentalists, literary critics, and policymakers alike as one of the most influential of the twentieth century.[35] Indeed, as a direct result of the book, President Kennedy set up a special panel of his Science Advisory Committee to study the problem of pesticides, which marked a first step toward new environmental policies.[36] In the words of environmental historian Ralph Lutts, "never before or since has a book been so successful in alerting the public to a major environmental pollutant, rooting the alert in a deeply ecological perception of the issues, and promoting major public, private, and governmental initiatives to correct the problem."[37] Why did Silent Spring have such an extraordinary impact?

While the issue of pesticide pollution was not new, Lutts observes, there was something remarkably different between the response to Carson's book in 1962 and the pesticide-control efforts in the first half of the century.[38] Carson's message reached a public increasingly concerned about the dangers of nuclear fallout following the various tests in the 1950s. As radioactive clouds from the Nevada test sites passed over populated areas of the country, causing mysterious illnesses and the sudden death of cattle, Americans' initial fascination with nuclear power gave way to growing public anxiety, in spite of the reassuring statements made by the Atomic Energy Commission. In 1954, fallout from the Bravo test of the American superbomb in the Pacific provoked the illness of twenty-two Japanese fishermen who were sailing in the vicinity of the testing site, and the death of their radio operator. Many of the fish they brought back were found to be contaminated, but not until after they had been sold. Even more disturbingly, radioactive fish were also found on eighty-five other Japanese tuna boats, as sea currents spread radiation from the Bikini test site through the Pacific.[39]

As newly invented radiotracers showed that fallout from nuclear tests was spreading globally, Americans became increasingly alarmed to discover that their own food was contaminated. It was no longer the lives of a handful of distant fishermen that were at stake, but their very own and those of their children. It was as though preexisting ideas about geographical space, wilderness, and bodies as bounded entities were eroding. During early tests, health physicists had assumed that, as in the lab or the factory, fallout radiation could be largely confined within the precinct of the proving site. Although they were aware that explosions would discharge radiation into the atmosphere, they assumed the atmosphere to be passive—"nothing more than a radiation sink."[40] The Bikini test and other tests, however, dramatically demonstrated that significant amounts of radiation invisibly traveled often long distances outside of the proving grounds, as clouds of radioactive material were pushed by sea currents and winds. Absorbed into ecological food chains, radiation ended up accumulating inside the human body. Technocratic control now leaked. Space suddenly emerged as global and interconnected. The human body suddenly became permeable.[41]

Aware of public sensitivities, Carson thus opened *Silent Spring* not with the unfamiliar dangers of DDT (the main subject of her book), but with the by then familiar threat of radioactive fallout. Strontium 90, one of the most toxic isotopes released into the air by nuclear explosions, she wrote, "comes to earth in rain or drifts down as fallout, lodges in soil, enters into the grass or corn or wheat grown there, and in time takes up its abode in the bones of a human being, there to remain until his death." Similarly, she continued, chemicals sprayed on croplands or forests or gardens lie long in soil, entering into living organisms, passing from one to another in a chain of poisoning and death. Or they pass mysteriously by underground streams until they emerge and, through the alchemy of air and sunlight, combine into new forms that kill vegetation, sicken cattle, and work unknown harm on those who drink from once pure wells.[42]

What were thought to be waterproof surfaces, Carson showed, were in reality porous fabrics. In highlighting the interconnectedness between different parts of the biosphere and their permeability—from the cracked soil to the human skin and vegetal fibers—Carson also blurred the boundaries between different types of toxic substances. The initial scene of the contaminated idyllic village was itself wrapped in ambiguity. Unopened hens' eggs, dying litters and fish, fruitless trees, roadsides filled with browned vegetation, and, not least, sudden human maladies were ascribed to a mysterious substance: "In the gutters under the eaves and between the shingles of the roofs, a white granular powder still showed a few patches; some weeks before it had fallen like snow upon the roofs and the lawns, the fields and streams."[43]

Whether that strange white "snow" was a lethal pesticide sprayed from an airplane, or radioactive material, is not entirely clear. What is certain is that it would have reminded Carson's readers of the fallout that had silently descended over the unfortunate Japanese fishing boat few years earlier.[44] Incidentally, 1962, the year of publication of Silent Spring, was the year with the largest number of nuclear explosions history ever witnessed.[45] At this time, science fiction movies featuring mutants and disasters caused by radioactive fallout were capturing the American popular imagination. On the Beach (1957), for example, portrayed an apocalyptic post–nuclear war scenery of soundless ghost cities reminiscent of Silent Spring's contaminated village, whereas Dr. No, released in the same year as Carson's book, initiated a tradition of James Bond movies populated by mad nuclear scientists and evil geniuses threatening to poison the world.[46] Americans were thus prepared to receive Carson's book, in which they read but an extension of an already familiar threat destined to populate their imagination for the years to come.[47]

Although the power of Silent Spring lies in Carson's unique ability to weave scientific evidence with poetic insight, the book's success owed to her ability to give visual shape to, and to popularize, a new conception of space— and of the earth's mantle. In order to communicate her holistic and organic view of space and the environment to her public, the biologist made wide use of textile metaphors. Her favorite was the "web," or "fabric," of life. No other phrase or expression captures Carson's vision more effectively. In The

Edge of the Sea (1955), for example, she writes that living things are "bound to this world by many threads, weaving the intricate design of the fabric of life," while in *Silent Spring* she notes how the myriad organisms populating the soil constitute a "web of interwoven lives," making it "capable of supporting the earth's green mantle."[48]

As with the "earth's green mantle," the web metaphor was not Carson's invention. Used by Shakespeare and Thoreau alike in their meditations on the nature of human life, webs and, more generally, the fabric trope had also long permeated the realm of the natural sciences.[49] The web of life, we have seen, dominated Humboldt's understanding of the landscape and of the cosmos. Unsurprisingly, similar metaphors were also employed by Darwin.[50] In a famous lyrical passage at the end of *The Origin of Species* (1859), for example, he described "an entangled bank, clothed with many plants of many kinds, with birds singing on the bushes, with various insects flitting about, and with worms crawling through the damp earth." Such view, he said, led him "to reflect that these elaborately constructed forms, so different from each other, and dependent upon each other in so complex a manner, have all been produced by laws acting around us."[51]

As an ecologist in Darwin's wake, Carson shared his appreciation of the beauty of symbiotic relationships and subscribed to his evolutionary theory. Rather than the "struggle for life" that had so much enthused Herbert Spencer and his followers, however, Carson emphasized the themes of connectedness and organic unity shining through *The Origin of Species* and through the works of early twentieth-century naturalists, such as Arthur Thomson (1861–1933). For Thomson, one of the fundamental biological concepts was that "every thread of life is intertwined with others in a complex web." This was a relatively common notion at his time. Thomson, however, pushed the metaphor further, so that, he claimed, when one sought order and solidarity in nature,

> the multitudinous unique threads of life become more and more interwoven; the warp and the woof of the web are hunger and love; we get glimpses of a changing pattern becoming even finer. The web seems to become increasingly coherent, though man often rends the fabric ruthlessly.[52]

Like Darwin, Carson marveled at nature's "web of complex relations." Like Thoreau, she experienced a sense of awe before its mystery. Like Thomson, however, she was also deeply aware of man's damaging action on the fabric.

Carson engaged with contemporary ecological ideas, such as Tansley's concept of "ecosystem."[53] She presented a holistic view of the biosphere as a connected ecosystem, in which humans were as tightly interwoven in its tapestry as any other creature, and yet were also capable of profoundly affect- ing it.[54] In the "Earth's Green Mantle" chapter, for example, she discusses "the intricate web of life whose interwoven strands lead from microbes to man" and the effects human disturbance to one "strand" can have on another. She thus resolves, "The earth's vegetation is part of a web of life in which there are intimate and essential relations between plants and the earth, between plants and other plants, between plants and animals. Sometimes we have no choice but to disturb these relationships, but we should do so thoughtfully, with full awareness that we do may have consequences remote in time and place."[55]

Featuring different types of interpenetrating layers, or ecosystems ("Sur- face Waters and Underground Seas," "Realms of the Soil," "The Earth's Green Mantle"), *Silent Spring* is reminiscent of the *hexaemera* woven by the Church Fathers to celebrate creation. Yet, unlike its ancient predecessors, Carson's splendid fabric of life is insidiously traversed by the black threads of chemical contamination. In a sense, hers is a sort of "anti-*hexaemeron*" showing how the precious fabric of nature is being progressively undone by humans. In each layer of the biosphere, Carson demonstrates, life and death are sinisterly, and inextricably, intertwined:

> Seldom if ever does Nature operate in closed and separate compartments, and she has not done so in distributing the earth's water supply. Rain, falling on the land, settles down through pores and cracks in soil and rock, penetrating deeper and deeper until eventually it reaches a zone where all the pores of the rock are filled with water, a dark, subsurface sea, rising under hills, sink- ing beneath valleys. This groundwater is always on the move. . . . It travels by unseen waterways until here and there it comes to the surface as a spring, or perhaps it is tapped to feed a well. But mostly it contributes to streams and so to rivers. . . . In a very real and frightening sense, pollution of the groundwater is pollution of water everywhere.[56]

The web of life is part of a broader repertoire of textile metaphors used by Carson to promote her biocentric vision of nature and to render human threats visible to her readers. In *Silent Spring*, Romantic poets' translucent veils of ice and mist are thus replaced by deadly coats of chemicals. The po- rosity of organic "living tissues" is juxtaposed to the indestructible film pro- duced by pesticides on the leaves of sprayed plants. Rains, Carson explains, are not able to wash away this poisonous film, so when the leaves fall on the ground and become soil, they retain its chemical components, which are then absorbed by earthworms, which are in turn eaten by the birds, causing their death. The biologist thus evocatively compares the deadly film to Medea's "robe," a mythological garment said to have produced the immediate death of its wearer.[57]

Along with textile metaphors, Carson uses an array of other literary strategies and rhetorical devices borrowed from Romantic nature writers but adapted to her Cold War context. For example, like Thoreau, she personifies animals and plants, from the salmon swimming back to the "thread of water" in which it spent its first months of life, to the sagebrush, the first heroic colonizer of the windswept high plains of the American West. In this way, she ascribes human feelings to a variety of otherwise unnoticed species and arouses the reader's sympathy toward them. Sympathy, however, morphs into a sense of dismay as the stories of the fish's poisoning, or of the sagebrush's systematic eradication, are unfolded. Readers suddenly realize that "their home also belongs to many other living beings, whose lives human action has completely disrupted."[58]

While Romantics used metaphors to create a pantheistic connection between humans and the cosmos, Carson employs them to sensitize her public about humans' violence on nature. The poetic vocabulary of the fabric of life is thus constantly juxtaposed to a militaristic vocabulary of aggression and exploitation. Humans, for example, are said to wage "wars" against weeds and unwanted insects. Pesticides are likened to weapons. An Iron Curtain falls over the scene, as Carson questions "what deals are struck behind the closed doors of the Department of Agriculture, of university research labs funded by the chemical industry, and in the board rooms of pesticide manufacturers."[59] Tropes common during the IGY, like "assault on nature," or technocratic phrases like "control of nature" are condemned and ridiculed in the face of nature's recidivism.[60]

Carson's image of the wounded fabric of life, or ravaged green mantle, provided inspiration to later mainstream environmental writers, ecologists, and artists. For example, in 1977 the Italian sculptor Jorio Vivarelli (1922–2008) designed a series of coins for the Republic of San Marino featuring *Silent Spring*'s key motifs. The first coins in the series celebrated terrestrial life. The smallest, which Vivarelli appropriately titled "The Earth's Green Mantle," features the globe nestled in the starry firmament and crossed by the equator, the polar circles, and the tropics (fig. 9.2)—a reminder of Carson's holistic vision and of an Edenic time in which humans inhabited a clean planet.

Other coins show the dark shadow of the human hand progressively advancing. The mantle is thus imprinted with the "Mark of Man's Poisons"; the "Blue Sea Depths" are polluted by the "Footprint of Man's Pesticides"; the "Sky's Clean Transparency" is compromised by "Man's Aerial Disinfestations." The drama culminates with an "Earth Wounded by the Useless Slaughter"

FIGURE 9.2 Jorio Vivarelli, "Il verde manto della Terra" ("The Earth's Green Mantle"). The smallest coin (1 lira) in a series inspired by Rachel Carson's *Silent Spring* and designed by Vivarelli for the Republic of San Marino (1977).
Author's private collection.

and "The Elixir of Death in the Skies," featuring a bird killed by human poisons.[61] Juxtaposing elements of creation and destruction, the artist aimed at creating a tactile engagement with ecology. Touching the coin, he explained, would give his audience "the precise sensation of the message it carries."[62]

Vivarelli's tactile narrative evokes the visceral contact with the cosmos that humans seemed to have lost. It bears witness to a new environmental movement reacting against an increasingly visible and spectacular pollution. Eminent contemporary environmental writers and conservation biologists likewise employed textile metaphors to convey this sense of tactility. "Why after millions of years of harmonious coexistence, have the relationships between living things and their earthly surroundings begun to collapse?" asks Barry Commoner in *The Closing Circle* (1971). "Where did the fabric of the ecosphere begin to unravel?" Like Carson, Commoner repeatedly uses textile metaphors, such as "ecological fabric," but he also compares air, water, and soil—and the biosphere itself—to thin "skins," as a way to stress the organic, haptic continuum between humans and their ravaged home.

Introducing their landmark volume *Conservation Biology* (1980), Michael Soulé and Bruce Wilcox revived the "wounded green mantle" metaphor in what they aptly called "an emotional call to arms":

> The green mantle of the Earth is now being ravaged and pillaged in a frenzy of exploitation by a mushrooming mass of humans and bulldozers. Never in the 500 million years of terrestrial evolution has this mantle we call the biosphere been under such a savage attack. Certainly, there have been so-called "crises" of extinction in the past, but the rate of decay of biological diversity during these crises was sluggish compared to the galloping pace of habitat destruction today.... There is simply no precedent for what is happening to the biological fabric of this planet.... This is the challenge of the millennium. For centuries to come, our descendants will damn or eulogize us, depending on our integrity and the integrity of the green mantle they inherit.[63]

Gaia's Blanket

If the origins of modern environmentalism are closely tied to the fallout from nuclear tests and the development of radiotracers in the 1950s, the apotheosis of the movement in the 1970s was marked by another major Cold War development: the race to the moon. The Apollo missions (1968–72), which happened to coincide with the establishment of Earth Day and with the run-up to the first Earth Summit in Stockholm, were returning from space with unprecedented data. These included photographic images of our planet that revolutionized previous scientific understandings and had a tremendous emotional impact on the public.[64] As British astronomer Fred Hoyle observed, it was as though upon seeing those images, everybody had become suddenly concerned to protect the natural environment. "Where has the idea come from?" asked Hoyle. "You could say from biologists, conservationists and ecologists. But they have been saying the same things now as they have been saying for many years." Something, Hoyle concluded, had happened to create a worldwide awareness of the earth as a unique and precious place. "It seems to me more than a coincidence that this awareness should have happened at exactly the moment man took his first step into space."[65]

The image that had the strongest and most lasting impact was certainly the dramatic moon-shot of the full earth taken by the astronauts of the last Apollo mission in 1972. Although an unmanned vessel, Lunar Orbiter 1, had already captured the image of the earth in 1966, the stunning beauty of the Apollo full-earth photograph and the human drama behind its creation captivated popular imagination as no other picture had done. Locating a perfectly circular earth within a square frame, the photograph mimicked medieval *mappae mundi*; it conveyed the image of the globe as an island, or as a living organism floating in a lifeless universe.[66] The black void against which the planet was set had the effect of intensifying its colored textures: the green and ochre of Africa and the Arabian peninsula, the vivid blues of the oceans, the purity of the snows of Antarctica, the white of the clouds swirling over the southern hemisphere in delightful patterns.

The sublime splendor and grave majesty of the globe suspended in the eternal silence of the dark universe were nonetheless underpinned by a sense of fragility. "For a moment, the whole globe seems to float with the delicacy and iridescence of a bubble."[67] Indeed, astronaut James Irwin, who flew with Apollo 15 in 1971, poetically compared the tiny blue planet to a precious "Christmas tree bauble."[68] This vulnerable beauty helped viewers crystallize the sense of environmental awareness triggered by biologists and ecologists over the previous decade. The image also crystallized a new sense of the human condition: as seen from the moon, humans became invisible, or rather, they became "one species among the myriad which the earth's green mantle of life has brought into existence—and yet, the only one with the power to see and respect, or to damage and degrade, the planet as a whole."[69]

The idea of a living planet found what is probably its most famous and detailed literary expression in James Lovelock's popular book *Gaia: A New Look at Life on Earth* (1979). The subtitle chosen by the British scientist was an apt one. Not only did it refer to a "new" conceptualization of the surface of the planet as a self-regulating entity characterized by biochemical homeostasis—what Lovelock called "the Gaia hypothesis"—but it also evoked a plain and yet hidden truth: the green mantle of life could only be seen and truly understood from outside. It indeed required a new "look."

Lovelock dates the origins of his theory to the years when he was working as an engineer at the Jet Propulsion Laboratory in Pasadena, California. In 1965, he says, his work led him to look at the earth's atmosphere from top down, from space. "Air is invisible, almost intangible, but if you look at it from above, from space," he explains, "you see it as something new, something unexpected. It is the perfect stained glass window of the world, but also it is a strange mixture of unstable, almost combustible gases."[70] It was precisely this view that led Lovelock to speculate that the biosphere might be more than the habitat of all living things; that it was indeed controlled by living organisms; that the unique composition of the earth's atmosphere was the product of biological processes; that these conditions were regulated through the complex feedback systems of the world's ecosystems. "I differed from [other scientists]," Lovelock insists, "because the view from space let me see the earth from the top down, not in the usual reductionist way from the bottom up."[71]

At a time in which complex and expensive missions to Mars were being programmed, Lovelock reminded his NASA colleagues of the obvious: life was only possible under certain biochemical conditions. The best solution to answer the question about extraterrestrial life was thus not to physically travel to Mars and other planets, but to aim toward them with modest instruments designed to determine whether or not the atmosphere was in a state of chemical equilibrium. Bruno Latour recently contrasted the outward movement of Galileo's telescope piercing the mantle of the sky into the distant depths of outer space to the opposite movement of Lovelock's eye, shifting from the infinite universe back to "the narrow limits of the blue planet."[72] These two opposed movements produced opposed worldviews and understandings of the earth: raising the telescope to the moon led Galileo to believe that all planets are alike; gazing at the earth *from* space made Lovelock believe that our planet is like no other. It was as though, Latour notes, "three and a half centuries later, Lovelock had taken into account certain features of that same earth that Galileo could not take into account, if he were going to consider it simply as a body in free fall amidst all the others."[73] These features included the color, odor, surface, and the texture of its mantle.

It was not, however, the "green mantle" per se that mostly captured the imagination of Lovelock and his followers, but the atmospheric bubble enshrouding it. The Apollo shots came after Lovelock's initial conjectures, and they simply reinforced his views. The most evocative feature in the astronauts' first photographs of the earth from space was that thin luminescent film wrapping its surface. It was its magic blue glow, its trembling beauty, its fragile slenderness that captivated Lovelock and other scientists (pl. 7). Images of the earth captured from space made the invisible atmospheric layer visible in all its transcendent beauty. Such images led Lovelock to conclude that the atmosphere was "but an extension of the biosphere." In *Gaia*, "green mantle" and "web of life" metaphors thus gave way to "gaseous blankets" warmly "clothing" the biosphere, and to "carpets" of microorganisms changing their color from light to black, according to the season and the solar radiation received through the thin atmospheric bubble. "Could these black mats, produced by a life form with a long ancestry, be living reminders of an ancient method of conserving warmth?"[74]

While Carson started from the visible detail—be it a poisoned bird, or a withered roadside bush—to reach the global, Lovelock proceeded in the opposite direction: his initial view was a view from space, which allowed him and his readers to see how "the beauty of our home contrasts sharply with the drab uniformity of our lifeless neighbors."[75] While Carson's heroes were the anthropomorphized plants and animals that made up the visible web of life, the characters of Lovelock's story—gases, algae, and bacteria—largely eluded human sight. In order to move public opinion, Lovelock thus personified the whole system. Like Zeus in Pherecydes's myth, the scientist initially endowed Gaia with the mantle of a living female creature, for example, by repeatedly using the pronoun "she," or human attributes (for instance, "Gaia is sick"). Personification, he explained, was instrumental: without recognizing the earth as living, "we will lack the will to change our way of life and to understand that we have made it our greatest enemy."[76] Yet, Lovelock later became careful to describe Gaia as an organic system that is responsive and regenerative, rather than as a living creature. Gaia was, after all, not a Greek goddess, as the novelist William Golding had told the author, but a primeval force— the mystery of life itself.

*

Different understandings of the environment are deeply tied to different understandings of space. Shifts in the use of the earth's green mantle metaphor and its variants point to shifting conceptualizations of nature and transformations in Western public environmental consciousness. The green mantles eulogized by Thoreau, Muir, and early twentieth-century botanists all express an understanding of nature as "wilderness," for example, as special areas to be protected by enclosing them within the sacred precincts of natural reserves and parks. The green mantle was understood by these authors as a well-defined heritage to be safeguarded by humans. The environmental vision of Carson and her followers, by contrast, shifted from self-bounded wilderness areas to interconnected global space. This shift corresponded to environmental movements' shift in focus from aesthetic motivations to the importance of ecological integrity to human health.[77] Lovelock pushed this view further. He transported his readers to outer space, called their attention to the sensitive and perishable envelope wrapping the planet, and made the green mantle fully alive as a self-regulating super-organism.

All these developments were products of the Cold War. If nuclear testing was its most aggressive expression, the space race was its most graphic outcome. The first produced fallout, which proved instrumental in Carson's vision of global interconnectedness; the latter crystallized this vision in the Apollo images. As Cosgrove noted, the two programs were deeply connected "in both technology and culture." They both depended on the application of physics to engineering, and they both expanded the scale of "what counted as nature"—from the Romantic landscape of Yosemite to subatomic and extraterrestrial scales and spaces.[78] However, in the 1970s, physics was displaced by the life sciences as the most intellectually stimulating and popular branch of the natural sciences. Partly mobilized by Carson, partly triggered by the Apollo images, the sudden public surge of interest in ecology and biology was paralleled by a reassessment of US scientific research priorities. In the 1970s and 1980s, the earth sciences became associated with "the green mantle" of environmentalism, as opposed to the "khaki uniform" of military sciences that had dominated the IGY.[79] The nuclear race, on its end, became increasingly associated with images of indiscriminate destruction, "not only of humans, but of all terrestrial life."[80] At a more profound level, the mantle ceased to be a veil to be torn. Attention was redirected to the fine textures of its complex fabric.

The next (and final) section of the book moves to the social texture of the earth's mantle, that is, from the planetary consciousness triggered by the IGY and environmental movements to the phenomenon of globalization.

PART IV

Weaving Worlds

Cartographic Embroideries

In the opening of *The Crying of Lot 49* (1966), the American writer Thomas Pynchon describes an epiphanic encounter between Oedipa Maas—a California housewife and the heroine of the novel—and a painting by the Spanish artist Remedios Varo titled *Bordando el manto terrestre* ("Embroidering the earth's mantle," 1961). In the painting, Pynchon explains, "were a number of frail girls with heart-shaped faces, huge eyes, spun-gold hair, prisoners in the top room of a circular tower, embroidering a kind of tapestry which spilled out the slit windows into a void, seeking hopelessly to fill the void: for all the other buildings and creatures, all the waves, ships and forests of the earth were contained in this tapestry, and the tapestry was the world."[1]

In Pynchon's novel, Varo's painting appears as a modern, or rather post-modern, incarnation of Pherecydes's myth. The tapestry is nothing less than the cartographic mantle Zeus placed upon Chthonia, transforming her into Gaia. Here, however, the frightening presence of Chthonia—the monstrous, shapeless matter—is replaced by something even more frightening: a non-presence, a disquieting "void." The cartographic mantle does not simply imprint legibility and order on reality; it becomes reality itself.

The omnipotent, patronizing figure of Zeus is likewise replaced by a group of fragile girls—and by the viewer herself. Indeed, as Pynchon explains, in contemplating the painting, Oedipa immediately identifies herself with those girls. She suddenly realizes that she too is locked in a tower—a tower that is "everywhere." Her ex-lover died, having named the woman executor of a secret underground postal delivery service, whose existence she set off to investigate. Oedipa's researches, however, eventually led her to paranoia. The heroine finds herself torn between believing in the existence of the secret service and believing that it is a hoax established by her ex-lover as a conspiracy toward her. Unable to distinguish between fact and fiction, Oedipa stands in front of the painting and cries. "No one had noticed; she wore dark green bubble shades. . . . She could carry the sadness of the moment with her that way forever, see the world refracted through those tears."[2]

Oedipa's world is a world of surfaces; it is a vast tapestry fabricated out of her imagination and refracted through the glass screen of her spectacles and through the liquid screen of her tears. But just as Oedipa recognizes her own condition in the painting, so literary critics have recognized in the novel the condition of American society in the 1960s: "a surface affluence, outwardly undimpled by discontent or want." According to literary critic David Coward, the dilemma of the young woman imprisoned in her mental tower with no way out symbolizes "the paralysis of a whole culture."[3] It symbolizes that same consumerist culture of glamorous surfaces that, in their own way, environmental writers like Carson sought to challenge and reverse. There is, however, a tiny detail in the actual painting that both Oedipa and Pynchon seem to have missed: one of the girls stitched her lover in the earth's mantle, which she will use as a means to escape from the tower. What the postmodern novelist and his paranoid heroine seem to have forgotten is that weaving the mantle of the earth is, first of all, a creative gesture; it is a powerful metaphor for imagining the world, *as well as* for shaping and transforming it.

The playful ambiguity of cartographic mantles has long captivated the fantasies of artists and writers alike. In an oft-cited passage, Lewis Carroll, for example, wrote of a folded 1:1 map, which was never spread out, as the farmers feared that it would cover the whole country, and shut out the sunlight. The map's uselessness allowed Carroll's character Mein Herr to sarcastically conclude: "We now use the country itself, as its own map, and I assure you it does nearly as well."[4] In a short fictional tale, Jorge Luis Borges likewise told the story of a perfect life-sized map that eventually ripped and weathered to shreds across the actual land it covered. "In the Deserts of the West, still today, there are Tattered Ruins of that Map, inhabited by Animals and Beggars; in all the Land there is no other Relic of the Disciplines of Geography."[5] As in Pynchon's novel, here the cartographic mantles feature as powerful epistemological metaphors, with the difference, though, that for Carroll and Borges the map seems to be a copy of territory, and is thus ridiculed as such, while for the postmodern novelist it appears to be territory (or reality) itself.

The textile medium adopted by Varo (and Pynchon) seems to lack the rigidity and the fragility of traditional paper maps. Unlike Borges's map, it seems to be immune to the damaging action of the elements and of time. It is a plastic, almost liquid, element that constantly spills, flows, and unfolds through the tower's slit windows. Like the human imagination, fabric has the

capacity to constantly remodel itself. Fabric contracts and expands; it fluctu-
ates and undulates, like the surface of a lake, or of the sea. This fluidity blurs
boundaries and distorts straight lines. It challenges fixed geometries and cartographic truths. As such, fabric has been appropriated by modern writers
and artists as a privileged medium for crafting "countermaps."[6] Pynchon and
Varo used representations of cartographic mantles to confuse the boundaries
between map and territory, between representation and reality. Over the following decades, conceptual artists literally manufactured their own cartographic mantles as tactile metaphors and tools for questioning extant territorial orders and conventional ways of seeing an increasingly globalized world.

   This chapter explores some of these crafts as a way into modern and contemporary perceptions of globalization and human interconnectedness. The
first section introduces the act of embroidering the earth's mantle as a metaphor for artistic creativity and imaginative processes through Alfred Tennyson's nineteenth-century poem *The Lady of Shalott* and Varo's painting. The
following sections focus respectively on the cartographic textiles of the Italian artist Alighiero Boetti (1940–94) and contemporary London-based artists
Mona Hatoum and Katy Beinart. Whether reflecting on troubled existential
conditions, denouncing the artificiality of territorial boundaries, or promoting cosmopolitan narratives, all these mantles are utterly human. Intertwined
in their fabrics run stories of exile and segregation, of mobility and diaspora,
of desperation and hope—alongside new and old textile metaphors, world
images, and imaginations.

Weaving Worlds

Weaving, embroidering, stitching, knitting are all creative acts. There is
something tactile, repetitive, and gentle about them that eludes words and
straight lines. Traditionally classed as feminine practices, they all belong to
the domain of the private and the domestic. But so does the geographical
imagination. In order to flourish and expand, imagination demands privacy.
In order to fashion worlds, the artist, the writer, and the poet need to separate
themselves from the world, if only temporarily. Varo's *Bordando el manto terrestre* (fig. 10.1) best expresses this tension between domestic self-enclosure
and the unbounded realm of the imagination. The young women, so vividly
described by Pynchon, embroider the tapestry of the world in the enclosed
space of a tower suspended in a primordial sea of darkness. Each girl works
alone embroidering images on the continuous, overflowing fabric, as if in a
medieval scriptorium. In their communal isolation, those frail girls literally
craft the world.

FIGURE 10.1 Embroidering the earth's mantle as a creative act: Remedios Varo, *Bordando el manto terrestre (Embroidering the Earth's Mantle)*, 1961. Oil on masonite, 39 ½ × 48 ½ inches (100 × 123 cm). © Remedios Varo, DACS/ VEGAP 2019. Courtesy Gallery Wendi Norris, San Francisco.

There are, of course, precedents for the scene. Between the 1770s and the first decades of the nineteenth century, before the sewing machine and indelible ink were invented, embroidering maps and globes made of silk was a common practice and learning exercise among British and American schoolgirls (fig. 10.2). Geographical samplers—which is how these embroidered maps were known—were widespread crafts in the material culture of women's education. Made of minute stitches in silk threads, they attest to mastery of both geography and needlework. The deeply tactile and repetitive act of stitching enabled young girls to memorize the outlines of countries and regions, the meandering courses of famous rivers, old and new political boundaries, familiar and exotic place names, and other geographical facts and features. Map samplers of Palestine, for example, were used in religious classes to teach Scripture and biblical geography, while samplers of England and the United States imparted contemporary geographical education, while at the same time boosting patriotism.[7] Typically executed in dame schools, boarding schools, and academies, cartographic samplers evoke the collective intimacy and careful handiwork portrayed in Varo's painting.

Weaving the world was also, and perhaps more famously in Victorian England, the subject of Alfred Tennyson's poem *The Lady of Shalott* (1832). Here, however, communal work is transmuted into a quintessentially solitary act. Collective intimacy is morphed into oppressive seclusion. Possibly inspired by a medieval Italian novel, the poem features the story of an unidentified woman trapped in a tower and condemned by a curse not to engage with the external world except for what she sees reflected in a mirror.[8] Day and night, the woman restlessly reproduces on her loom the mirror's "magic sights." Like a cartographer of old in his cabinet, she endeavors to create a perpetually unfolding "map" of the world around her. Yet, the two are closely intertwined. The landscape outside the tower seems to blend into her own craft; like the fabric woven on her loom, it does have a fine texture. It is, we are told, a landscape of "long fields of barley and rye, / That clothe the world and meet the sky"; it is "a space of flowers" reminiscent of the earthly mantles in vogue amongst early nineteenth-century poets.[9]

The whole story is articulated through a complex play of reflections. Isolated in a twilight zone of shadows and illusions, the Lady of Shalott is said to

FIGURE 10.2 Girl stitching globe, a common practice and learning exercise among British and American schoolgirls between the late eighteenth and early nineteenth centuries. Drawing by Edward Shenton. Esther Duke Archives, Westtown School, West Chester, Pennsylvania.

"delight herself" in weaving in her web the mirror's sight of a funeral procession moving through the night; a few lines later, however, as the image of a young, newlywed couple illumined by the moonlight materializes in the mirror, she complains of being "half-sick of shadows."[10] This melancholic longing for life eventually leads her to rebellion and, ultimately, to death. As the figure of Lancelot flashes into the crystal mirror and his distant voice breaks the silence of her seclusion, the woman is eventually compelled to turn her back on images and gaze out of the window, into the real world:

> She left the web, she left the loom
> She made three paces thro' the room
> She saw the water-flower bloom,
> She saw the helmet and the plume,
>   She look'd down to Camelot.
> Out flew the web and floated wide;
> The mirror crack'd from side to side;
> "The curse is come upon me," cried
>   The Lady of Shalott.[11]

Robed in white like a bride, the lady abandons the tower and steps onto a boat moored in the river underneath. Yet, the moment she leaves her loom, the flowery mantle of the earth reveals nothing but gloom. It turns into a gray landscape of pale yellow woods, stormy low skies, and copious autumnal rains.[12] As the current of the river transports the boat to the kingdom of Camelot, the lady slowly dies.

Separated from and yet connected to the world though the thin thread of representation, the Lady of Shalott has been interpreted by literary critics as an allegory of the poet. Like the poet, the lady lives in a world of images, which she crafts and upon which she depends.[13] Art historians have likewise recognized in her the artist who, "entrusted with a mission, fails to deliver it, by succumbing to personal desires."[14] Either way, the lady's magic mirror and loom seem to reveal a deep truth, a truth more real than reality itself.

This unsettling hyperrealism compelled pre-Raphaelite painters to make the story one of their favorite subjects. In his own rendition (1886–1905), William Holman Hunt, for example, famously captures the moment of rapture, when the lady turns to the window and the loom falls down on the floor (fig. 10.3). The scene seems to freeze time, suspended as it is between the lady's golden hair floating in midair and an intricate pattern of bundled colored skeins of wool amid which she now appears to be trapped. The disordered threads act as a visual metaphor for the woman's distraught state of mind.[15] "The coherent web she was weaving has to break, and illusion with it."[16]

FIGURE 10.3 William Holman Hunt, *The Lady of Shalott*, ca. 1886–1905. The scene captures the moment of rapture, as the lady turns her back to the world of reflections in which she is trapped, and her loom falls on the floor.
Manchester Art Gallery, UK / Bridgeman Images.

Sidney Meteyard, by contrast, preferred the quietness of weaving to the drama of rapture. In *I Am Half-Sick of Shadows* (1905; pl. 8), the lady is immortalized as she gently falls asleep before her web and the nocturnal view of the two lovers reflected in the mirror. As in Hunt's painting, here the flow of time and of her weaving seems to have stopped, this time in the calmness of a moonlit night. Lancelot's half-completed figure on the web foreshadows the next turn of the story, as if the lady had already started to weave her own destiny. The sensuality of the scene reflected in the mirror extends to the entire scene. The painting is imbued with a mysterious silky texture. The lady's thin blue dress, her exquisitely soft hair, the delicately colored silk skeins, the shiny stitches on the loom, the velvety curtain revealing the mirror: all conjure up an intimate tactility, a magical realism. The viewer is invited to caress the loom, to touch the lady's silk dress, to toy with the skeins. Representation becomes palpable to the eye, while reality becomes a mere reflection in the mirror, a distant dark shadow.

Even more than an epistemological allegory built on the opposition between illusion and truth, however, Tennyson's original allegory, it has been suggested, was a social one built on oppositions between the feminine and the masculine, between the private and the public, between the self-enclosed domestic space of the tower and the open landscape outside it.[17] In her boldness to cross the boundary between these spaces, the Lady of Shalott prefigures Varo's heroine—and the life of Varo herself.

The daughter of a devout Catholic woman and an agnostic civil engineer, Maria de los Remedios Varo Uranga (1908–1963) was born in a town north of Barcelona. She received her education in a Catholic convent school and subsequently in the Academy of Fine Arts of Madrid.[18] Throughout her life she came to cross the boundaries between many different worlds. Her rebellious temperament, ill suited to convent schools as much as to parental control, led her to an early marriage with a fellow artist from the academy. Contracted mainly to gain independence from her family, the marriage lasted less than two years, spent amid Barcelona's vanguard circles. During the Spanish Civil War, Varo fled to Paris with her new partner. As the Nazi army occupied the French capital, she was then newly forced into exile. Having traversed the worlds of Catholic Madrid, Republican Barcelona, and Surrealist Paris, she thus eventually landed in Mexico City, where she achieved economic stability and artistic fame.[19]

If pre-Raphaelite paintings are striking for their hyperrealism, Varo's Surrealist work presents a fluid world of reverie, alchemy, and metamorphoses in which boundaries and edges are constantly done and undone. Her world is, in Janet Kaplan's words, "a world of permeable membranes in which chair backs can open to reveal human faces, hands may reach out from behind walls, and tabletops peel back to expose organic roots." It is a dream world in which "the inanimate moves, the hard is made pliable, the soft stiffens, bodies and their shadows become interchangeable."[20] It is a world populated by heroines crossing thresholds into "a *transmundo* beyond the natural and the supernatural," of mystical seekers and eternal pilgrims.[21] Whether in series or in individual paintings, Varo's female characters move between different states and different spaces toward secret destinations, which mark the apotheosis of an inner rebirth.

Painted in Mexico shortly before her death, *Bordando el manto terrestre* is the central panel of a triptych that has been recognized as Varo's most autobiographical work. It reflects "on the restrictive atmosphere of those early years in Spain and on her rebellious schoolgirl yearning for escape."[22] The first painting of the triptych, *Toward the Tower* (1961), shows her self-portrait character in a group of identical uniformed girls in an orderly group bicycling away from a beehive-like building behind their mother superior. The building represents the space of Varo's childhood, a space she perceived as austere and oppressive: a world of common meals and classes, of group sewing and group confessions, an old world "enclosed by walls of virtue" in which, for Varo, "improvisation found no place, for all was pre-arranged and predictable."[23] A girl in the group nonetheless rebels, her gaze reaching out disobediently, resisting what the artist called "hypnosis." The same girl reappears in the central panel, this time embroidering her escape route in the earth's mantle, and then again, in the last panel, where she eventually flees with her lover to a primordial, "Chthonic" landscape of mountains and caves.

The earth's mantle in the central panel (fig. 10.1) is striking for its insistent physical presence. It is a solid, three-dimensional fabric amid whose folds houses, trees, streams, ponds, seas, boats, animals, and humans are nestled. The tactility of this giant mantle conveys the artist's "exceptional love for all that could be experienced through the senses," as a friend remembered her:

> Her touch passed and re-passed over the warm surface of wood or the coolness and solidity of a rock. She could be absorbed for hours in the weave of a cloth or the play of light on a window pane. She saw in everything the life latent from within; she observed the most diverse objects, delighting in all their details, the infinite hues, the textures, colors, and forms.[24]

As Kaplan comments, in *Bordando*, a "masterful variant on the myth of creation, Varo has used this most genteel of domestic handicrafts to create her own hoped-for escape."[25] Female creativity and imagination will free Varo's blond heroine—and Varo too. While the mantle of the earth is woven according to the instructions dictated by a taskmaster and from threads spun from the same alchemical vessel, the girl shows that she has a mind of her own. Unlike the Lady of Shalott, Varo's young heroine imprisoned in the tower is not a mere metaphor for confinement, but the agent of her own life, of her own dreams, and, eventually, of her own liberation. While the Lady of Shalott is but a shadow of the world condemned to die for trespassing the boundary of representation, Varo's heroine is an active part, indeed a shaper, of the world. Freeing herself from anonymity, she reclaims her life by literally "weaving" her own destiny.[26]

In Varo's real life, however, this escape merely resolved in an escape from one tower to another—the tower of her own passions. And the walls of this tower proved far thicker and more impenetrable than those of her childhood's convent school. A highly superstitious and very anxious woman, whose life carried the deep scar of forced dislocation and exile, Varo eventually fell prisoner to her own disquiets. "She feared disease, she feared aging, and her fertile imagination created other fears she could never explain."[27] It is ironic that, after a life of continuous search—whether through her interest in alchemy or in Eastern mysticism and science, or through her relationships with multiple lovers—the artist longed to spend the last years of her life in a Carmelite convent in Aguilar de la Frontera, a secluded, penitential order inaccessible to the rest of the outside world, except for a screen through which the nuns communicated with visitors.[28] Yet, the scissors of death moved faster than her ever expanding embroidery of fears and longings. She died of a heart attack at her home in Mexico City at the age of fifty-four.

> Threads of death, of life, of time.
> The weft is woven and unwoven,
> The unreality that we call life, the unreality that we call death . . .
> Only the canvas is real.
> Remedios the anti-Moira.[29]

Mappe's Shifting World Orders

Tales of confinement and imagination, of travel and segregation, of visibilities and invisibilities weave through what are probably the most magnificent, iconic, and controversial of all twentieth-century cartographic mantles: the approximately 150 hand-embroidered world maps that form Alighiero Boetti's *Mappa* ("Map") series (1971–94; fig. 10.4). The works were commissioned by the Italian artist from women in workshops and families in Kabul, Afghanistan, and, later, in refugee camps in Peshawar, Pakistan.[30] Ranging from the size of a tablecloth to that of an entire wall, they all feature a brightly colored mosaic of national flags filling the outlines of each country. Spacious continental landmasses and tiny, jewel-like islands are set against a vast monochrome (usually blue) ocean and framed by an inscription in Italian, English, Farsi, or Dari (the variant of Persian spoken in Afghanistan). Sometimes the

FIGURE 10.4 Alighiero Boetti, *Mappa* (1979–83), one of about 150 exemplars of hand-embroidered world maps on canvas in the series.
Private collection, Switzerland. Photo by the author.

inscription spells out the date and place of production, accompanied by a greeting; at other times it contains some cryptic statement by the artist; at still others, it features verses from Sufi poems.

Boetti's are certainly not the first cartographic mantles. Embroidered and woven world images are found throughout history and all over the world, as we have seen—from Queen Kypros's textile gift to Gaius Caesar to the above-mentioned nineteenth-century British and American samplers; from Byzantium to Mexico; from central Asia to the Far East. Christian Jacob, for example, writes of "a woman of the Chinese imperial court who produced a map on silk in which the warp and the woof of the fabric formed a cartographic grid."[31] During World War II, Allied soldiers had maps of escape routes embroidered on silk handkerchiefs jealously treasured in their pocket or tied around their neck to elude enemy scrutiny, while women of the British Women's Voluntary Service produced three-dimensional aerial views for the army on hessian mats, stitching cloths to make trees and hedges.[32] What sets Boetti's maps apart from such ingenious predecessors, however, is the temporal scale they consciously embrace. Commissioned almost uninterruptedly for a period of more than two decades, the works reflect shifts in borders and regime changes from the height of the Cold War to its end. During this time span, no fewer than five hundred women—that is, one thousand hands—were employed in the task. But how did the *mappe* come into being?

Boetti's initial encounter with Afghanistan, we are told by his first wife and biographer Annemarie Sauzeau, happened in 1971 "almost by chance," and yet, it left a powerful mark on the artist, to the point that he decided to open a hotel in Kabul and make the city his second home.[33] In that Afghanistan— the Afghanistan of the early 1970s—the artist discovered a country locked in a peaceful feudal past, a past "as solid as the mountains of the Hindu Kush and the Pamir that dominate the horizon to the North." He discovered "a bare, pink landscape bathed in a crystal clear light that made distances seem to shrink."[34] It is against this unfamiliar background of light and rock, be- tween its open landscapes and the secluded enclosure of *purdahs*,[35] between the squalor of refugee camps and the homely space of the artist's studio in Rome, that Boetti's *mappe* are to be situated.

On his second trip to Kabul, Boetti carried with him a white piece of linen upon which the design and colors of the first *mappa* had been carefully drawn. He took the linen to the embroidery school of a certain Mrs Kandi, where the work was to be completed.[36] Over the years, the Italian artist dis- tributed his increasingly numerous commissions among different workshops and families. Designs and colored yarns were either carried to the produc- tion sites or shipped by post, and the completed *mappe* were shipped back to the artist's studio in Rome. With the Soviet invasion of 1979, Boetti's regular trips to Afghanistan came to an end, as it was no longer possible to enter the country. Production was interrupted. It slowly resumed three years later in camps housing Afghan refugees in Peshawar, where it continued until Boetti's premature death in 1994.[37]

Mappe intrigue because of their unusual origin and hybrid nature. On the one hand, we are faced with a familiar, modern, deeply Eurocentric world im- age, an image that many westerners have probably learned to take for granted. On the other, we are faced with an unconventional "premodern" medium, hand embroidery. Sometimes we are also faced with exotic strips of Arabic calligraphy running around the edge of the map—and with the orientalist tales surrounding the whole venture.[38] This tension between expected and unexpected disturbs the sensibilities of the Western viewer. It makes the embroidered maps media for "*mettere al mondo il mondo*" ("giving birth to the world"), as the inscription framing a 1972 *mappa* states. In other words, these cartographic textiles enable the viewer to see the world "as if for the first time." They at once surprise and invite scrutiny. In their familiar unfa- miliarity, they become instruments for unveiling, or rather, they "reveal an empirical and phenomenological questioning, like that of someone who is getting to know the world day by day, and whose end products show us the result of this gradual unveiling."[39]

Both individually as patiently crafted artifacts and collectively as a series, the textiles speak *longue durée* stories. Their crowded, seemingly infinite stitches and carefully executed details bear witness to the slow process of their manufacturing.[40] Shifting boundaries and changing flag patterns bear witness to the slow passage of time and shifting world orders. The first *mappe* present the frozen image of a split world dominated by the Western and Soviet blocs. Magnified by the Mercator projection, the two rival blocs face each other, occupying most of the northern hemisphere: on the one hand, there are stars and stripes stretching across America and Alaska, Canada's huge maple leaf, the giant Danish cross inscribed upon Greenland, and the lively patchwork of European countries; on the other, there is a vast and uniform sea of red covering most of Eurasia, from Leningrad to Peking, from Karelia to Kamchatka. Year after year, *mappa* after *mappa*, newly independent African colonies reclaim their own flags, while contested territories linger in ambiguity. Israel, for example, is often absent from the map, while the territories it occupied during the Six-Day War continued to be portrayed in the same colors as the Egyptian flag. Likewise, the flag of South Africa was not sewn over Namibia, and, after the Soviet invasion, Afghanistan was represented with a flag bearing the name of a local political faction, rather than in red.[41]

As a whole, the series can be read as "an expression of humanity's temporal relationship with the world." Set in sequence, the cartographic embroideries unfold the unpredictable story of humankind—"*a tempo in tempo col tempo il temporale*" ("in time, on time, with time, the temporal"), as an inscription around a *mappa* recites.[42] *Mappe* express the fragility of global politics and, in general, the absurdity of any attempt at imposing artificial boundaries and abstract human concepts upon the natural world. Countries and flags raise and fall. Political alliances are built and broken. Fabrics are made, unmade, and remade.

While the textile medium helps convey this sense of fluidity, it also helps articulate ancient Stoic narratives of global unity in diversity, as the Farsi inscription gracing another *mappa* seems to suggest: "Of one essence is the human race."[43] In spite of the political fragmentation conveyed by its flags, in spite of the multiple colors and consistencies of its threads, a *mappa*, like the world, is one and the same fabric. To Boetti and his Western audience, it has been suggested, *mappe* "seemed to embody what was then, in the immediate wake of the Cold War, a newly emerging sense of the world as a more holistic, interactive and intercommunicative entity."[44] In stressing interconnectedness and fragility by way of the textile medium, *mappe* bodied forth a cultural counterpart to the ecological visions promoted by Carson's followers in those decades. At the same time, their ability to cross boundaries between culturally distant worlds made them ideal vehicles for armchair travel. As with any map, and perhaps more than any map, *mappe* enlarged the geographical imagination of their viewers. As Sauzeau affectionately recalls,

> Now, in every part of the world flags change design and color after wars. At that point Alighiero's maps had to be redesigned. Even the color of the seas can change, according to the language of the people (black sea, white, red, yellow . . .). It may also alter according to the hours of day and night, according to the predilection of poets for silver, amethyst or jade . . . or the choice made by the women embroiderers among the skeins of thread in the bazaar. . . . These were the gentle stories whispered in the ear of little Agata by her father when, on returning from Kabul, he emptied out his bag and pulled out the embroidered *Maps of the World*, still smelling of goat, curry and other spices. After a while, in the Umbrian grass, the tapestries would float like distant oceans, like mountains higher than the Apennines. These flying carpets were offered to our little girl.[45]

Whether floating on the grass, or hanging in the sanitized white space of a contemporary art gallery, what immediately strikes the viewer is the material texture of the *mappe* (pl. 9). A close look reveals a patchwork far more complex than the flags. Regardless of whether we scour their polychrome continental spaces, or navigate their vast oceans, looking at the surface texture of these works, we discover different shades, consistencies, and textures. Some sections reveal lighter or darker colored thread; in others, the stitches follow different directions, like sea currents. Occasional stretched marks and stains bear physical witness to the relentless passing of time.

Boetti worked predominantly with two-dimensional surfaces, yet his engagement with them was deeply tactile and three-dimensional; when he drew, we are told by Sauzeau, he "traced," "massaged," or else "filled in with dense hatching."[46] In intensifying this haptic experience, the textile medium, nonetheless, also unveils nonpresences—the invisible hands of its Afghan embroiderers. It is in the rugged surface of the *mappe*, in their technical mishaps, in their unexpected occurrences and errors that we discover the traces of these anonymous women. The changing direction of the stitches, or the use of different kinds of thread, bears witness to the separate production of different parts; misplaced boundaries or distorted outlines speak of the manufacturers' geographical illiteracy. Secluded in appositely designated female spaces and generally lacking formal education, most of these women, it has been observed, treated countries and flags as just abstract patterns, except perhaps for Afghanistan, whose familiar outline could be easily seen on trucks, carts, and stamps.[47] Unlike Varo's heroine, the Afghan women embroidered places they were unlikely ever to see, or perhaps even to hear of. Their tower was not psychological, but physical.

Boetti complacently proclaimed that for the *Mappa* he "did nothing, chose nothing, in the sense that the world is made as it is." He claimed to be a simple spectator of a self-generating process he merely set in motion.[48] The Italian artist did not have direct contact with the embroiderers. He generally interacted with them through a male intermediary, usually the family leader, who would take the work to the women of his clan and arrange the shipping of the final product to Rome.[49] In the best orientalist tradition, the stories of production of the *mappe* are articulated through a series of veils, curtains, and barriers separating the artist from the women: those of the closed domestic space of their *purdah*, of female workshops, of refugee camps, of the *hijab*.[50] As a result, the women who manufactured the *mappe* become an anonymous mass of invisible hands; we do not know their names, we do not know their stories. As an inscription framing one of the embroideries states, the *mappe* were simply *"embroidered to the order by unknown Afghan women."*[51]

While Boetti's cartographic embroideries have been traditionally celebrated as epitomes of cosmopolitanism and globalization, or even international cooperation, for contemporary feminist and postcolonial critics they are ultimately products of a culture experiencing multiple phases of cultural and economic exploitation. Their colored patterns ironically veil what Nigel Lendon identified as the "dark side," or negative effects, of modernity and globalism: ethically questionable production practices, deeply asymmetrical gender and power relations, alienated labor. In his words, these works "exude an aura of excessive abstract labour." Through the evidence of endless repetition, stitch by stitch, they create "a particularly contradictory mode of representation whereby their Afghan makers are simultaneously present through the embedded labour and yet made absent through the nature of Boetti's iconography."[52] Cartographic transparency obscures the fact that these artifacts were manufactured on fraught terrain.

As with any map, or indeed any mantle, Boetti's cartographic embroideries reveal and conceal all at once. As with any image, they are open to the scrutiny of multiple gazes. Some people, Agata Boetti notes, only look at their own country; others admire the beauty and strangeness of symbols on the flags; others note the vast presence of the oceans; others contemplate the beauty of the work of the earth.[53] And then there are others—the postcolonial critics—who, reversing their imposing fabric, expose the messy and unappealing tangle of the threads constituting the ordered image it presents to the world. For little Agata, *mappe* were flying carpets floating on the grass and taking her to distant lands. Hanging in an art gallery in London, to critical theorist Charlotte Kent, the *mappa* appeared "a silent and sad geopolitical artifact." But then,

> A mother and daughter arrive, smile at its beauty, tracing with their fingers the route from the mother's birthplace to London. Others stop by and find their own tracks in Boetti's map. Soon it is full of life, traveling again.[54]

Life Threads and Cosmopolitan Textures

The metaphorical power of cartographic textiles and their ability to stir the geographical imagination have continued to inspire contemporary artists well into our present. The ancient textile metaphor seems to have become especially apt and poignant in a world that is, on the one hand, globalized and interconnected as never before and, on the other, punctuated with conflicts and traversed by ceaseless threads of migrants and refugees. Textile metaphors and cartographic mantles have helped artists give visual expression to the complexities of such a world. If Boetti's *mappe* are caught in ambiguity, however, the new twenty-first-century cartographic mantles seem to have assumed openly political and participatory overtones. Rather than simply commenting on changing world orders, these mantles are meant as powerful interventions in the very fabric of a globalized Western society.

Reflecting on the postcolonial condition, *Bukhara (maroon)* (2007) by the British Palestinian artist Mona Hatoum, for example, features a world map created through the removal of fibers in a traditional red, Persian-style carpet (fig. 10.5). *Bukhara (maroon)* is a sort of "anti-*mappa*." The traditional Mercator projection used in most of Boetti's embroideries here is replaced by the more equalitarian Gall-Peters projection, which reveals the true proportions of distributed land mass.[55] Yet, this more "equal world," in which Europe and North America are redimensioned, and Africa and South America are given back their actual prominence, is just a shadow. Unlike in Boetti's *mappe*, here the continents are not colorfully and carefully embroidered, but violently carved out of the fabric, as if moth-eaten, or eroded by the continuous trampling of humankind. The world appears as a dark space constructed in negative. Where does this vision come from?

As with Varo and the Afghani embroiderers of Boetti's last maps, Hatoum's life bears the scar of exile. The artist was born into a Palestinian family that was uprooted from Haifa during the Arab-Israeli War in 1948 and settled in

FIGURE 10.5 Mona Hatoum, *Bukhara (maroon)*, 2007. Wool, 148 × 261 cm. The world worn out by the continuous trampling of humankind.

© Mona Hatoum. Courtesy Galleria Continua, San Gimignano / Beijing / Les Moulins. Photo: Ela Bialkowska.

Beirut. During a visit to London in 1975, civil war broke out in Lebanon. Unable to return to her family for several years, she decided to stay in Britain.[56] Much of her work reflects on the experience of displacement; it speaks of the sense of irremediable loss and separation triggered by the war.[57] Hatoum used maps in various previous installations to convey the transiency of territorial orders and of territorial promises: from a soap map of the territories agreed to be returned to Palestinian control in the Oslo Accords of 1993, to world maps made of glass marbles set on a wooden floor.[58] Ready to be washed away or dissolved, these ephemeral cartographies forced viewers to reassess their place in the contemporary world and their relationship to its numerous ongoing conflicts.

Bukhara (maroon) shares in this narrative of disruption and transiency. The metaphorical power of the textile medium and its cultural legacy, however, add new connotations to this work. This is an oriental rug; as such, to the Western eye, it acts as a synecdoche for the land and culture from which the artist has been displaced. Its consumed surface brings to mind biblical verses and prophecies likening God's creation to a garment that shall "wax old," be "fretted away" or "worn out," and eventually depart at the end of times.[59] With its beautiful symmetric patterns and fine textures, the artifact evokes Byzantine cosmographic mantles. It enshrines a sort of sacred quality. At the same time, however, the carpet is also an everyday object. It is similar to those that furnished the artist's childhood home. Its familiarity as a domestic item and as a traditional symbol of comfort conveys an uncanny sense of intimacy, which makes territorial disintegration all the more dramatic. As with other artworks by Hatoum, here "familiarity and strangeness are locked together in the oddest way, adjacent and irreconcilable at the same time."[60] The unmaking of the rug's precious fabric undoes certitudes and disrupts affective bonds. It destabilizes the surface of the earth; it turns it into shaky ground.

In disrupting fabrics, Hatoum reverses the ancient weaving metaphor. Domestic or political, profoundly ritualized, weaving, as Scheid and Svenbro observe, has for over two millennia brought into play "an ensemble of notions capable of being inscribed in the collective memory, gestures that allow one to grasp, to touch, social organization."[61] Just as the act of weaving consists in interlacing and binding threads together, so do social and urban fabrics interlace and bind the multiple threads of human lives and their diverse colors. To weave these fabrics is to unite, to entwine, to connect different people and different imaginative scales. Weaving "offers a simple model to the mind seeking ideas about the nature of social cohesion: how is it that the human group, the family alliance, and the city can hold together?"[62]

While Hatoum focuses on the unmaking, on the tattering of these human fabrics, British artist Katy Beinart uses the metaphor to highlight the opposite process: their making, the creative process of crafting, of coming together. Her installation *My Life Is But a Weaving* (2016) is a communal blanket embroidered by migrant women of different religious faiths and ethnic backgrounds in Ealing, a culturally diverse suburb of West London (fig. 10.6). The blanket consists of a dark blue velvet base upon which the women stitched a map of white lines of prayer and other embroidered pieces representative of their faith identities, as well as an appliquéd river.[63] The cloth has been exhibited in various London venues on a table surrounded by kneelers, prayer mats, and other devotional objects.

The concept was consciously inspired by Boetti's and Hatoum's cartographic textiles. Key to Beinart's "tablecloth map," however, is the process of making, rather than the final output. Collectively embroidered and assembled in a local church, the blanket is the reflection, or rather the crystallization, of a communal experience, "a shared space of hospitality and exchange," more than anything else. The tablecloth evokes the act of coming together and gathering around a table to share meals and stories—stories and memories

FIGURE 10.6 Katy Beinart and women from the Fabric of Faith Group, *My Life Is But a Weaving*, 2016. Detail from communal textile artwork embroidered by women of different religions and ethnicities who live in the Ealing (West London) area.
Photo by the author.

of migration, diaspora, exile.[64] Rather than mastermind and maker (as in the case of Hatoum), or detached observer (as in the case of Boetti), here the artist consciously takes on the role of catalyst for communal participation and interactivity; she becomes "a conduit, at times facilitator, of events or environments."[65] Ultimately, her job is to weave life threads together.

The title of the artwork, *My Life Is But a Weaving*, is "a popular prayer which uses the metaphor of woven threads, some visible, others hidden, to describe the role of spirituality and faith in ordinary lives."[66] Beinart selected the theme of prayer for the tablecloth map "as a shared practice across the different faith communities and as a practice linking both private and communal spaces, much like the experience of sewing."[67] Stitching is a repetitive act. Like praying the rosary or reciting a psalm, stitching has its own quiet cadence. It entails the rhythmic crossing of a surface.[68] This simple, peaceful gesture has the power to trigger stories, to bring long-buried memories to surface, and to expand domesticity beyond the walls of the household.

It helps create a safe, almost therapeutic space in which women share their pasts, an intergenerational space in which life threads and stories intersect and are passed on to new cohorts.[69]

The communal blanket is a fabric made of stitches and family histories intertwined with micro-stories of home, migration, and faith. Like Ealing itself, it is a crossroad of life trajectories: the Lebanese refugee escaping the bombings of the civil war; the Indian migrant who arrived in London's winter in her cotton dress; the migrant daughter who moved from the West Indies with her late mother's skeins; the Armenian expatriate who left a pile of ruins behind her.

> There is a town called Marash,
> She tells me.
> It was called Kermanig before,
> Back when it was Armenian, like us.
>
> Everything else is gone, now.
> Razed in 1915.
> Nothing remains of us,
> There,
> Save for this stitch.
> . . .
> The stitch covers
> Her grandmother's velvet cushion
> In fine interlaced crosses.
> It is a map
> Of a scattered people
> Crisscrossing the globe
> In search of new homes
> To replace the one lost to them.
> Kermanig.[70]

The communal blanket enshrines a religious quality in the literal sense of the word—*religare*, "to bind together." It is a metaphor for harmonious coexistence rooted in the Stoic tradition of Alexandria's cosmopolitan *chlamys*, in Marcus Aurelius's web of human lives, and, perhaps more evocatively, in the collective ritual weaving of textiles by women from ancient rival *poleis* to mark the end of war.[71] In contemporary London as in ancient Greece, weaving (or stitching) binds communities. "All joined together to do something," an embroiderer of Beinart's tablecloth map observes. "It was very interesting, the unity; there was no fighting."[72] While Hatoum engages with the unmaking of geopolitical texture of the world, the embroiderers of Beinart's communal blanket attempt to weave the strands back together:

> We know diaspora is a scattering
> The ripping up of the fabric of the family quilt.
> We try to piece them together into new shapes
> But some are lost and the edges are raw.[73]

✶

Fabrics are closely connected with humans and with the space they inhabit. From birth to death, fabrics wrap human bodies and human lives. They provide warmth and shelter. As such, it is not surprising that their materials and the processes through which they are crafted supply one of the richest reservoirs of metaphors to describe life processes and the workings of society. Life threads, urban fabrics, social networks, and many other textile metaphors weave through everyday speech, as they have been doing for centuries—from destiny-spinning Moires to Alexandria's cosmopolitan *chlamys*. Reacting against the misplaced concreteness of cultural groups, poststructuralist anthropologists have often used textile metaphors to describe a new cultural world "of not only fragments and *bricolage*, but also of rifts, threads, weaves, sinews, knots, looms and tangles."[74]

For Latour, modern societies can only be described as "having a fibrous, thread-like, wiry, stringy, ropy, capillary character that is never captured by the notions of levels, layers, territories, spheres, categories, structure, systems."[75] Resorting to yet another textile metaphor, the American sociologist Elijah Anderson coined the concept of "cosmopolitan canopy," an urban public space in which diverse people come together and practice peaceful coexistence; a giant mantle sheltering difference and multicultural interaction, like the biblical tent of heaven hanging over the world (Ps. 104:2).[76]

If diaspora scatters people around, the mantle brings them back together in new configurations. Cartographic mantles give tangible expression to the metaphor and to the complex workings of human society. Echoing ancient myths and metaphors, modern cartographic textiles speak of interconnectedness and multiculturalism, as much as of shifting geopolitical orders, precariousness, and instability. More significantly, they weave together the macro-scale of the world map with the micro-scale of the domestic enclosure of their sites of production, and not least, the threads of the disrupted lives of their makers. The next chapter continues to explore the theme of globalization and precariousness by extending the weaving metaphor to the realm of the digital.

The Digital Skin

Weaving metaphors, and textile metaphors in general, are not simply effective devices for describing human interconnectedness and the workings of society in a globalized world. They are part of the complex digital reality that over the past two decades or so has come to wrap the planet by way of satellites, electromagnetic waves, coaxial cables, optical fibers, and omnipresent plasma screens. Today, "the web" and "the net" are deeply embedded in the daily lives of more than half of the planet's population and more than 80 percent of the inhabitants of the developed world.[1] Increasing numbers of people routinely access the internet to check email and pay bills, to manage personal bank accounts and shop, to search for jobs and properties, to read the news and play video games, to download music and TV programs, to book flights and hotels, to check public transport schedules, pay taxes, share personal experiences with other users, or simply chat with family and friends. Like a giant fabric, the internet links the behind-the-scenes infrastructures of everyday life, threading through administration, financial transactions, governmental policy, and transport systems. Consciously or unconsciously, for a rapidly growing number of people around the world, real and virtual have become thoroughly intertwined, to the point that "access to cyberspace is no longer a luxury, but a necessity."[2]

At the same time, software and networks have come to saturate the urban landscape. Since the late 1990s, cities have become more and more reliant on digital systems. Traditional urban infrastructures have been augmented with networked sensors, transponders, cameras, and a plethora of other electronic devices. Embedded in tickets, credit cards, identification cards, and all sorts of mundane objects, digital codes enable us to pass through turnstiles and catch the tube or the metro or the subway, to buy groceries at the local supermarket, to check out books from the library, among myriad other quotidian activities. Day and night, hundreds of networked electronic eyes silently monitor streets, squares, parks, and alleys, while giant digital screens wallpapering entire buildings with their luminescent surfaces compete to grab attention. Increasingly embedded in the surfaces of our physical surroundings, screens shape our interactions with and perceptions of them. "As screens become ubiquitous in urban environments, they network into maps of information control and consumption."[3] And as they become increasingly portable and digitized, they further modify our relationships with the city, with the world, and with space in general.

Textile metaphors are powerful indicators of these shifting spatial relationships, and of changing perceptions of the digital by and large. Back in the late 1990s, geographers used to compare the internet to a "net" laid over territory and yet somehow disconnected from it. Stress was placed on its capacity to ignore distances and boundaries, thus challenging the traditional geographies of the nation-state. Today, this metaphor is no longer tenable. It is as though the net has become intimately woven into the texture of territory and, more disturbingly, into the textures of many human lives. Geographers talk about "augmented places," "digital envelopes," and "ecologies of screens." Some prefer to use "skin" rather than "net" metaphors. Similarly, while in the 1990s sociologists and science fiction writers used to refer to the virtual as a realm paralleling the real, recent literature has done away with such binarisms. Not only has the digital mantle ended up wrapping the earth (or at least, a good portion of it), but it has become part of its same texture, to the point that the one has become indistinguishable from the other.

Textile metaphors speak of different aspects of contemporary digital communication technologies. More specifically, webs, nets, nodes, and threads highlight their connective power and complex workings. Such metaphors provide an essential vocabulary for visualizing their invisible infrastructures and making sense of their rhizome-like geographies. Echoing the Stoic tradition of cosmopolitan textile metaphors, weavings and fabrics hold a promissory potential of global connectivity that was central to the vision of internet pioneers. At the same time, such metaphors also speak of the pervasive nature of digital infrastructure in contemporary everyday life. Tapestry, carpet, and palimpsest metaphors, by contrast, speak of new augmented realities as more or less homogeneous layers of digital objects and information piling over the material landscape. Ecologies, tissues, skins, membranes, and other biological metaphors, for their part, suggest a new and disturbing reality, in which the digital is no longer a kind of covering, but an organ of perception, something that can be hardly separated from the human body. This chapter explores some of these metaphors, from the weaving of the World Wide Web in the early 1990s to its weaving into the earth's mantle, and the transformation of the world into a map of itself.

Weaving the Web

Maps of worldwide internet connections have become familiar images over the past decade. With their intricate patterns of ethereal white or fluorescent threads crisscrossing an ink-black void, they hold a spectral quality. They are reminiscent of NASA composite satellite views of "the earth by night" and their city lights flickering in the darkness. Internet maps also resemble those airline route maps we usually find in glossy in-flight magazines. Internet maps, however, hold a unique quality that sets them apart from both NASA nocturnal views and airline route maps: on internet maps there are no landmasses, no geographical features; continental outlines are not physical presences over-layered with connecting lines, but rather the result of those very connections. "What really struck me," writes Paul Butler, the author of a popular Facebook connection map in 2010 (pl. 10), "was knowing that the lines didn't represent coasts or rivers or political borders, but real human relationships. Each line might represent a friendship made while traveling, a family member abroad, or an old college friend pulled away from various forces of life."[4]

On these maps, the earth is not a physical presence; its surface is not a mantle made of solid rock, water, sand, and vegetation, but pure geometrical space. It is a textile pattern woven out of human connections and relationships, like a very crowded airline route map. More characteristically, it is a web that is continuously in the making. We are looking at a representation not of the land, but of space-time. Early maps of world internet traffic, such as those designed by the American computer scientist Chris Harrison (fig. 11.1), resemble spiderwebs under construction. Their ethereal filaments traverse the black void, weaving a thick stretch of fabric across the Atlantic, that is, between the two most industrialized regions of the planet. Like a spiderweb hanging in the corner of a dark attic, this stretch of fabric seems precariously anchored to distant spots (mostly in the global South and Russia) by way of thin radial threads.

On Butler's more recent Facebook connection maps, by contrast, the outlines of most continental landmasses and islands are clearly visible. Connection patterns are not abstract forms, but carefully woven samplers that accurately reproduce geographical patterns. Yet, these weavings too are in the making: China and Russia lack definition, while Brazil and Saharan Africa present large holes. On more recent versions of the same map, some of these

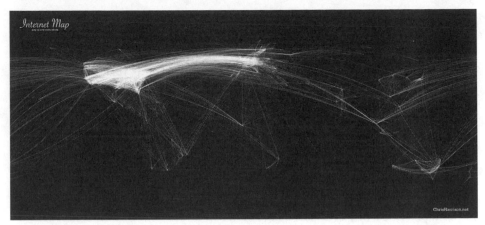

FIGURE 11.1 A spiderweb still under construction: Chris Harrison, internet map of world city-to-city connections, 2007.
By permission of Chris Harrison, Carnegie Mellon University.

black spots are starting to be covered by the expanding web, while in previously well-covered areas, such as Europe and India, the texture has become so thick and intricate as to approximate a uniform white surface.[5] Will the same maps still bear a vague resemblance to spiderwebs in fifty years' time, or will they have become more similar to an impenetrable silkworm's cocoon wrapping the globe in its entirety?

In spite of their differing patterns, these visualizations, and probably those to come, resonate with the vast array of textile metaphors that have underpinned the history of the internet since its inception—as the very word "inter-*net*" suggests. Internet is an abbreviation of Inter-networking, that is, "'working in network,' of millions of wired computers physically located all around the world, thanks to a special communication protocol which enables computers running different software to exchange information in real time."[6] The word brings to mind a fisherman's net made of fibers interlocked in regular patterns. Just as its threads are patiently knotted together, so are the millions of computer terminals gracing labs, offices, commercial premises, and private homes around the world. Yet, this metaphor is just one out of many. The language of the internet is punctuated with metaphors, and "without them we could not really talk about it."[7] Nodes, servers, flows, streams, and webs, for example, provide users with "scaffolding" for grasping its functions and interactions. Without the aid of these metaphors, it would be extremely difficult, perhaps even impossible, to understand such a dynamic and abstract infrastructure. Internet metaphors thus play a didactic role, as they enable common users to make sense of and talk about highly complex realities.[8]

Textile metaphors are especially effective not only in illustrating the working principles of the internet (and thus fulfilling the didactic role), but also in making a largely invisible and abstract realm of information somewhat more tangible, and therefore within users' reach. The reason is that, in both our mind and experience, textiles are intimately associated with touch; "traditionally, a cloth would have passed through many different hands, representing the epitome of handmade."[9] "Weaving" metaphors, for their part, convey a temporal quality; they speak of dynamism and open-endedness of those systems and their workings. Characteristically, *Weaving the Web* (1999) is the title that Tim Berners-Lee, the inventor of the World Wide Web (WWW), chose for his book.[10] The book charts the origins and evolution of the system and presents some of Berners-Lee's views on issues such as censorship, privacy, the increasing power of software companies, and the future of the web.

Although the terms are often used interchangeably, Berners-Lee reminds us, the internet and the Web are by no means the same thing. The former is a general communication infrastructure that links computers together. The latter is one of the principal internet services that enable users to navigate and access a huge amount of content by way of hyperlinks. The internet is therefore the structure on top of which the WWW operates; without the backbone of internet technology, there would be no web. While the web is a creation of the 1990s, the origins of the internet are to be sought in the Cold War. More precisely, they are to be sought in the work of the US Defense Department's Advanced Research Projects Agency (ARPA), a lavishly funded research body established in response to the launch of Sputnik and the threat posed by Soviet military power.[11] ARPANET, the embryo of what would later become the internet, was conceived as a way to allow computers to communicate with one another in the event of a nuclear attack. "Earlier network topologies had a central hub; if the hub was destroyed, the whole network would go down. For this reason, defense system developers considered it necessary to create a decentralized network, so that if anyone knocked out any one portion of it, the rest of the network would continue to work."[12]

As Berners-Lee explains, by the 1970s the internet was up and running, but transferring information was still extremely difficult for a non–computer expert. "The system needed a simple way for people to represent links in their documents, and to navigate across links."[13] The British engineer thus envisaged the new system—what he later called the World Wide Web—as a dynamic, changing object from the outset. And it would continue to grow, thanks to its completely decentralized nature. "I wanted the act of adding a new link to be trivial," Berners-Lee explains. "If it was, then a web of links could spread evenly across the globe."[14]

The web metaphor is apt, in the sense that it describes a dynamic, self-generated, and open system. When she spins her web, the spider "uses her back legs to draw out silk from her abdomen, using gusts of air to cast strong, radial threads to points farther away than she can step. Sometimes she secretes thread with a special glue on the ends. Sometimes she eats part of her web and weaves it anew."[15] The lines of her web are thus an organic extension of the spider into the world. The spider's web is not a closed, self-contained object; it is not an entity, but "a bundle or tissue of strands, tightly drawn together here but trailing loose ends there, which tangle with other strands from other bundles."[16] Likewise, the World Wide Web is an open structure with links constantly woven and unwoven by its users, and yet, like the spider's web, one that needs to be supported by an existing structure—whether it be the twigs of bushes and the ceiling of an attic, or the backbone of the internet.[17]

As Berners-Lee makes clear, however, the name he chose for the new system did not come from a spider, but from maths. In other words, the web was already a metaphor, which the engineer simply adopted. Mathematicians used the word to denote "a collection of nodes and links in which the node can be linked to any other. The name [World Wide Web thus] reflected the distributed nature of the people and computers that the system could link. It offered the promise of a potentially global system."[18]

Throughout the book, the web metaphor is used to illustrate the connective and social dimension of the new system. As such, it assumes an uncanny organic quality. From a technological device, the web becomes a "web of life" evocative of Darwin's and Carson's—yet, in this case, "a web of life that continues to grow."[19] For Berners-Lee, the web is but a reflection of a broader worldview; it is an extension of the principle of human knowledge, and of human life itself:

> In an extreme view, the world can be seen as only connections, nothing else. . . . I like the idea that a piece of information is really defined only by what it's related to, and how it's related. There really is little else to meaning. The structure is everything. There are billions of neurons in our brains, but what are neurons? Just cells. The brain has no knowledge until connections are made between neurons. All we know, all that we are, come from the way our neurons are connected.[20]

The organic fabric of neurons and synapses is projected onto the global scale. The World Wide Web becomes a sort of "global brain," an analogy the author finds appealing, as "both involve huge numbers of elements—neurons and web pages—and a mixture of structure and randomness."[21]

Echoing Stoic cosmopolitan tropes, Berners-Lee's World Wide Web also becomes "the web of people," that is, "more a social creation than a technical one." As the scientist explains, he designed the web "for a social effect—to help people work together—and not as a technical toy. The ultimate goal of the Web is to support and improve our web-like existence in the world. We clump into families, associations, and companies. . . . The essence of working together in a web-like way is that we function in groups—groups of two, twenty and twenty million."[22]

Berners-Lee's vision is religious in the original sense of the term (*religare*, "to bind together"). An enthusiastic adherent of Unitarian Universalism, a "liberal religion" asserting no single creed but weaving together beliefs and philosophical strands from different faiths, Berners-Lee is a believer in interconnectedness as the foundation of society, indeed of humanity: "The Web's universality leads to a thriving richness and diversity. If a company claims to give access to the world of information, then it presents a filtered view and the Web loses its credibility."[23] The web is thus meant to be a collective global enterprise: "Link by link we build paths of understanding across the web of humanity. We are threads holding the world together. . . . [And] hope in life comes from interconnections among all people in the world."[24]

From the Net to the Digital Skin

Metaphors help us make sense of unfamiliar things and concepts, but also structure the way in which we respond to them. While Berners-Lee's web metaphor speaks of the principles of connectedness and outward expansion underpinning his vision of the World Wide Web, a plethora of early metaphors dealing with the internet highlight ways in which the system was perceived by the public in the 1990s. Al Gore's well-known "information superhighway" metaphor, for example, conveyed the exhilaration with speed caused by the new medium, while "frontier" and "ocean" metaphors (along with derivatives such as "exploring," "navigating," and "surfing") conveyed a perception of the internet as a largely unknown and potentially treacherous realm, a vast *terra incognita* open to exploration and discovery (from which came the names of popular early browsers, such as Internet Explorer, or Netscape Navigator).[25]

Ironically, as the internet was likened to infrastructures and environments of the "offline" world, the "offline world" was understood in terms of the internet. In the mid- to late 1990s, digital networks and network metaphors thus came to inform new understandings of modern society.[26] The Spanish sociologist Manuel Castells, for example, coined the phrase "network society" to emphasize process, connectivity, and mobility, as opposed to a former focus on boundedness, hierarchy, and form. While networks are an old form of social organization and the human experience, argued Castells, digital technologies, characteristic of what he called the "Information Age," powered social organizational networks in novel ways. Because digital networks transcend the borders of the nation-state, "the network society constituted itself a global system, ushering in the new form of globalization characteristic of our time."[27]

Among the key concerns of Castells and other social scientists were the patterns of social and economic inequality generated by a global network connecting financial hubs disconnected from their surrounding "urban texture."[28] Urban geographer Stephen Graham talked about nets "thrown over spaces."[29] Largely divorced from the territory they traversed, such networks, scholars believed, constituted a new type of space, a space "invisible to our direct senses, a space which might become more important than physical space itself, [and which is] layered on top of, within and between the fabric of traditional geographical space."[30]

The streams of data and capital flowing through these networks also bypassed the nation-state, causing, according to many, its crisis. While social movements and geopolitical strategies became increasingly global, "the institutions of the nation-state inherited from the Modern Age and from industrial society gradually lost their capacity to control and regulate global flows of wealth and information."[31] Economists and business management scholars, such as Kenichi Ohmae, went as far as to proclaim "the death of the nation state," or talk about "cartographic illusion."[32] The colorful territorial containers embroidered on Boetti's *mappe*, they believed, were destined to become a nostalgic fantasy.

Imagining a digital net "cast over" (and largely disregardful of) territory implied envisaging the online and offline worlds as separate, parallel spatial realms, each with its own geography and responding to its own logic. As opposed to solid physical reality, in the 1990s cyberspace (or "virtual space") was usually perceived as an ethereal alternate dimension, a space in which geographical constraints could be transcended.[33] As such, the virtual fascinated and stirred the popular imagination. Sometimes this "space" was believed to intersect with reality; more frequently, it was perceived to be in competition with it, causing alienation and estrangement.[34] By the end of the decade, however, "the daily dose of cyberspace (using the internet to make plane reservations, check email using a Hotmail account, or download MP3 files) became so much the norm that the original wonder of cyberspace was almost completely lost."[35] Even if the virtual became domesticated, it nonetheless continued to be imagined as a separate "space" one could "access" and "navigate" to retrieve information and perform a range of activities.

Since the beginning of the new century, the perceived gap between cyberspace and physical space has progressively narrowed. "Research agendas, media attention, and practical applications have come to focus on a new agenda—the physical—that is, physical space filled with electronic and visual information."[36] Even though metaphors such as "the cloud" might lead us to believe that our data are floating somewhere in an ethereal space, for many users today the separation between the physical and the digital in two separate spheres is no longer tenable. Improved bandwidth and mobile technologies have caused a dramatic increase in the number of users. Above all, they have enabled continuous access, to the point that teenagers in countries like Japan no longer perceive a disconnection between physical and digital spaces, because their first online experiences were through smartphones.[37]

According to media scholars, we have moved from a "desktop" paradigm of personal computers anchored in the workspace or domestic space, and accessed exclusively in those spaces at certain times, toward a new paradigm in which the internet has been embedded in outdoor everyday activities and delivered to a mobile user.[38] The key feature of the new paradigm, however, is not so much mobility as coverage and perpetual connectivity. As the internet has gone mobile and "always on," digital and physical have blurred in new "hybrid spaces," "mixed realities," and "augmented places."[39] Rather than a sparsely knit net of connected lines, in the minds and experience of many, the internet has come to form an organic continuum—and, as such, to demand new metaphors.

In particular, attention seems to have shifted from nets, nodes, flows, streams, and lines toward textures, screens, and surfaces in general. Today, the promise of global connectivity and the speed that caused so much excitement in the early 1990s are taken for granted. What seems to matter, and to intrigue contemporary designers, users, and commentators, is the blend between physical and digital, the continuum between online and offline. Attention to this aspect is reflected in a new profusion of textile metaphors, this time conveying continuity, rather than dynamism or expansion. Webs have become "seamless," and networks akin to "nervous systems" or "skins" (that is, organic tissues). As the mobile internet becomes "intertwined" with people's everyday practices, digital and real are "enfolded" into one another, "woven together," or "knotted together."[40] Far from being a frontier land or a fisherman's net with a wide mesh, the internet has become "the fabric of our lives"[41]—a fabric so tightly woven as to approximate, at least in appearance, a smooth surface, or a screen.

A new mantle has wrapped the earth. Thrift called it "an ecology of screens," a vast "web of perception" that crowds our lives, "communicating, informing, entertaining, affecting life, or simply being there providing ground."[42] No matter where we go, or what we do, we cannot escape the mighty spread of this new digital mantle. Computer screens await us at work. ATM screens and self-checkout machine screens help us fulfill our daily tasks. At the airport, screens direct our travels; embedded in the passenger seats of planes and trains, they inform us on our journey's progress, or simply help us forget about it. Safely stored in bags and pockets, mobile phone and iPad screens follow us virtually everywhere. Light to carry and immune from annoying light reflections, Kindle reader screens have become loyal holiday companions to many. Traditional screens, like TV screens, for their part, have migrated out of the private space of the living room and bedroom into pubs, waiting rooms, and even onto the sides of buildings, often acting "more like electronic ad-soaked wallpaper than conventional television."[43] As our lives become increasingly "woven" into this giant fabric of plasma, it has been suggested that we might soon come to inhabit "a world with more screens than people"—a "postdigital" world.[44]

Although screens are but the visible manifestation of network "nodes"— every terminal is in essence a visual interface, a "screen-based" artifact—they have a presence and a consistency of their own. As such, digital screens transform environments and mediate our relationship with them in their own peculiar ways.[45] Although built environments have always been covered with texts (for instance, shop signs) and images (fresco paintings, icons, sculptures), the multimedia information that screens embed in these environments is new in that it is dynamic.[46] Electronic screens add luminescence and movement to static images. In so doing, they amplify the ability of images to capture human attention. No matter what its size, every screen cries for a look, and every gaze we direct at a screen is one we remove from the physical environment. Onscreen and offscreen spaces thus constantly "merge, blend, and even battle for ocular time."[47] Sometimes screens compete for attention even among themselves, as someone, for example, uses their smartphone while watching TV—a behavior encouraged by certain shows featuring hashtags designed for audience interaction on social media.

Mimicking the space of a page, or of a painting, screens usually come in a rectangular shape. Their sizes vary from the minuscule (the screen of a smartwatch, for example) to the gigantic (the mega-screens in Time Square, New York, or in London's Piccadilly Circus). The irony is that most of the time we are not aware of the physical presence of the screens. These objects have become so embedded in our lives, and we have grown so used to their presence, that we no longer take notice of them, until their surface cracks, or they fail to switch on. The simple fact that many people still use their phones after they have smashed their screens stands as proof that we do not see the screen; we only see its contents, as if the new digital mantle of the earth were a transparent film set between us and the world—or rather, a giant skin.[48]

Skin, we are told, "is not primarily a membrane of separation between subject and object, inner and outer, the self and the world, but a medium of inter-subjective connection."[49] Our skin touches the world, and the world touches our skin. Interactive screens demand constant tactile interaction: they ask us to tap, cross, and caress their surfaces with our fingertips, to hold them with our hands.[50] As we switch off the screen of our mobile phone or tablet, finger marks and other unwanted traces of our tactile interactions with them suddenly emerge, spoiling their surfaces of their glitter. Specialized cleaning products work to restore their original appearance and wipe out the invisible carpet of germs covering them after intensive use. At the same time, many of these devices are activated through sensors able to recognize their owners' fingerprints. On these surfaces, "organic and inorganic co-mingle."[51] As with a skin, the new ecology of screens has come to operate as a natural interface between our inner and outer worlds, or, in Roland Barthes's words, as an "umbilical cord" linking places, people, and objects to the gaze—and shaping our gaze on places, people, and objects.[52]

Screens do not simply mediate; they also flatten. Their smooth plasma surfaces flatten landscapes, people, experiences, and relationships. Coupled with social media networks, digital screens turn events and encounters into bi-dimensional images, stories into "tweets," places and holidays into time-less views, people into endless pages of contacts. And as networks of contacts expand—the average Facebook user has 229 "friends"—our capacity to sustain close ties weakens.[53] Likewise, the infinite sequences of flickering images that constantly traverse the shiny surface of the screen produce an inevitably "detached, disembodied, non-committal relation with the world, a zapping domain where everything is watchable rather than graspable."[54] On searchable digital maps, places, distances, and prospects turn into decontextualized landmarks. We no longer navigate the map to find our destination; the map navigates to find us. GPS works with cellular towers and software platforms to locate us, or more accurately, our phones. We just enter the address of our destination on our smartphone, and its location (and perhaps a picture of it) will promptly pop up on the screen. Mediated through the digital skin of the screen, our perception of a city, or of the world itself, thus morphs into a collection of fragments, of scattered icons and attractions.

Simultaneously displaying multiple views, screens disrupt the monofocal view of Renaissance linear perspective and destabilize the notion of a stable, grounded horizon.[55] The result is a feeling of continuous distraction and disorientation: "the lines of the horizon shatter, twirl around and superimpose in a multiplication and de-linearization of horizons and perspectives."[56] Hence, programs like Google Earth heavily emphasize the ground, although this ground is unstable and always contingent. It slips under our fingertips.

By definition, screens at once reveal and conceal. While screens are used to hide unwanted sights, human bodies and objects are "screened" to check whether there is anything wrong with them, often there where the naked eye cannot reach. Duplicity is a quality digital screens not only share with their nondigital cognates, but they also increase and intensify. While the images or words they display might appear bright and neat, their technologies are in reality opaque and impenetrable to most users. As such, digital screens embed a strange paradox: the more tightly they are woven into the fabric of the earth's mantle and into our everyday routines, the more disconnected we become from their workings. It seems that the key enabling their layer of superficial multiplicity and fluidity is "packed away in the black boxes (sometimes literally) of [our] computers and cell phones." Critical attention has therefore recently turned to what is concealed "beneath the smooth surface in continuous motion on which digitally mediated social life is played out"— that mysterious software-mediated layer that "invites, even encourages, us to ignore it."[57] Metaphors of "revealing innards," of "unmasking deeper, more basic understandings of the work that the digital does in the world," thus add to the long history of geographical and anatomical "unveiling" metaphors initiated in the Renaissance.[58]

At the same time, new anatomical and textile metaphors are being used to shed light, or at least to speculate, on the nature of pervasive digital devices and their effects on contemporary human life. For contemporary urban sociologists, for example, our nervous system is no longer contained within the body's limits, but is extended through the urban fabric by way of myriad networked sensors, cameras, hotspots, and other *invisibilia*.[59] Like the spider's web spun from the interior of her body, digital networks and interfaces form an organic continuum with the human body. Yet, in this case the spinning movement is two-way. Often designed to encourage more screen time and addictive behavior, digital technologies are provoking changes in our brain structure, including shorter attention spans.[60] It is as though the digital net is penetrating the depths and folds of cerebral tissues and silently weaving into the human nervous system. Consciously or unconsciously, we become a visceral part of the web's digital texture.

Similarly, Thrift sees in the future of information technologies able to track movement "a redefinition of the world of persons and objects as constituent elements of . . . a fabric that is constantly being spun over and over again as position becomes mobile, sometimes producing new patterns."[61] Others talk about "envelopes" wrapping the users of screen-based technologies. These envelopes, they argue, momentarily shape users' perceptions of space and time.[62] The result is, nonetheless, always the same: an increasingly fragmented Self inhabiting many worlds all at once; a multitasking brain, no longer able to gain pleasure from focusing on a single activity; an increasingly "distracted present," in which "human capacities for imagination, long-term thinking and careful reflection are being broken down."[63]

The World as a 1:1 Map

> As the most beautiful allegory of simulation, [Borges's] fable has now come to full circle for us, and possesses nothing but the discrete charm of second-order simulacra. Today abstraction is no longer that of the map. . . . The territory no longer precedes the map, nor does it survive it. It is nevertheless the map that precedes the territory—precession of simulacra—that engenders the territory, and if one must return to the fable, today it is territory whose shreds slowly rot across the extent of the map. It is the real, not the map, whose vestiges persist here and there in the deserts that are no longer those of the Empire, but ours. The desert of the real itself.[64]

Not only do digital technologies shape our relationship with the world, they also shape the world—in the most physical sense. The spiderweb has in fact become a "*Geo*-web," a digital web deeply intertwined into the fabrics of Gaia; a hybrid mantle of soil and plasma, of georeferenced data, augmented reality overlays, and sand. Or rather, the earth has become a map of itself. At the origin of this amalgamation is "geotagging," or the addition of a geographical location to digital information. Geotagged information comes in various digital media; it encompasses both texts and images, including videos, websites, SMS messages, and QR codes. This information is assigned latitude and longitude coordinates (and optionally altitude, compass bearing, and other fields). In this way, diverse and often disparate data are pinpointed on the map and visualized by way of interfaces such as Google Earth, so that the map becomes a vast searchable archive to be used alongside a plethora of other contextual information available from the internet.[65] At the same time, physical places are "overlaid" with digital information and, more recently, literally embedded with digital objects (such as QR codes, for example).

Each of these technologies creates a different experience of place, and the metaphors that are used to talk about them reflect these experiences. For example, an early locative media art piece titled *Urban Tapestries* (2002–4) enabled people to digitally annotate the space of their neighborhoods on a digital map and thus "weave" their own stories into it. The goal of the project, Frith

comments, "was to collect the largely invisible pathways left by urban occu-
pants in order to better understand the identity and specificities of place," to
enable them to "embed social knowledge into the wireless landscape of the city."
In this way, the city was made legible in new ways and "its texture became open
to multiple interpretations."[66] The result was a digital urban map covered with
a dense web of intersecting "threads," as well as textual and visual annotations.

Since the release of Google Earth in 2005 and its expansion via Google
Maps and other programs, a larger and larger amount of online information
has been routinely attached to physical places. The resulting layer of online
information covering places has become increasingly crucial to contemporary usages of online data, as well as to physical interactions with those "augmented" places.[67] Unlike *Urban Tapestries*, contemporary augmented reality
technologies allow users to access such places through transparent surfaces,
rather than through the mediation of an annotated map. A development of
previous GPS and locative mapping applications, recent augmented reality
map applications, for example, create overlays on the surrounding environment when viewed through a smartphone camera pointed at it.

Yelp Monocle, for instance, superimposes names and ratings on physical
shops, restaurants, and other businesses. *Lookator* shows the location of Wi-
Fi hotspots, while *Theodolite*, a multifunction viewfinder, overlays bearing,
heading, altitude, and elevation on the surrounding environment, among
other spatial information. Its uses range from land measurements to outdoor
sports, sightseeing, navigation, and wayfinding.[68] In 2017, London's Gatwick
Airport graced its terminals with an augmented reality wayfinding tool, so
that "passengers can be shown directions in the camera view of their mobile
device, making it easier for them to locate check-in areas, departure gates,
baggage belts etc." (fig. 11.2).[69] *Pokemon Go* and augmented reality versions
of *Geocaching* offer playful counterparts to these applications, adding yet another dimension to augmentation and mixed realities.

Thanks to these applications, the neighborhood, the city, the wilderness,
and the airport all become three-dimensional 1:1 maps through which the
traveler, the vacationer, the geologist, the runner, the passenger, and the game
player are invited to move, assisted by digital paths, signs, and other pointers.
"Imagine being able to enter an airport and see a virtual red carpet leading
you right to your gate," media scholars Adriana de Souza e Silva and Jordan
Frith comment.[70] As we walk on this virtual red carpet, we become akin to the
King of Gaul in the sixteenth-century woven tapestry of Beauvais, as he strolls
upon a real-size map of his kingdom, signposted with toponyms and terms
of orientation (fig. 11.3). In both cases, "the map is the territory, and mastery
over space means walking among the signs, not on the earthly ground."[71]

FIGURE 11.2 From the web to the digital skin: Gatwick Airport augmented reality app, 2017.
Source: https://www.diorama.com/2017/05/26/gatwick-installs-2000-indoor-navigation-beacons-enabling
-augmented-reality-wayfinding/.

FIGURE 11.3 "The map is the territory": Beauvais tapestry featuring the "Fabulous History of the King
of Gaul," ca. 1530.
Photo: François de Dijon.

Tapestries and overlays share an all-encompassing quality. Unlike a
sparsely knotted net, they promise full coverage, leaving no gaps on the surface. Yet, the two words are by no means the same. "Tapestry" evokes thickness and richness; it conjures up tangles of colored threads, hosts of woven
(or, in its larger usage, embroidered) figurines and decorative patterns. The
word "overlay," by contrast, indicates a thin covering of something (for example, a golden overlay covering a wooden surface, or a transparent sheet placed
over a piece of artwork for protection, or over a map or an aerial photograph

to provide additional information, such as a route, or a boundary). While *Urban Tapestries* emphasized the collective and creative process of "making," of drawing the threads together (like Beinart's communal tablecloth), overlays and the augmented reality technologies that produce them aim and claim to make the map (and data) equal to territory.

The result, however, is never a totally seamless tapestry, or continuous smooth surface. It is, rather, a fabric with its own topographies characterized by different densities and textures—like Boetti's maps. While certain spots, for example, are crowded with information, like a densely stitched patch, in other spots the layer of geotagged information is thinner, sometimes even nonexistent. Wireless signals are stronger in certain areas and out of reach in others, forming "holes," or "worn-out patches" in the digital tapestry. "Mobile networks have to negotiate the architecture of spaces that they attempt to inhabit," as the radio waves connecting mobile devices are reflected, refracted, and absorbed by the same obstacles, creating "not a seamless network, but a series of ebbs and flows."[72] In this sense, digital topographies are materially intertwined into the topographies of the irregular surfaces they claim to cover. At the same time, in order to survive, the network needs constant updates, like an ancient tapestry exposed to the action of time and requiring the periodical intervention of restorers and conservationists.

As smartphones and tablets pave the way to wearable technologies, fabric and overlay metaphors blend into actual fabrics and into the fabrics of the human body. New technologies include cloths embedding self-tracking devices, which allow users to collect biometric data from their everyday routines, such as skin conductance, blood oxygen level, and performance, whether mental or physical. In this way, the digitally augmented body is turned into "a laboratory of the self."[73] New technologies also include e.paper (or e.ink), "a very thin electronic display on a sheet of plastic, which can be flexed to different shapes and which displays information that is received wirelessly."[74]

Unlike mobile phones, computer screens, and other traditional rigid interfaces, these new technologies share the flexibility and fluidity of textiles. Google Glass, for its part, "points towards a possible future in which augmented reality becomes an ordinary part of human vision, rather than an interface that users can only access by holding a smartphone or tablet in front of them."[75] Filtered through screens and special glasses, reality takes on a new dimension. Or rather, it loses a dimension. As Baudrillard once claimed, "something has disappeared. The sovereign difference, between the [map] and the [land], that constituted the charm of abstraction. Because it is difference that constitutes the poetry of the map and the charm of the territory, the magic of the concept and the charm of the real."[76]

The map and the land, however, blend to different extents and in different ways. Not only are digital layers of geocoded content intertwined in our experience of place, but they also contribute to the physical shaping of the land. For example, shortly after the release of Google Earth, architects, urban planners, and developers started to turn their attention to the appearance of buildings, urban developments, and even artificial landforms, as seen from above, through the eye of a satellite. Ironically, such attention to the flattening satellite view was part of a broader movement to digitize planning processes in 3D, including three-dimensional mappings of the earth above and below ground, and 3D mapping simulations (for example, to establish where a new railway could fit under the congested subterranean space of a large city). Over the past decade, an increasing number of sites, ranging in size from the rooftop of a house to massive human-made islands, have been purposely engineered to appear aesthetically appealing when beheld through Google Earth, or simply odd enough to catch viewers' attention.

Traditionally neglected, rooftops—what the American photographer Alex Maclean called "the fifth façade"—have thus turned into urban icons and sites for artistic experimentation.[77] In 2006, for instance, an article in the *Los Angeles Times* featured a project to cover the new de Young Museum rooftop in San Francisco with long, thin skylights and copper panels (then viewable under construction in Google Earth) and a satellite image centered on the museum's observation tower. This was described as "a wry twist on the idea of the observation tower in an age of digital technology and pervasive surveillance—a reminder that while you are looking down on the world, the world is also looking down on you."[78]

Other architectural trends have since included "green roofs," and even "green roof parks." Floating high above the noise and the traffic, these aerial green islands have become so popular over the past decade that a whole luxury "roofscape" has emerged in cities like New York.[79] In Athens, by contrast, tarpaulins were draped over rooftops in 2009, when the government sought to locate wealthy tax evaders by using Google Earth to find their swimming pools.[80] Likewise, when two years earlier the ground plan of a 1950s US Navy building on Coronado Island in San Diego was shown from Google Earth images to have the form of a swastika, the whole site was modified at a cost of $600,000.[81]

Artists, activists, and private businesses have similarly used rooftops, and indeed the earth's surface at large, as a giant canvas waiting to be inscribed and looked at from above. Hence, "while US Secretary of Energy Stephen Chu would like us all to paint our rooftops white—sending excess heat and blinding glares back into space," artists such as New Yorker Molly Dilworth

FIGURE 11.4 "The world is the map": Dubai's Palm Jumeirah and The World, as seen through Google Earth.
Source: Google Earth.

have been producing colorful rooftop paintings that can be seen from outer space.[82] Likewise, soon after the release of Google Earth, American businesses and brands such as KFC and Target emblazoned the tops of buildings and open spaces with their giant logos, as a way to capture customers' attention, while in 2011 a company called Blue Marble, offered, among its services, giant QR codes to be painted on its customers' rooftops. "The idea is that when viewers are searching on Google Maps or Google Earth, the codes will automatically be deciphered by Google," and "the image, logo or website details that are embedded will display as you scan across the building."[83] In this way, the earth's surface becomes a vast media surface open to manipulation, not just virtually on the Google Earth interface, but literally.[84] In Jean Baudrillard's words, "territory no longer precedes the map."

More spectacularly, GPS technologies and the new visual regimes triggered by Google Earth and other digital interfaces have also encouraged, or indeed enabled, the emergence of new landforms. Completed in 2006 and covering an area equal to 800 football pitches, Dubai's Palm Jumeirah, for example, was the first of a series of giant artificial residential islands in the Gulf, designed in complex, eye-catching shapes, including an archipelago of private islands forming the world's image (fig. 11.4).

Whether coming in curious fractal shapes or in the shape of a map, these new artificial islands share an iconic quality. They defy the logic of the grid, the traditional symbol of modernity and efficiency. While the grid enabled rational control over America's expanding cities and cultivations, producing an open, predictable, replicable, if not monotonous, landscape, the new artificial islands subordinate practicality to aesthetics, and predictability to wonder.[85] They are giant curios engineered to capture the global viewer's attention from above. They are new cultural icons for aspiring global cities striving to get to the center of the world scene. Yet, unlike the modern gridded city, these artificial islands remain self-enclosed private spaces seen by many around the world, but physically accessible only to few.[86]

Like the new rooftops and the giant commercial logos or QR codes inscribed on them, the new artificial islands are invisible from ground level. They are conceived to be viewed from above and, more specifically, through a digital interface, such as Google Earth or Google Maps. Made of sand pumped from the ocean floor, their often complex designs are likewise accomplished following instructions dictated "from above," by way of GPS devices. Like the screens on which they appear and through which they take shape, the new artificial islands are by nature duplicitous. While developers and real estate agents market them as ornaments inscribed on an empty sea surface, or as embroideries expanding and beautifying extant urban fabrics, environmentalists have harshly condemned the disruption they caused to preexisting marine ecosystems. In other words, their fancifully shaped surfaces conceal a dark side invisible to the eyes of the most viewers.

Even so, since the realization of Palm Jumeirah, the earth's mantle has been embroidered (or has been planned to be embroidered) with several of these landforms. By 2010, almost twenty countries were claiming one. Some of these mega-projects have been realized; many others, however, have remained mere fantasies on the screen.[87] Either way, these hybrid landforms mark the ultimate fulfillment of Baudrillard's prophecy—the land copying the map, rather than the other way around.

*

Metaphors dealing with the digital are symptomatic of its developments. The proliferation of metaphors for the internet in the 1990s reflected the need to come to grips with its novelty and with a new perception of reality.[88] Textile metaphors such as "net" or "web" highlighted global interconnectedness and the new medium's perceived ability to transcend the national boundaries and lie of the land. "Skins" and "ecologies of screens" signaled the naturalization of the technology and its pervasiveness in everyday life. Today, "overlays" and augmented space speak of yet another stage, in which physical places, digital objects, and interfaces have become inseparable parts of the same seamless texture.

If modernity started with rectilinear inscriptions on the land by way of railroads, highways, canals, geometrical boundaries, and urban gridded patterns, and late modernity started with the invisible digital network underpinning (or indeed, rising above) those inscriptions, today augmented reality overlays, iconic rooftops, and, not least, artificial islands complete and yet at the same time transcend this project. Not only do they ask places to become more hybrid and unfixed, but they are visually accessible only through the mediation of the same technologies by means of which they have been crafted. The world has become the map—and the map the world.

Epilogue

Cloak, garment, vernicle. Stage curtain, drape, veil. Tablecloth, carpet, surface. Web, tapestry, skin. Multiple incarnations of the earth's mantle metaphor speak of the unbounded imaginative power of the human mind and its continuous attempt to comprehend the planet we inhabit. Spanning three millennia of human history, all these metaphors share a basic matter of fact: our experience and knowledge of the planet are by necessity superficial. We dwell and move, struggle and thrive, live and die on the terrestrial crust or, perhaps more precisely, within the narrow bounds of the biosphere. Each variant of the mantle metaphor nonetheless reveals a different type of engagement with the terrestrial crust (or the biosphere): awe and respect in front of its mystery; a desire for its penetration and mastery; integration and interconnectedness. If we imagine all these metaphors as colored patches sewn together in a single fabric—the fabric of history—and take a look at this giant drape from afar, different patterns will slowly emerge on its polychrome surface.

First of all, we will note patterns of recurrence. In particular, we will note two basic kinds of metaphors running across our patchwork: weaving metaphors and unveiling metaphors. The former have been consistently used to speak of interconnectedness; the latter have surfaced at times of discovery. What is perhaps more intriguing is that we have not moved from the one to the other, as a linear understanding of Western history might suggest. Rather, these metaphors seem to alternate and return to us again and again as in a spiral, sometimes claiming visibility, other times receding to the background, to emerge once again unexpectedly.

In ancient Greece, weaving Gē's mantle meant bringing order to creation; it literally meant giving Chthonia a readable face. Textile and web metaphors in patristic writing likewise spoke of the harmony of creation; they gave visual shape to the idea that everything, from the boundless ocean to the tiniest grain of sand, is wisely interconnected in a single whole, as a manifestation of God's creative power. For centuries, the contemplation of this *atechnon hyphasma* ("web unwrought by art") redirected the eye of the soul to the inscrutable depths of divine wisdom—for, as Western medieval *mappae mundi* remind us, the face of the earth was a vernicle imprinted with the face of the Creator. The same metaphors were appropriated and popularized by Romantic poets, naturalists, and philosophers to express similar concepts. Humboldt and Darwin, for example, both employed "web of life" metaphors to give account of organic integration—whether at the macro-scale of the cosmos, or at the micro-scale of a riverbank. Thoreau, Muir, and other late nineteenth-century nature writers used similar metaphors to reconnect an alienated modern subject to the transcendent grandeur of nature. However, while the Church Fathers had challenged their audience to see creation through the eyes of God—to see it, that is, from the end for which it was created, for the union of all things in the person of his incarnate Son—for these transcendent pantheists, the web of nature itself became the object of worship.

Modern human geographers, such as Ratzel and Vidal de la Blache, adopted web and fabric metaphors to call attention on the "textures" of their object of study—the earth's surface—and the complex patterns of interrelationship between humans and the land they inhabit. Web metaphors were once again evoked by Rachel Carson at the height of the Cold War, but this time to show the threats that irresponsible human action posed to the planet and to humans themselves, as an integrating part of the same "web of life."

Weaving and web metaphors also speak of human society and its workings. Plato, for example, used the weaving metaphor to describe the good statesman making a tightly woven fabric out of citizens with different characters and dispositions. Ritual weavings by women from rival *poleis* marked the end of hostilities. Alexandria of Egypt, a miniature *chlamys*-shaped world, became the paradigm of cosmopolitan coexistence. Marcus Aurelius envisaged human lives as threads of the same great web. Over the past few decades, these metaphors seem to have gained increasing currency among sociologists and anthropologists, as they attempt to give account of complex phenomena such as globalization, multiculturalism, and diaspora. Modern and contemporary artists, for their part, have given the metaphor tangible shape by way of "world tapestries," "communal tablecloths," and other textile crafts made of stitches and intersecting "life threads." Perhaps more glamorously, computer scientists and engineers have woven "webs" and "nets" that have become so thick and so deeply intertwined in contemporary society as to form an organic tissue, or skin.

If web and weaving metaphors direct our attention to the surface and to the fragile complexity of its texture, the second type of metaphor—unveiling— demands us to go beyond it. Although nature has been veiled since the times of Heraclitus, it was in the Renaissance—that is, during the Age of Discovery and the opening of oceanic space—that the unveiling metaphor gained currency and expanded to the planetary scale. As Baconian experimental science established itself in the West as the primary mode of knowledge, and the telescope and the microscope empowered the human eye to penetrate into a dazzling multiplicity of new worlds, opening stage curtains and parted veils multiplied both in text and in image. Knowledge of the world and of nature was articulated as a dialectic between concealment and revelation. Human progress came to be understood as a perpetual act of unveiling, of chasing new horizons, of challenging new frontiers—to the point that, by the end of the eighteenth century, the veil had become diaphanous. To the Romantic eye, including Humboldt's, distant horizons, and the landscapes framing them, were perpetually wrapped in ethereal curtains of mist and light.

Surviving through the heroic tales of nineteenth-century polar exploration, unveiling metaphors re-emerged with new impetus at the height of the Cold War, as "ideological geopolitics" penetrated the untapped domains of Antarctica, outer space, and the underground. Veils of ice were torn; the curtain of heaven was for the first time physically violated; the skin of the earth was pierced to unprecedented depths. In this panorama of hubristic races and competition, web metaphors appear as correctives, or perhaps reactions, to the violence of unveiling metaphors—and to the threats of atomic power. Carson and other environmental writers redirected public attention to the fabric of the surface, just as Romantics had reacted against the Newtonian mechanistic worldview and the dissecting gaze of its supporters.

In a sense, web metaphors and unveiling metaphors are reflections of the opening and closing movements of the earth's mantle: the opening of the self-enclosed order of the medieval *mappa mundi* and its closure with the end of geographical discoveries; the opening of new vertical frontiers during the Cold War race to space and deep drilling, and the "closing" of the biosphere with the awakening of modern environmental consciousness; the opening of the internet frontier and its closing into an almost omnipresent digital skin.

The horizontal expansion of space alternates with its vertical expansion. The mantle thickened as human imagination and action were directed upward and downward; as Humboldt climbed Chimborazo; as Whittlesey imagined new vertical worlds during the Air Age; as atomic tests were conducted in the upper layers of the atmosphere and in the profundities of the earth; as the horizontal axis of Renaissance linear perspective turned ninety degrees upward and downward, producing new "perpendicular sublimes"; as humanity saw the thin iridescent film of the atmosphere through the eyes of the first astronauts; as digital cables and networks continue to weave and expand under our feet, and the invisible threads of electromagnetic waves and satellite orbits wrap the planet with increasing density.

The history of this expanding and contracting mantle is underpinned by a constant tension between the macro-scale of creation and the micro-scale of the human body: the body of Pherecydes and the cosmos he endeavored to describe; the body of Christ and the body of creation; the body of ancient emperors wrapped in cosmographic cloaks, and the body politic of their global domains; Hélinand's mantle of worms and fire wrapping the soul enslaved by the passions of flesh, and the biblical garment of creation "waxing old" and "fretting away" at the end of times; the opening space of the anatomical body and the opening space of the earth and of the cosmos in the Renaissance; the suffering body of Carson and the suffering body of the earth; the "augmented body" of the self-tracker and the digital hypercortex spreading its axons throughout the planet.

In between these two scales, one can recognize yet other recurring patterns on our fabric: a pattern of beauty shining through its glowing multicolored texture, and a dark thread, a pattern of destruction threateningly underpinning it. In this sense, Alain de Lille's medieval image of Natura's splendid garment defiled by human perversion is no less dramatic and timely than Carson's and Vivarelli's wounded earth's mantle, Hoskins's injured English landscapes, or Mona Hatoum's "anti-mappe." Many other patterns could be discerned in the texture of our polychrome mantle, and a study encompassing other cultures and traditions would probably reveal much more than this book has been able to do.

As with any geographical metaphor, the earth's mantle mediates between theory and the unexpected eruptions of the terrain. It suggests a fusion of metaphorical and material space, to the point that the two can become indistinguishable. As I am writing these last pages, heavy showers are bringing some relief to the land, after an anomalously hot and dry summer. Scorching temperatures and sun-burned fields are a rare occurrence in northern Europe. They are a powerful and disquieting reminder of the magnitude and consequences of climate change and, more broadly, of what scientists and the media call the Anthropocene, the geological epoch in which human action has become the dominant force modifying the earth's mantle.

While climate change reinforces the image of an earth wrapped, or rather trapped, in a mantle of human-produced gases, the land itself is revealing previously unseen shapes. As the summer sun beat down on the British Isles,

over the past two months ghosts of forgotten pasts have mysteriously surfaced in the yellowing fields under the mechanical eyes of drones. Normally
kept hidden by lush grasses and crops, old and prehistoric features have become suddenly visible beneath the unusually thinned and emaciated mantle
of vegetation. A neolithic henge emerged in Newgrange, Ireland; the imprints
of a grand eighteenth-century mansion in Nottinghamshire; the traces of a
World War II RAF station in Lasham, East Hampshire.[1] It is hard to look at
these images of ghostly ruins surfacing from a worn-out, yellowed mantle
with dispassionate eyes, without asking what we are doing to the planet.

Mantles and metaphors are not neutral. At the end of this journey, it is
therefore important to interrogate ourselves on their power and their effects.
What do these metaphors do? What are their ethical implications? Imagining the earth's surface as a veil to be parted can have (and indeed did have)
consequences dramatically different from its appreciation as a precious yet
fragile fabric. Likewise, as Carson insisted, imagining the biosphere as a web
of which we humans are an integrating part, rather than external spectators,
can change the destiny of our planet. Or again, we might ask, what power
do metaphors have in producing imaginaries of the internet for multiple
stakeholders? What work do these metaphors do in shaping imaginaries of
local-global relations, technological affordances, and futures? Perhaps more
important, what impact does our belief in the metaphors have on our experiences of everyday life? And finally, what is the geographer's task? Weaving
images reflected in a magic mirror, like the Lady of Shalott? Rising high above
the earth's surface, seeking to attain a God's-eye view? Attempting to pierce
the veil of appearance? Reversing its tapestry? Or simply, trying not to lose
sight of the mantle upon which "human joys and sorrows unfold"?[2]

Just as the Lady of Shalott's shuttle moves back and forth as it lays down
the weft, so does the geographer's pen move up and down, leaving behind
it a thread of dark ink on the page and on the old face of the earth. Like the
shuttle of a skilled weaver, the geographer's pen crafts new mantles of words.
Warp and weft cross again and again, as the fabric of human history silently
unfolds. Which thread we should follow is up to us.

Acknowledgments

Many people and institutions have contributed, in different ways, to the making of this book. First and foremost, I thank Fr. Maximos Constas, the late Denis Cosgrove, and the late David Lowenthal for being endless sources of intellectual and human inspiration, both in life and in memory.

I am also especially grateful to Jeremy Brown and Mike Duggan for their expert insights and valuable suggestions on Renaissance veils and on digital nets; to Dan Crawford for many inspiring conversations; to Claudio Piani and Diego Baratono, whose work sparked my original interest in cosmographic mantles; and to Pete Adey, Klaus Dodds, Sasha Engelmann, Giorgio Mangani, Ken Olwig, Tania Rossetto, and Alessandro Scafi for the precious threads they compelled me to follow and embroider in this book.

My thanks go to the two anonymous University of Chicago Press readers for their constructive comments and excellent suggestions, and to all the institutions and individuals that granted their permission to reproduce the images in this book.

I am grateful to Phil Crang for being such a generous and supportive head of department, and to my students at Royal Holloway for their energy and enthusiasm.

My thanks and appreciation go to Felix Driver for his always wise advice, for his friendship, and for all the support he has given me over the years, especially in emotionally difficult times—that has meant a lot to me.

I am likewise deeply thankful to Mary Laur for trusting in this project and inviting me to submit a full manuscript; to Rachel Kelly Unger and Michaela Luckey for their exemplary editorial guidance and tireless assistance; and to the rest of the team at Chicago for their excellent support during editing and production. I am also grateful to Lys Weiss, who besides being a fantastic copyeditor is also a tapestry artist-weaver—what better combination for a book on mantles? Working with all of them has been a true privilege.

The initial chapters of the book were written during my sabbatical at the Academy of Athens in 2017. The stay was funded through an Onassis Fellowship. I am most grateful to the Onassis Foundation for making this possible; to Frédérique Hadgiantoniou for all her assistance; and to all the wonderful colleagues in the academy's Research Center for Byzantine and Post-Byzantine Art for their warm hospitality and generosity. I am especially thankful to Chryssa Maltezou, the supervisor of the center; to Ioanna Bitha for letting me use her office and enjoy stunning views of the Acropolis; and to Ioanna Christoforaki and Ioannis Vitaliotis for their friendship and for making my stay so special. I would also like to thank the staff of the Gennadius Library and director Maria Georgopoulou for their usual kindness and for all their assistance during my time in Athens.

The research for this project was funded through a British Academy/Leverhulme Senior Research Fellowship. I would like to express my deepest gratitude to both institutions for this unexpected and most precious gift, which freed me from teaching responsibilities and allowed me to focus on the book. I am likewise indebted to the Stanford University History Department and Humanities Center for sponsoring and hosting a visit that enabled me to benefit from the rich resources of the David Rumsey Historical Map Collection and from their thriving scholarly community.

Some of the ideas that informed this book were presented in my inaugural lecture at Royal Holloway, at the department of geography of the University of the Aegean in Mytilene, Greece, and at the conference "Time in Space: Representing the Past in Maps," at the David Rumsey Map Center, Stanford University, in 2017. I am thankful to the organizers of these events and to my audiences for their precious input and feedback, which greatly helped me develop the book. In particular, I thank Theano Terkenli, Kären Wigen, and Caroline Winterer for making my visits to their respective institutions possible and for being such generous and welcoming hosts.

Finally, I would like to thank the Fathers of the Holy Monastery of Docheiariou, Mount Athos, Greece, especially Abbot Amphilochios, for their loving support and for their prayers. As I was starting to write this book, Fr. Apolló, my spiritual father, was diagnosed with an incurable disease, which he fought with faith, dignity, and altruism until the very end. The first reader's report arrived a couple of days before he reposed in the Lord. It is to him, the person who illumined and changed my life, that I dedicate this book.

Abbreviations

Alberti	Leon Battista Alberti; *De pict.* = *De pictura*
Am.	Ambrose, Bishop of Milan; *Ep.* = *Epistula*; *Hex.* = *Hexaemeron*
A. R.	Apollonius Rhodius
Arist.	Aristotle; *Metaph.* = *Metaphysica*; *Po.* = *Poetica*
Artemid.	Artemidorus Tarsensis; v. *Anthologia Graeca*
Ast.	Asterius of Amaseia; *Hom.* = *Homilies*
Ath.	Athanasius, Bishop of Alexandria
Au.	Augustine, Bishop of Hippo
Bas.	Basil the Great; *Hex.* = *Hexaemeron*
Chry.	John Chrysostom; *Hom.* = *Homilies*; *Hom. Gen.* = *Homilies on Genesis*
Clem. Alex.	Clemens Alexandrinus; *Strom.* = *Stromateis*; *Cypr.* = *The Cypria*
Cyr. Jer.	Cyril of Jerusalem; *Cath. Lec.* = *Catechetical Lectures*
Democr.	Democritus, *Fragments*
Diod.	Diodorus; v. *Anthologia Graeca*
DK	H. Diels and W. Kranz, *Fragmente der Vorsokratiker*, 6th ed. (1952)
Enc. Trec.	Istituto Giovanni Treccani, *Enciclopedia italiana di scienze, lettere e arti* (Rome, 1929)
Eriugena	Johannes Scottus Eriugena; *Periph.* = *Periphyseon*
Eur.	Euripides; *Hip.* = *Hippolytus*
Eus.	Eusebius of Caesarea; *In laudem Const.* = *In laudem Constantini*
Gr. Naz.	Gregory of Nazianzus; *Hom.* = *Homily*
Gr. Nys.	Gregory of Nyssa; *Ep.* = *Epistula*; *Hom.* = *Homily*
Hdt.	Herodotus; *Hist.* = *Historia*
Il.	Homer, *Iliad*
Isoc.	Isocrates; *Ep.* = *Epistulae*
Jo. D.	John of Damascus; *Exp.* = *An Exact Exposition of the Orthodox Faith*; *Hom.* = *Homily*

LS	Henry George Liddell and Robert Scott, *A Greek-English Lexicon* (Oxford, 1940)
M. Aur.	Marcus Aurelius; *Med.* = *Meditationes*
Max. Conf.	Maximus the Confessor; *Amb.* = *Ambigua*
Mēnaia	*Mēnaia tēs Apostolikēs Diakonias Ekklēsias tēs Ellados* (Athens, 1959–66)
MWD	*Merriam-Webster Dictionary* (Dallas, 1995)
Od.	Homer, *Odyssey*
ODB	Alexander Petrovich Kazhdan, *The Oxford Dictionary of Byzantium* (New York: Oxford University Press, 1991)
OEtymD	Online Etymology Dictionary
Op.	Hesiod, *Opera et dies*
Or.	Origenes
Orph.	Orphica; *H.* = *Hymni*
Ov.	Ovid; *Met.* = *Metamorphoses*
PG	J. P. Migne, ed., *Patrologiae cursus completus: Series Graeca*, 166 vols. (Paris, 1857–66)
Pi.	Pindar; *I.* = *Isthmian Odes*
Pl.	Plato; *Cra.* = *Cratylus*; *Lg.* = *Leges*; *Phd.* = *Phaedo*; *R.* = *Respublica*
Plin.	Pliny the Elder; *Nat.* = *Naturalis Historia*
Plu.	Plutarch; *Alex.* = *Alexander*; *Moralia*; *Pel.* = *Pelopidas*
Proclus	Proclus, Archbishop of Constantinople; *Hom.* = *Homily*
Ptol.	Ptolemy; *Geog.* = *Geographia*
Str.	Strabo; *Geog.* = *Geographia*
Te.	Tertullian
Verg.	Vergil; *G.* = *Georgics*

Notes

Introduction

1. *MWD.*

2. Eugene Robertson, "The Interior of the Earth," *USGS,* http://pubs.usgs.gov/gip/interior/ (accessed 28 June 2019).

3. David Oldroyd, "Geophysics and Geochemistry," in *Cambridge History of Science,* vol. 6, ed. Peter Bowler and John Pickstone (Cambridge: Cambridge University Press, 2003), 404n29. Wiechert proposed a model of the earth with an iron core (*Kern*) and a stony shell (*Steinmantel,* or *Mantel*). Wiechert was the first scientist to attribute "both physical and chemical significance to the two-part earth model and to argue that the mantle-core boundary corresponds to a discontinuous change from stone to iron, as well as a jump in density" (Stephen Brush, "Discovery of the Earth's Core," *American Journal of Physics* 48 [1980]: 706). In 1897, following observations of an exceptionally violent earthquake in Northeast India, Richard Dixon Oldham, superintendent of the Geological Survey in India, clearly identified compressional (P), shear (S), and surface waves. Based on their speeds and paths through the earth's mass, he concluded that the planet had a core probably made of iron, surrounded by magma (Brush, "Discovery," 710). Oldham nonetheless makes no reference to Wiechert and uses the phrase "outer shell" to designate the mantle (Richard Dixon Oldham, "The Constitution of the Interior of the Earth, as Revealed by Earthquakes," *Quarterly Journal of the Geological Society* 62 [1906]: 456). For a comprehensive account of nineteenth-century theories and models of the interior of the earth, see Stephen Brush, "Nineteenth Century Debates about the Inside of the Earth: Solid, Liquid or Gas?" *Annals of Science* 36 (1979): 225–54. For an earlier history, including premodern myths and speculations, see David Oldroyd, *Thinking about the Earth: A History of Ideas in Geology* (London: Athlone, 1996), 7–41; see also Robert Wood, *The Dark Side of the Earth* (London: George Allen & Unwin, 1986).

4. *OEtymD;* Oldroyd, *Thinking about the Earth,* 341n13. An English definition of "earth's mantle" is provided by the Harvard geology professor Reginald Daly: "The rest of the planet [between crust and core], beneath ocean and atmosphere, reacts like a solid to the stresses set up by the passage of earthquake waves, and may be distinguished as a whole by the name 'mantle'" (Reginald Daly, *Strength and Structure of the Earth* [New York: Prentice-Hall, 1940], 1).

5. With local variants tied to the fashion of the time. The *Vocabolario degli Accademici della Crusca* (1612), the first Italian dictionary, for example, mentions a collar (*bavero*). The

Enciclopedia italiana Treccani (1929) emphasizes "draping" and refers to a circle of fabric with a central hole for the head and an opening along the radius as the most common type of mantle, "typical of the traditional dresses of some Italian regions." Nathan Bailey's *Universal Etymological English Dictionary* (1775), by contrast, defines it as "a kind of cloak or long robe."

6. *LS*; Antoine Furetière, *Dictionnaire universel, contenant generalement tous les mots fran-çois, tant vieux que moderns, & les termes des sciences & des arts*, 4th ed., vol. 1 (The Hague: Chez Pierre Husson et al., 1727).

7. According to Robert Sullivan's *Dictionary of Derivations* (1860), "poets have applied the term to the vine, from its spreading or extending itself; to a blush, because it spreads or suf-fuses itself over the cheeks; and to a goblet covered with froth or overflowing." The meaning of "to embrace kindly" is mentioned in Bailey's *Universal Etymological English Dictionary* (1775). While in Italian there is no equivalent of the verb "to mantle," the noun can figuratively refer to "an exterior appearance assumed for opportunism, in order to deceive" (*Enc. Trec.*).

8. The figurative sense of "covering up" is from the mid-fifteenth century. The intransitive sense of "to become covered with a coating" (of liquids) is from the 1620s (*OEtymD*).

9. Examples here range from the mantle hiding Fenisa's astute machinations in Lope de Vega's play *La Discreta enamorada* (1618) to the mantle Enrico uses to conceal his illegitimate son in Giraud's *Ajo nell'imbarazzo* (1807) and the exchange of mantles in Lorenzo Daponte's version of *Don Giovanni* (1787). In fiction the phrase "cloak and dagger" refers to espionage and secretive crimes (it suggests murder from hidden sources).

10. In Italian "mantle" also designates the fur of some mammal species in relationship to their color ("*il manto pezzato delle mucche*;" "*il m. maculato del leopardo*;" "*il m. dei cavalli*" [*Enc. Trec.*]).

11. Nikolaij Vasil'evich Gogol, *The Overcoat* (London: Bristol Classical Press, 1991); Dino Buzzati, "Il Mantello," *Il Dramma* (June 1960): 37–47.

12. D. J. Pangburn, "Cloak è l'app per evitare le persone che odi," *Motherboard*, 19 March 2014, http://motherboard.vice.com/it/read/cloak-app-per-evitare-le-persone-che-odi (accessed 10 June 2016). This can be linked to the long tradition of invisibility cloaks. These devices are found in ancient Greek and Welsh mythology, in medieval Italian poems, in nineteenth-century German fairy tales, and in contemporary science fiction alike—from the Grimm brothers to *Star Trek* and the Harry Potter novels (see "Mantello," in Remo Ceserani, Mario Domenichelli, and Pino Fasano, *Dizionario dei temi letterari* [Turin: UTET, 2007], 1402–3). A magical item em-ployed by treacherous characters, cast over a spaceship to make it disappear, or worn by a hero to accomplish his mission, in today's culture of digital surveillance and simulacra the invisibility cloak holds special appeal. On October 19, 2006, a team led by scientists from Duke University produced a cloak that routed microwaves of a certain frequency around a copper cylinder in such a way that it rendered the object almost invisible to the human eye. Since then, the pos-sibility of creating an "invisibility cloak" has become the object of sustained scientific inquiry and of periodic attention in the media. Every now and then, science magazines and the popular press feature articles reporting new small findings toward the achievement of a mantle enabling total invisibility, along with its possible applications. These range from espionage and military camouflage to more mundane uses (William Harris and Robert Lamb, "How Invisibility Cloaks Work," http://science.howstuffworks.com/invisibility-cloak.htm [accessed 10 June 2016]). In a way, the invisibility cloak is the literal "embodiment" of Bentham's panopticon, or perhaps just the ultimate fulfillment of the modern escapist dream—seeing without being seen, being there without really being there.

13. Joyce Chaplin, *Round about the Earth: Circumnavigation from Magellan to Orbit* (New York: Simon & Schuster, 2014).

14. Clarence Glacken, *Traces on the Rhodian Shore: Nature and Culture in Western Thought from Ancient Times to the End of the Eighteenth Century* (Berkeley: University of California Press, 1967); Keith Thomas, *Man and the Natural World: Changing Attitudes in England 1500–1800* (New York: Penguin Books, 1984). These seminal accounts end with the eighteenth and nineteenth centuries, respectively.

15. "Well then, my friend, first of all the true earth, if one views it from above, is said to look like those twelve-piece leather balls, variegated, a patchwork of colors, of which our colors here are, as it were, samples that painters use. There the whole earth is of such colors, indeed colors far brighter still and purer than these: one portion is purple, marvelous for its beauty, another is golden, and all that is white is whiter than chalk or snow; and the earth is composed of the other colors likewise, indeed of colors more numerous and beautiful than any we have seen" (Pl., *Phd.* 110–15). The Latins called the imaginative dreaming associated with the rising over the earth "*somnium.*" Examples of this long tradition span from Cicero's *Somnium Scipionis* to Dante Alighieri's *Paradise* (22.151–53). The *somnium* demanded an imaginary viewpoint external to the earth. Appropriate elevation was usually achieved through interstellar flight and spiritual ascent, and it provided the human mind with a perspective over the world denied to the physical eye. The resulting synoptic view implied "rising over the mundane" and was therefore "a mark of the exceptional being, the call to heroic destiny of the paradigmatic human" (Denis Cosgrove, *Apollo's Eye: A Cartographic Genealogy of the Earth in the Western Imagination* [Baltimore: Johns Hopkins University Press, 2001], 27).

16. Cosgrove, *Apollo's Eye*, 3.

17. Abraham Ortelius, quoted in Svetlana Alpers, "The Mapping Impulse in Dutch Art," in *Art and Cartography: Six Historical Essays*, ed. David Woodward (Chicago: University of Chicago Press, 1987), 88.

18. Or what geographer Gunnar Olsson called the "abyss" between "the mindscape of pure meaning" and the "rockscape of pure materiality," or "the realm of things" and "the affections of the soul" (Gunnar Olsson, *Abysmal: A Critique of Cartographic Reason* [Chicago: University of Chicago Press, 2007], 121).

19. Christian Jacob, *The Sovereign Map: Theoretical Approaches in Cartography throughout History* (Chicago: University of Chicago Press, 2006), 18.

20. Jacob, *Sovereign Map*, 11.

21. Jacob, *Sovereign Map*, 11.

22. Jacob, *Sovereign Map*, 31; Olsson, *Abysmal*, 364.

23. Jacob, *Sovereign Map*, 30.

24. Franco Farinelli, *Geografia. Un'introduzione ai modelli del mondo* (Turin: Einaudi, 2003).

25. Mark Jackson and Veronica della Dora, "From Landscaping to 'Terraforming': Gulf Mega-Projects, Cartographic Visions and Urban Imaginaries," in *Landscapes, Identities and Development: Europe and Beyond*, ed. Zoran Roca, Paul Claval, and John Agnew (London: Ashgate, 2011), 95–113.

26. Horacio Capel, *Filosofia e scienza nella geografia contemporanea* (Milan: Unicopli, 1987).

27. Pierre Hadot, *The Veil of Isis: An Essay on the History of the Idea of Nature* (Cambridge, MA: Harvard University Press, 2006), 96.

28. Hadot, *The Veil of Isis*, 93.

29. Francis Bacon, quoted in Hadot, *The Veil of Isis*, 93. Violence reverberates in Georges Cuvier's famous remark: "The observer listens to nature; the experimenter interrogates it and

forces it to unveil itself," quoted in Hans Blumenberg, *Paradigms for a Metaphorology* (Ithaca, NY: Cornell University Press, 2010), 27.

30. Blumenberg, *Paradigms*, 5.

31. "By providing a point of orientation, the content of absolute metaphors determines a particular attitude or conduct; they give structure to a world, representing the non-experienceable, non-apprehensible totality of the real. To the historically trained eye, they therefore indicate the fundamental certainties, conjectures, and judgments in relation to which the attitudes and expectations, actions and inactions, longing and disappointments, interests and indifferences, of an epoch are regulated" (Blumenberg, *Paradigms*, 14). On metaphors as cognitive structures, see Mark Johnson, *The Body in the Mind: The Bodily Basis of Meaning, Imagination and Reason* (Chicago: University of Chicago Press, 1990); George Lakoff and Mark Johnson, *Metaphors We Live By* (Chicago: University of Chicago Press, 2003). See also Trevor Barnes and James Duncan, eds., *Writing Worlds: Discourse, Text and Metaphor in the Representation of Landscape* (New York: Routledge, 1992), 1–17.

32. Indeed, Livingstone explores the history of geography as a succession of metaphorical visions and investigates the "social circumstances in which particular metaphors arise, survive, or decline." David Livingstone, *The Geographical Tradition* (Oxford: Blackwell, 1992), 28.

33. Here biology met environmental determinism: like every living organism, "a state must struggle against the environment to survive. This required that it acquires space and resources to feed its healthy growth" (John Agnew, *Geopolitics: Re-visioning World Politics* [New York: Routledge, 1998], 101).

34. William Morris Davis, "The Physical Geography of the Lands," *Popular Science Monthly* 57 (1900): 169.

35. Livingstone, *Geographical Tradition*, 327. Like metaphorology, the history of geography thus appears as an old palimpsest, that is, as a thick textual mantle beneath which other texts (or, in this case, sometimes uncomfortable discourses and paradigms) lie dormant (see Clive Barnett's critical commentary on Livingstone's book "Awakening the Dead: Who Needs the History of Geography?" *Transactions of the Institute of British Geographers* 20 [1995]: 417–19). Similarly, Robert Mayhew's genealogical approach to the history of geographical thought suggests a family tree, or a giant root penetrating the quiet depths of the terrain pushing through its geological strata (Robert Mayhew, "Geography's Genealogies," in *The SAGE Handbook of Geographical Knowledge*, ed. John Agnew and David Livingstone [London: SAGE Publications, 2011], 21–38). Both approaches imply a vertical, top-down movement; a metaphorical excavation into a hidden Freudian subconscious. The compatibility (or incompatibility) of geographical metaphors with metaphors used in other disciplines has been regarded by some as a presupposition for the possibility (or impossibility) of dialogue and unified intellectual agendas (see, for example, David Demeritt, "The Nature of Metaphors in Cultural Geography and Environmental History," *Progress in Human Geography* 18 [1994]: 163–85).

36. Richard Boyd, "Metaphor and Theory Change: What Is 'Metaphor' a Metaphor for?" in *Metaphor and Thought*, ed. Andrew Ortony (Cambridge: Cambridge University Press, 1993), 487.

37. Boyd, "Metaphor," 487. Thomas Kuhn also acknowledged the importance of metaphors. Metaphors, he claimed, "play an essential role in establishing links between scientific language and the world. Those links are not, however, given once and for all. Theory change, in particular, is accompanied by a change in some of the relevant metaphors and in the corresponding parts of the network of similarities through which terms attach to nature" (Thomas Kuhn, "Metaphors in Science," in *Metaphor and Thought*, ed. Ortony, 539).

38. Boyd, "Metaphor," 489–90. Boyd's theory-constitutive metaphors are similar to Blumenberg's "absolute metaphors" in that they have their own existence. They constitute, at least for a time, "an irreplaceable part of the linguistic machinery of a scientific theory;" in other words, they are "constitutive of the theories they express, rather than merely exegetical" (Boyd, "Metaphor," 486). As with Blumenberg's absolute metaphors, shifts in theory-constitutive metaphors mark paradigmatic shifts. Indeed, some scholars deem scientific models as ultimately sophisticated metaphors: "in their endeavors to come to grips with some aspect of reality hitherto unexplained, scientists and social scientists look around for similar processes" (Livingstone, *Geographical Tradition*, 19).

39. Gaston Bachelard, *The Poetics of Reverie: Childhood, Language, and the Cosmos* (Boston: Beacon Press, 1971), 3.

40. Giuseppe Dematteis, *Le metafore della terra. La geografia umana tra mito e scienza* (Milan: Feltrinelli, 1985), 23. Here Dematteis borrows the metaphor from Yves Lacoste. A similar argument is made by Neil Smith and Cindy Katz, "Grounding Metaphor: Towards a Spatialized Politics," in *Place and the Politics of Identity*, ed. Michael Keith and Steve Pile (London: Routledge, 1993), 67–84.

41. Dematteis, *Metafore della Terra*, 126–27.

42. Christie Wampole, *Rootedness: The Ramifications of a Metaphor* (Chicago: University of Chicago Press, 2016), 17.

43. Stephen Norwick, *The History of Metaphors of Nature: Science and Literature from Homer to Al Gore*, vol. 1 (Lewiston, NY: Edwin Mellen Press, 2006), 346.

44. Vaughan Cornish, "On Kumatology: The Study of the Waves and Wave Structures of the Atmosphere, Hydrosphere and Lithosphere," *Geographical Journal* 13 (1899): 624–28.

45. Yi-Fu Tuan, "Surface Phenomena and Aesthetic Experience," *Annals of the Association of American Geographers* 79 (1989): 233.

Chapter One

1. Released in 2007, the commercial advertises the Citroen C-crossover: http://adsoftheworld .com/media/tv/ccrosser (accessed 25 June 2016). It is discussed in Stefania Bonfiglioli, "Quando le mappe si increspano," *E/C* (2009): 1–20.

2. "'Αναξίμανδρος ὁ Μιλήσιος ἀκουστὴς Θαλέω πρῶτος ἐτόλμησε τὴν οἰκουμένην ἐν πίνακι γράψαι" (DK 12A6). The passage is discussed by Alex Purves, *Space and Time in Ancient Greek Narrative* (Cambridge: Cambridge University Press, 2010), 110. On Anaximander, see J. B. Harley and David Woodward, "The Foundations of Theoretical Cartography in Archaic and Classical Greece," in *The History of Cartography*, ed. J. B. Harley and David Woodward (Chicago: University of Chicago Press, 1987), 1:134–35.

3. Claudio Piani and Diego Baratono, "Teofanie cosmografiche, ovvero l'origine del sacro manto geografico," http://www.mastromarcopugacioff.it/Articoli/Teofanie2.htm, 2011 (accessed 30 June 2016).

4. According to Olsson, the Babylonians are also responsible for the oldest creation epic, *Enuma elish*. The mythical account bears resemblance with Pherecydes's tale. Tiamat, the wife of Apsu (the Abyss), is defeated by Marduk through her capture in a giant net, which Olsson envisages as the cartographic grid. "Marduk enters the stage disguised in new clothes, no longer dressed in the warrior's coat of mail but in the uniform of the land surveyor. . . . The magic net [is] no longer a tool for capturing monstrous rivals but a device for ordering the thing-like entities of stars, towns and people" (Olsson, *Abysmal*, 21).

5. Ewa Kuryluk, *Veronica and Her Cloth: History, Symbolism, and Structure of a "True" Image* (Oxford: Blackwell, 1991), 167.

6. Quran, surat al-Hijr 15:19; see B. W. Higman, *Flatness* (London: Reaktion, 2017), 58.

7. Florence Holbrook, *Book of Nature Myths* (Boston: Houghton, Mifflin, 1902).

8. Kuryluk, *Veronica and Her Cloth*, 179.

9. The passage is translated and discussed in James Scott, *Geography in Early Judaism and Christianity* (Cambridge: Cambridge University Press, 2002), 5–22.

10. Glacken, *Traces on the Rhodian Shore*, 8–9; "Anaximander," in *Chambers' Encyclopaedia*, vol. 1 (London: George Newnes, 1961), 403. Proclus writes that "Pherecydes used to say that Zeus changed into Eros when about to create, for the reason that, having created the world from opposites, he led it into agreement and peace and sowed sameness in all things, and unity that interpenetrates the universe" (Geoffrey Kirk, John Raven, and Malcolm Schofield, *The Presocratic Philosophers* [Cambridge: Cambridge University Press, 2003], 62).

11. Purves, *Space and Time*, 18.

12. Karl Ritter, quoted in *Cos'è il mondo? È un globo di cartone. Insegnare geografia fra Otto e Novecento*, ed. Marcella Schmidt (Milan: Unicopli, 2010), 33–34.

13. The word φᾶρος, later also φάρος, indicated a large piece of cloth, commonly, a wide cloak or mantle without sleeves worn by both men and women, drawn over the head. It was also used as a shroud, as well as a bedspread (LS).

14. Hadot, *Veil of Isis*, 9.

15. The custom continued in ancient Rome. In a poem by Tibullus, Mars is said to wrap the heads of the dying with the darkness of death, which was both an article of clothing, a web or veil, and a cloud. See Richard Onians, *The Origins of European Thought about the Body, the Mind, the Soul, the World, Time and Fate* (Cambridge: Cambridge University Press, 1955), 428–29. Likewise, the mantle as a metaphor for concealment took different forms: for example, in the *Ora Maritima*, a Spanish geographical work dating to the fourth century (but probably derived from a text from the sixth century BCE), a dark fog is said to "enshroud the air as in a kind of cloak and the clouds hide the face of the deep always."

16. Tibullus, *Odes* 1.8.23–24; Ov., *Met.* 2.254–55. See Brigitte Postl, "Die Bedeutung des Nil in der römischen Literatur. Mit besonderer Berücksichtigung der wichtigsten griechischen Autoren" (PhD diss., University of Vienna, 1970), 34, 205–8; Alessandro Scafi, "Mapping the Nile," in *The Nile in Medieval Thought*, ed. Pavel Blazek, Charles Burnett, and Alessandro Scafi (London: Warburg Institute, forthcoming). In ancient Egyptian cosmography the sources of the "true" Nile were believed to be located in the underworld (Glacken, *Traces on the Rhodian Shore*, 38). Representations of the Nile with his head wrapped in a veil, as an indication of the river's unknown sources, include Bernini's famous fountain in Rome. See also the frontispiece of Comte de Caylus, *Recueil d'antiquités*, vol. 5 (Paris: Tilliard, 1762); and the Italian baroque painter Paolo de Matteis's headless representation of the river, reproduced in Livio Pestilli, *Paolo de Matteis: Neapolitan Painting and Cultural History in Baroque Europe* (Farnham, UK: Ashgate, 2013), 69. I am thankful to Alessandro Scafi for bringing these references and images to my attention.

17. Chthonic divinities included Hecate (A. R., iv.148; Orph., *H.* 35.9), Nyx (Orph., *H.* 2.8), and Melinoë (Orph., *H.* 70.1), but especially of Demeter (Hdt., 2.123; Orph., *H.* 39.12; Artemid., 2.35; A. R., 4.987).

18. Purves, *Space and Time*, 101.

19. Gaston Bachelard, *On Poetic Imagination and Reverie* (Putnam, CT: Spring Publications, 2005), 51.

20. "Ζὰς μὲν καὶ Χρόνος ἦσαν ἀεὶ καὶ Χθονίη. Χθονίη δὲ ὄνομα ἐγένετο Γῆ, ἐπειδὴ αὐτῇ Ζὰς γῆν γέρας διδοῖ" ("Zeus, Chronos and Chthonia always were. But Chthonia was named Gē, when Zeus gave her the earth as a gift"), F14 I.229, in Hermann Schibli, *Pherekydes of Syros* (Oxford: Clarendon Press, 1990), 51.

21. F68 II (Schibli, *Pherekydes*, 50), trans. Purves, *Space and Time*, 103. The mantle therefore becomes also a point of contact between the realm of the humans and that of the gods. As Schibli observes, in the marriage of Zeus and Chthonia "the divine world touches upon the human world. The institutions and customs of men are traced back to the gods . . . marriages are literally made in heaven as each marriage re-enacts the first divine marriage" (Schibli, *Pherekydes*, 67).

22. Franco Farinelli, *L'Invenzione della Terra* (Palermo: Sellerio, 2007), 49. On the ritual of the *anakalyptria*, see John Scheid and Jesper Svenbro, *The Craft of Zeus* (Cambridge, MA: Harvard University Press, 2001), 64.

23. Farinelli, *Invenzione della Terra*, 51–52.

24. Democr. 177, trans. Hadot, *Veil of Isis*, 47.

25. Intriguingly, later commentators such as Maximus of Tyre and Clement of Alexandria (second century CE) mention a mysterious cosmic tree alongside the mantle (τὸ δένδρον καὶ τὸν πέπλον). This image can be linked to other cosmographic accounts (such as Hesiod), whereby the cosmic tree is formed by a trunk that constitutes the main body of earth and sea. Its roots extend into the invisible underworld (the realm of the dead) and its branches stretch to heaven (the realm of the gods). The mantle represents only Earth and Ocean, underneath and above which are respectively the invisible realms of Tartaros and Ouranos. Hence the visible mantle acts as a boundary between these invisibilities. See Schibli, *Pherekydes*, 70–71.

26. Mesomedes, quoted in Hadot, *Veil of Isis*, 27.

27. Quoted in Hadot, *Veil of Isis*, 53-54.

28. Edmund Leach, *Genesis as Myth and Other Essays* (London: Jonathan Cape, 1969), 8.

29. Scheid and Svenbro, *Craft of Zeus*, 12.

30. Scheid and Svenbro, *Craft of Zeus*, 12.

31. Scheid and Svenbro, *Craft of Zeus*, 65.

32. The word comes from *cum* ("together") + *nubere* ("to wed," but also "to cover," from which the word *nubes*, "cloud") (Charlton Lewis and Charles Short, *A Latin Dictionary. Founded on Andrews' Edition of Freund's Latin Dictionary* [Oxford: Clarendon Press, 1879]). The *chlaina* also symbolized the coming together of the newlywed under the same roof, hence the close relationship between *stegasma* ("roof") and *skepasma* ("fabric"). The *chlaina* was usually provided by the groom's household, hence, Pherecydes's special emphasis on *oikia* and *pharos* (see Scheid and Svenbro, *Craft of Zeus*, 66).

33. Schibli, *Pherekydes*, 3.

34. Scott, *Geography*, 5.

35. Scheid and Svenbro provide a rich account of this use of the metaphor, ranging from the text as a point of encounter between writer and reader to the material texture of the inscribed surface, the papyrus, itself reminiscent of a true fabric. The metaphor was also used in a musical context, whereby the harp became the loom. In the *Odyssey*, Calypso sang "most sweetly" while weaving in her cave (*Od.* 5.58–62). On the relationship between singing and weaving in Homer and in different cultures, see Anthony Tuck, "Singing the Rug: Patterned Textiles and the Origins of Indo-European Metrical Poetry," *American Journal of Archaeology* 110 (2006): 539–50.

36. Purves, *Space and Time*, 32.

37. Scheid and Svenbro, *Craft of Zeus*, 105.

38. Aristotle called this "eusynoptic view." The phrase and his commentary on the *Iliad* are discussed in Purves, *Space and Time*, 24–64.

39. Indeed, Clement of Alexandria describes Achilles's shield in order to compare it with the embroidered robe in Pherecydes's tale (Clem. Alex., *Strom.* 6.2). On Achilles's shield as a narrative device and as a cosmological representation, see respectively Purves, *Space and Time*, 46–55; and Cosgrove, *Apollo's Eye*, 31.

40. Schibli, *Pherekydes*, 54.

41. Aristotle used the word to describe eusynoptic plots (see Purves, *Space and Time*, 25, 54). Other examples include: "αἱ περὶ τὴν λέξιν π." (Isoc., 5.27); "ὀνόμασι π." ("adorned with beautiful words;" Menex., 235); "ποικίλλειν δὲ ἔξεστι ταῖς συλλαβαῖς, ὥστε δόξαι ἂν τῷ ἰδιωτικῶς ἔχοντι ἕτερα εἶναι ἀλλήλων τὰ αὐτὰ ὄντα: ὥσπερ ἡμῖν τὰ τῶν ἰατρῶν φάρμακα χρώμασιν καὶ ὀσμαῖς πεποικιλμένα ἄλλα φαίνεται τὰ αὐτὰ ὄντα" ("But variety in the syllables is admissible, so that names which are the same appear different to the uninitiated, just as the physicians' drugs, when prepared with various colors and perfumes;" Pl., *Cra.* 394); "ἡ π. τῆς λύρας" (Pl., *Lg.* 812d) (LS; and Lorenzo Rocci, *Vocabolario Greco—Italiano* [Rome: Società Editrice Dante Alighieri, 1989]).

42. "Νῦν δ' αὖ μετὰ χειμέριον ποικίλων μηνῶν ζόφον/ χθὼν ὥτε φοινικέοισιν ἄνθησεν ῥόδοις/ δαιμόνων βουλαῖς" (Pi., 1.4.18–20). Of stars, "ἡ περὶ τὸν οὐρανὸν π." (Pl., *R.* 529d; Pl., *Phd.* 110d); "Σικελικὴν π. ὄψου" (Pl., *R.* 404d).

43. *Cypr.* 2.1–7.

44. Purves, *Space and Time*, 103.

45. Schibli, *Pherekydes*, 56.

46. Schibli, *Pherekydes*, 56–57.

47. From the fourth century BCE to the second century CE, a number of authors, including Plato, Aristotle, Aeschines, and Lucian of Samosata, repeatedly used the extravagant figure of Xerxes as a personification of *hubris*. His barbarian irrational instinct led him to rebel against things as they are. For example, when a great storm destroyed a pontoon he had built across the Hellespont, the Persian king grew so furious that he commanded his men to punish the sea: "He told those who laid on the lashes to say these words, of violent arrogance, worthy of a barbarian: 'You bitter water, our master lays this punishment because you have wronged him, though he never did you any wrong. King Xerxes will cross you whether you like it or not; it is with justice that no one sacrifices to you, who are a muddy and briny river.' So he commanded that the sea be punished and ordered the beheading of the supervisors of the building of the bridge" (Hdt., *Hist.* 7.35).

48. Hdt., *Hist.* 109.9.1–3; trans. A. D. Godley (Cambridge, MA: Harvard University Press, 1920).

49. Kathleen Wilson-Chevalier, "Alexander the Great at Fontainebleau," in *Alexander the Great in European Art*, ed. Nikos Hadjinikolaou (Thessaloniki: Institute for Mediterranean Studies and Foundation for Research and Technology Hellas, 1997), 25.

50. Cosgrove, *Apollo's Eye*, 43.

51. According to Plutarch, Alexander would have said to his architect Dinocrates: "Let Athos remain as it is. It is enough that it be the memorial of the arrogance of one king [that is, Xerxes]" (*Moralia* 335.17–20).

52. "There was no chalk at hand, so they took barley-meal and marked out with it on the dark soil a rounded area, to whose inner arc straight lines extended so as to produce the figure of a *chlamys*, or military cloak, the lines beginning from the skirts (as one may say), and narrowing the breadth of the area uniformly. The king was delighted with the design" (Plu., *Alex.* 26.5;

trans. Bernadotte Perrin, *The Parallel Lives*, vol. 7 [Cambridge, MA: Harvard University Press, 1919], 301). The *chlamys*-like outline of the city is also mentioned by Diodorus of Sicily (Diod., 17.52), Strabo (*Geog.* C.793), and Pliny the Elder (*Nat.* 5.62).

53. On the shape of the *chlamys*, see Frank Bigelow Tarbell, "The Form of the *Chlamys*," *Classical Philology* 1 (1906): 283–89.

54. Maria Papadopoulos, "The *Chlamys* City: Urban Landscapes and the Formation of Identity in Hellenistic Egypt," *Ενδυματολογικά* 5 (2015): 122–27, at 123. *Bēmatistai* (or bematists) were surveyors trained to measure distances by counting their steps. Bematists accompanied Alexander the Great on his campaign in Asia, measuring the distances traveled by his army. On Alexander's *chlamys*, see Andrew Collins, "The Royal Costume and Insignia of Alexander the Great," *American Journal of Philology* 133 (2012): 371–402.

55. *Encyclopedia of Fashion*, http://www.fashionencyclopedia.com/fashion_costume_cul ture/The-Ancient-World-Greece/Chlamys.html#ixzz4UAlRpWzL (accessed 10 June 2016).

56. Schein and Svenbro, *Craft of Zeus*, 9.

57. Schein and Svenbro, *Craft of Zeus*, 28.

58. Schein and Svenbro, *Craft of Zeus*, 103.

59. Social order was reinforced by the physical structure of the *poleis* and their insular, or rather archipelagic, character (see John Gillis, *Islands of the Mind: How the Human Imagination Created the Atlantic World* [New York: Palgrave Macmillan, 2004]). In a world of city-states, in which what lay outside of them was often perceived as potentially unsafe and hostile, "the analogy between garments that protect the body from the harsh weather conditions and the defensive walls was quite obvious" (Papadopoulos, "The *Chlamys* City," 123).

60. The rise of Alexandria represented a radical shift in the political geography of the classical world. "Alexander's military conquests had transformed the Greek world from a group of small insular Greek city states into a series of imperial dynasties spread across the Mediterranean and Asia" (Jerry Brotton, *The History of the World in Twelve Maps* [New York: Penguin Books, 2013], 17).

61. Plu., *Moralia* 4.1.6.

62. Peter Fraser, *Ptolemaic Alexandria* (Oxford: Clarendon Press, 1971).

63. Christian Jacob and François de Polignac, eds., *Alexandria, Third Century BC: The Knowledge of the World in a Single City* (Alexandria: Harpocrates Publishing, 2000), 14.

64. Edward Parsons, *The Alexandrian Library, Glory of the Hellenic World: Its Rise, Antiquities and Destructions* (London: Cleaver-Hume Press, 1952), 57.

65. Christian Jacob, "Mapping in the Mind: The Earth from Ancient Alexandria," in *Mappings*, ed. Denis Cosgrove (London: Reaktion, 1999), 30.

66. Giorgio Mangani, *Cartografia morale* (Modena: Cosimo Panini, 2006), 23.

67. Claude Nicolet, *Space, Geography and Politics in the Early Roman Empire* (Ann Arbor: University of Michigan Press, 1991), 61.

68. Nicolet, *Space, Geography and Politics*, 61.

69. Jacob, "Mappings of the Mind," 31.

70. Str., *Geog.* 2.5.6, 14 and 18.

71. "So let us presuppose that the island lies in the aforesaid quadrilateral. We must then take as its size the figure that is obvious to our sense, which is obtained by abstracting from the entire size of the earth our hemisphere, then from this area its half, and in turn from this half the quadrilateral in which we say the inhabited world lies and it is by an analogous process that we must form our conception of the shape of the island, accommodating the obvious shape to

our hypotheses. But since the segment of the northern hemisphere that lies between the equator and the circle drawn parallel to it next to the pole is a spinning-whorl in shape, and since the circle that passes through the pole, by cutting the northern hemisphere in two, also cuts the spinning-whorl in two and thus forms the quadrilateral, it will be clear that the quadrilateral in which the Atlantic Sea lies is half of a spinning-whorl's surface; and that the inhabited world is a *chlamys*-shaped island in this quadrilateral, since it is less in size than half of the quadrilateral. This latter fact is clear from geometry, and also from the great extent of the enveloping sea which covers the extremities of the continents both in the East and West and contracts them to a tapering shape; and, in the third place, it is clear from the maximum length and breadth" (Str., *Geog*. 2.5.6; trans. Leonard Jones, *The Geography of Strabo*, vol. 1. [Cambridge, MA: Harvard University Press, 1917], 435–36).

72. "The shape of the area of the city is like a *chlamys*; the long sides of it are those that are washed by the two waters, having a diameter of about thirty stadia, and the short sides are the isthmuses, each being seven or eight stadia wide and pinched in on one side by the sea and on the other by the lake" (Str., *Geog*. 17.8; trans. Leonard Jones, *The Geography of Strabo*, vol. 8. [Cambridge, MA: Harvard University Press, 1932], 33).

73. Str., *Geog*. 17.8–10 (trans. Jones, *The Geography of Strabo*, 33–43). Interestingly, Strabo also insists on the interconnectedness of the buildings and different parts of the city, as if to emphasize the finely detailed weaving of the urban tissue.

74. Pierre Nora, *Realms of Memory*, vol.1 (New York: Columbia University Press, 1996).

75. Piotr Grotowski, *Arms and Armour of the Warrior Saints: Tradition and Innovation in Byzantine Iconography, 843–1261* (Leiden: Brill, 2010), 255.

76. Str., *Geog*. 1.1.18. The political aspect of Strabo's *Geography* is discussed in Christiaan van Passen, *The Classical Tradition of Geography* (Groningen: J .B. Wolters, 1957), 9–10.

77. "For it is the ultimate and most sublime contemplation of mankind to demonstrate by mathematical theorem an understanding of the heavenly firmament revolving about us and likewise of the earth itself which, since it cannot be physically encompassed by one man, can at least be moulded into an image of itself" (Ptol., *Geog*. 1.1.9). The original title of Ptolemy's work is *Geographikē yphēgēsis*, literally meaning "geographical guidebook," but it became known simply as *Geography*.

78. As he wrote, "a greater similarity to the likeness of inscription on a sphere as opposed to previous methods is at once obvious" (Ptol., *Geog*. 1.2.23). On the two Ptolemaic projections, see O. A. W. Dilke, "The Culmination of Greek Cartography in Ptolemy," in *The History of Cartography*, ed. J. B. Harley and David Woodward (Chicago: University of Chicago Press, 1987), 1:177–200.

79. Hipparchus, cited in D. R. Dicks, *The Geographical Fragments of Hipparchus* (London: Athlone Press, 1960), 53.

80. Lakoff and Johnson, *Metaphors We Live By*, 185.

81. Arist., *Po*. 1459a.

82. Interestingly, Strabo compares Providence to "a broider and an artificer of countless works" (Str., *Geog*. 17.1.36).

83. Brotton, *History of the World in Twelve Maps*, 53.

84. A graphic expression of the politicized "*chlamys*-shaped" *oikoumenē* would have been Agrippa's map, whose length, according to Isidorus, was double its width (Nicolet, *Space, Geography and Politics*, 104). The map was publicly displayed in the colonnade named after the emperor, Porticus Vipsania, and showed the full extension of the Roman empire after the battle

of Actium in 31 BCE. As Dilke comments, "there can be little doubt that by the late Republican period Roman rulers and their advisers had come to recognize the value of geographical maps in both administration and propaganda" (O. A. W. Dilke, "Maps in the Service of the State: Roman Cartography to the End of the Augustan Era," in *The History of Cartography*, ed. J. B. Harley and David Woodward [Chicago: University of Chicago Press, 1987], 1:207).

85. On Strabo's symbolic mimesis, see Stephen Halliwell, *The Aesthetics of Mimesis: Ancient Texts and Modern Problems* (Princeton, NJ: Princeton University Press, 2002), 273–74. Allegorical or symbolic mimesis has also been associated with Crates of Mallos (second century BCE), another thinker influenced by Stoicism, and to his characterization of Achilles's shield as a "*mimēma* of the cosmos."

86. For example, he talks about the "nature" (φύσις) of Europe as its ability to house many people with different ways of life, whom the Romans harmoniously brought into contact with one another. See Van Passen, *Classical Tradition*, 19–21.

87. One of the most characteristic traits of Strabo's work was the diagrammatization of reality. For example, the shapes of countries and geographical features were associated with geometrical forms and objects (a mouth for the Black Sea, the hide of an ox for the Iberian peninsula, a plane-tree leaf for the Peloponnese, and so on); in other words, they were made visible by way of metaphor (see van Passen, *Classical Tradition*, 6–7).

88. Strabo, *Geog.* 1.1.23.

89. According to Zeno, "all the inhabitants of this world of ours should not live differentiated by their respective rules of justices which separate cities and communities, but we should consider all men to be of one community, and we should have a common life and order to common us all" (Plu., *Alex.* 329A–B). Philo of Alexandria (25 BCE–50 CE) likewise compares the cosmos to a great city with one law (the passage is discussed by Glacken, *Traces on the Rhodian Shore*, 110). In his *De Officiis* Cicero (106–43 BCE) explicitly argued for the potential integration of all the world's peoples within a single civic order governed by Roman law; see Denis Cosgrove, "Globalism and Tolerance in Early Modern Geography," *Annals of the Association of American Geographers* 93 (2003): 860.

90. M. Aur., *Med.* 4.40, 4.26, 4.34. These passages are discussed in Hans Urs von Balthasar, *Theo-Drama: Theological Dramatic Theory*, vol. 1: *Prolegomena* (San Francisco: Ignatius Press, 1988), 143. The "stage of life" metaphor in Marcus Aurelius's meditation and in Stoic philosophy is discussed in Linda Christian, *Theatrum Mundi: The History of an Idea* (New York: Garland Publishing, 1987), 11–23.

91. Glacken, *Traces*, 57.

92. "Pherecydes the wise man, who was put to death by the Lacedaemonians, and whose skin was preserved by their kings, in accordance with some oracle" (Plu., *Pel.* 21.2, trans. Bernardotte Perrin, *The Parallel Lives by Plutarch* [Cambridge, MA: Harvard University Press, 1919], 393). This is just one version of Pherecydes's death. Diogenes Laertios provides four other accounts of his death (see Schibli, *Pherekydes*, 7). Plutarch's passage is also discussed by Mark Munn, *The Mother of the Gods, Athens, and the Tyranny of Asia* (Berkeley: University of California Press, 2006), 48.

Chapter Two

1. Texture and styling usually reflect the social and spiritual status of biblical characters and the physical environment in which they are found: "But what went ye out to see? A man clothed

in soft raiment? Behold, they that wear soft clothing are in kings' houses" (Matt. 11:8; compare Luke 7:25).

2. Kuryluk, *Veronica and Her Cloth*, 186.

3. Proclus, *Hom.* 32.8, in Nicholas Constas, *Proclus of Constantinople and the Cult of the Virgin in Late Antiquity: Homilies 1–5, Texts and Translations* (Leiden: Brill, 2003), 318. Ephraim the Syrian (306–73 CE) likewise traces the life of Christ in terms of clothing, assigning a specific piece of clothing to every stage of Christ's earthly sojourn: having clothed himself in the body of mortal Adam, "He was wrapped in swaddling clothes, and offerings were offered him. He put on garments in youth, and from them there came forth helps. He put on the waters of baptism, and from them there shone forth beams. He put on linen cloths in death, and in them were shown triumphs. . . . All these are the changes of raiment, which Mercy put off and put on, when he strove to put on clothes as Adam with leaves; and clad garments instead of skins. He was baptized for Adam's sin, and buried for Adam's death. He rose and raised Adam into Glory. Blessed be he who came down and clothed him and went up!" (*Hymn on the Nativity* 16.12–13, quoted in Kuryluk, *Veronica and Her Cloth*, 188). The same motif is found in John Chrysostom's exegesis of the biblical narrative, although with an emphasis on God's mercy and didactic goal. In *Hom. Gen.* 16, Chrysostom repeatedly contrasts the "garment of glory" God originally bestowed upon Adam and Eve with the garments of fig leaves they ended up making for themselves. A few pages thereafter, he explains the verse "The Lord God made garments of skin for Adam and his wife and clad them in them" (Gen. 3:21) as an ultimate act of mercy, rather than punishment: "Take the case of a kindly father with a son of his own who was brought up with every care, who enjoyed every indulgence, had the run of a fine house, was clad in a silken tunic, and had free access to his father's substance and wealth; later, when he saw him tumble headlong from this indulgence into an abyss of wickedness, he stripped him of all those assets, subjected him to his own authority and, divesting him of his clothes, clad him in a lowly garment usually worn by slaves lest he be completely naked and indecent. Well, in just the same way the loving God, when they rendered themselves unworthy of that gleaming and resplendent vesture in which they were adorned and which ensured they were prepared against bodily needs, stripped them of all that glory and the enjoyment they were partakers of before suffering that terrible fall. He showed them great pity and had mercy on their fall: seeing them covered in confusion and ignorant of what to do to avoid being naked and feeling ashamed, he makes garments of skin for them and clothes them in them" (*Hom. Gen.* 18).

4. Mary Cunningham, "The Mother of God and the Natural World," *Analogia* 1 (2016): 41–51. See also Paul Blowers, "'Entering This Sublime and Blessed Amphitheatre': Contemplation of Nature and Interpretation of the Bible in the Patristic Period," in *Nature and Scripture in the Abrahamic Religions: Up to 1700*, ed. Scott Mandelbrote and Jitse van der Meer, 2 vols. (Leiden: Brill, 2009), 148–76.

5. Constas, *Proclus*, 315.

6. Compare 2 Cor. 5:1–2: "If the earthly tent we live in is destroyed, we have a building from God, a house not made with hands, eternal in the heavens. For in this tent [of the world] we groan, longing to be clothed with our heavenly dwelling."

7. Alfred Biese, *The Development of the Feeling for Nature in the Middle Ages and Modern Times* (London: George Routledge & Sons, 1905), 38.

8. The Byzantine empire lasted from 330 to 1453 CE, that is, from the foundation to the fall of its capital Constantinople (previously named Byzantium). From the time of the division of the late Roman empire into eastern and western, Byzantine rulers styled themselves "the Emperors

of the Eastern Romans" and viewed Byzantine citizens as Roman citizens and as the Christian inheritors of the ancient Greeks.

9. Almost concomitant to the founding of Constantinople in the fourth century, Byzantium established a luxury Constantinopolitan imperial silk-weaving industry. Throughout the succeeding millennium, and indeed beyond the fall of Byzantium to the Turks in 1453, Byzantine silks developed into symbols of the power and prestige of the Byzantine empire. The link between the silks and Byzantine identity is explored by Anna Muthesius in *Studies in Silk in Byzantium* (London: Pindar Press, 2004).

10. Désirée Koslin, "Byzantine Textiles," http://fashion-history.lovetoknow.com/fabrics-fibers /byzantine-textiles (accessed 30 January 2017). Interestingly, Augustine of Hippo uses the garment as a metaphor for perfection: "For there is an end of food, and an end of a garment; of food when it is consumed by the eating; of a garment when it is perfected in the weaving. Both the one and the other have an end; but the one is an end of consumption, the other of perfection" (*Sermon 3 on the New Testament* 53.6).

11. Ast., *Hom.* 1, Migne, *PG* 40:165–68.

12. See Kuryluk, *Veronica and Her Cloth*, 76. In a fifth-century homily Theodoret of Cyrrhus described the arts of weaving and embroidery as gifts from God (Theodoret of Cyrrhus, *De providentia oratio* 4, PG83, 617–20. The passage is discussed in Muthesius, *Silk in Byzantium*, 23–24.).

13. Gr. Nys., *Ep.* 15, "To Adelphios the Lawyer."

14. Eus., *In Praise of Constantine* 6.6; trans. H. A. Drake (Berkeley: University of California Press, 1976), 92.

15. The Byzantines inherited the *chlamys* as a regal insignia from ancient Roman emperors. It formed part of the ceremonial vestment worn by the imperial couple during the most solemn court rituals and was employed in official portraiture "to propagate the image of imperial power within and beyond the frontiers of the empire" (Maria Parani, *Reconstructing the Reality of Images: Byzantine Material Culture and Religious Iconography, 11th–15th Centuries* [Leiden: Brill, 2003], 11). See also Muthesius, *Silk in Byzantium*, 92–96.

16. Muthesius links the belief that the Byzantine emperor was God's representative on earth to earlier pagan panegyric ritually linking Roman emperors to the gods. As early as the seventh century, she observes, Christian rites were added to celebrations of imperial victory and Byzantine emperors were increasingly styled "Christian warriors." In a ninth century manuscript of the *Homilies* of Gregory of Nazianzus, for example, Constantine the Great is shown in the purple imperial military tunic, "the very embodiment of the Christian warrior" (Muthesius, *Silk in Byzantium*, 26).

17. Eus., *In laudem Const.* 16.6–7. "For thus the mutual concord and harmony of all nations coincided in point of time with the extension of our Saviour's doctrine and preaching in all the world: a concurrence of events predicted in long ages past by the prophets of God" (17.12).

18. Eus., *In laudem Const.* 10.6.

19. For a discussion of these different interpretations, see Kuryluk, *Veronica and Her Cloth*, 189–90.

20. "Βλαστησάτω ἡ γῆ. Νόησόν μοι ἐκ μικρᾶς φωνῆς, καὶ προστάγματος οὕτω βραχέος, τὴν κατεψυγμένην καὶ ἄγονον ὠδίνουσαν ἀθρόως καὶ πρὸς καρπογονίαν συγκινουμένην, ὥσπερ τινὰ σκυθρωπὴν καὶ πενθήρη ἀπορρίψασαν *περιβολήν*, μεταμφιεννυμένην τὴν φαιδροτέραν καὶ τοῖς οἰκείοις κόσμοις ἀγαλλομένην, καὶ τὰ μυρία γένη τῶν φυομένων προβάλλουσαν" (Bas., *Hex.* 1.2; my emphasis).

21. Col. 1:17; Susan Bowden-Pickstock, *Quiet Gardens: The Roots of Faith?* (London: Blooms-bury, 2009), 151.

22. Eus., *In laudem Const.* 12.1.

23. Bas., *Hex.* 1.8; compare Jo. D., *Exp.* 2.6. Interesting parallels can be traced with Ambrose's (340-97 CE) *Hexaemeron*, where the saint compares God with a weaver who leaves traces of his skill in the textile even when it is completed, "so that testimony is presented of the craftsman's own work" and "while we observe the work, the worker is brought before us" (Am., *Hex.* 1.5.17).

24. Chry., *Hom. Gen.* 5.13. The passage is further unpacked in the following homily (6.12): "When you see the earth adorned with flowers like some multicoloured garment, the foliage of plants enveloping it all over, don't think this is due spontaneously to the earth's power, or to the energy of the sun, or the moon; realize instead, as wisdom suggests, that for the creation of these things he simply spoke the word, 'Let the earth put forth a crop of vegetation,' and forwith the whole face of the earth was brilliant color." For a later exegesis, see Jo. D., *Exp.* 2.10.

25. Gr. Naz., *The Second Theological Oration* 28.26, emphasis added.

26. In his *hexaemeron*, Basil makes the same argument: what God finds beautiful (*kalon*) are not the material forms, "as it is with us." Rather, "beauty is that which is brought to perfection and which conforms to the usefulness of its end. God, therefore, established a clear aim for his works, and approved them individually as fulfilling his aim. In fact, a hand by itself, or an eye alone, or any other pieces of a statue lying about in fragments, would not appear beautiful to the viewer. But, when set in their proper place, they exhibit the beauty of relationship, the artist having organized them by directing his judgment to their final aim" (Bas., *Hex.* 3.10).

27. In the previous passages, Gregory directly discusses the art of weaving and mentions spiders interlacing "their intricate webs by such light and almost airy threads stretched in various ways, and this from almost invisible beginnings, to be at once a precious dwelling, and a trap for weaker creatures with a view to enjoyment of food. What Euclid ever imitated these, while pursuing philosophical inquiries with lines that have no real existence, and wearying himself with demonstrations?" (Gr. Naz., *Second Theological Oration* 28.25). Compare Bas., *Hex.* 1.2, 6.6, 6.7. Irenaeus of Lyons (130-202 CE), in contrast, talks about philosophers ignorant of God sewing together a "motley garment" out of a "heap of miserable rags" (*Against Heresies* 2.14.2).

28. Eus., *In laudem Const.* 12.10, 12.14; Bas., *Hex.* 3.10, 10.32; Chry., *De incomprehensibili* 10.32. Strabo refers to his *Geography* as a *kolossourgia*, that is, "a colossal work" (*Geog.* I.1.23). Ptolemy defines chorography as "the description of the individual parts [of the earth], as if one were to draw merely an ear or an eye," and geography as providing "a view of the whole, as, for example, when one draws the whole head" (*Geog.* I.3). Origenes (185-254 CE), likewise, writes: "Although the whole world is arranged into offices of different kinds, its condition, nevertheless, is not to be supposed as one of internal discrepancies and discordances; but as our one body is provided with many members, and is held together by one soul, so I am of opinion that the whole world also ought to be regarded as some huge and immense animal, which is kept together by the power and reason of God as by one soul. . . . To the same effect also are the words of Paul, in his address to the Athenians, when he says, In him we live, and move, and have our being. For how do we live, and move, and have our being in God, except by his comprehending and holding together the whole world by his power?" (Or., *De Principiis* 2.1.3).

29. Eus., *In laudem Const.* 12.11, 12.16, 14.5; Gr. Naz., *The Second Theological Oration* 28.24.

30. Ath., *Oratio contra Gentes* 34, cited in Glacken, *Traces*, 204. This metaphor is also used by John Chrysostom, Augustine, and Maximos the Confessor: "Creation is a Bible, just as the Bible is a kind of cosmos" (see Blowers, "Entering This Sublime and Blessed Amphitheatre," 156).

Maximos combines the book metaphor with the web metaphor: "The natural law . . . is to the highest possible degree evenly directed by reason through the marvellous physical phenomena that we see, which are naturally interconnected, so that the harmonious web of the universe [*to enarmonion tou pantos hyphasma*] is contained within it like the various elements in a book" (Max. Conf., *Amb.* 10.31). In the same *Ambiguum*, the author compares the words of Scripture to "garments," meaning that "their inner meanings are the 'fleshes' of the Word, and thus by means of the former we conceal, and by the means of the latter we reveal. In the same way, we can say that the forms and shapes of created things that appear within our vision are also 'garments,' the 'fleshes' of which are the principles according to which they were created."

31. Interestingly, the naming of some Byzantine garments evoked elements of creation. *Thalassai* took the name from the color (that is, the sea porphyry employed to dye them), while *arachnōdeis, aeria yphasmata*, and *anemitsia* were so called because of the thinness of their fabric (which was compared to a spider's web, to the air, or to the wind). See Phaidōn Kokoulēs, Βυζαντινῶν βίος και πολιτισμός, vol. 6 (Athens: Academy of Athens, 1955), 267–76.

32. Henri Maguire, "The Mantle of the Earth," *Illinois Classical Studies* 12 (1987): 228.

33. Scott, *Geography in Early Judaism*, 5.

34. Even though Ptolemy's *Geography* was known from late antiquity and in the Byzantine world, there is no evidence of inclusion of maps in the treatise before the end of the thirteenth century. See Stella Chryssochoou, "The Cartographical Tradition of Ptolemy's Γεωγραφικὴ Ὑφήγησις in the Palaeologan Period and the Renaissance (13th–16th Century)" (PhD diss., Royal Holloway University of London, 2010).

35. Almost fifty Greek manuscripts of Ptolemy's *Geography* survive to this day, though not all of them include maps. Most of them date to the fourteenth and fifteenth centuries. The earliest ones illustrated with maps, however, date no earlier than the thirteenth century, which is when Planudes was collecting manuscripts for the library of the Chōra monastery. To the end of this century belong three codices, Urb. gr. 82, Const. Seragl. 57, and Fabr. gr. 43, which include the text of the Γεωγραφικὴ Ὑφήγησις, Books I–VII closing with the map of *oikoumenē*, plus Book VIII with twenty-six regional maps (ten maps of Europe, four maps of Africa, and twelve maps of Asia). For a detailed discussion of these texts, see Chryssochoou, *The Cartographical Tradition*. A list of Greek manuscripts of Ptolemy's *Yphēgēsis* is provided by O. A. W. Dilke, "Cartography in the Byzantine Empire," in *History of Cartography*, ed. J. B. Harley and David Woodward (Chicago: Chicago University Press, 1987), 1:272–75.

36. See, for example, Urb. gr. 82 (thirteenth century) and BL Burney 111 (fifteenth century). The world map from Const. Seragl. 57, in contrast, is done according to Ptolemy's second projection.

37. Another example of a Byzantine Ptolemaic map conflating different temporal dimensions is Marc. Gr. 516 (fourteenth century), which features a second projection of the *oikoumenē* separated from the terrestrial paradise by an abyss.

38. Maximos Planudes, *Heroic Verses on Claudius Ptolemy's Map* (Τοῦ σοφωτάτου μοναχοῦ κυροῦ Μαξίμου τοῦ Πλανούδη στίχοι ἡρωϊκοὶ εἰς τὸν πίνακα τῆς Γεωγραφίας Κλαυδίου Πτολεμαίου), v. 9, trans. Chryssochoou, *The Cartographical Tradition*, 86.

39. *Heroic Verses* vv. 12–15. The fact that after the description of the *oikoumenē* (vv. 5–7) he mentions the "geography which has just now appeared" suggests that the poem must have been composed soon after the finding of the lost text and possibly after the construction of the map of the *oikoumenē* (Chryssochoou, *The Cartographical Tradition*, 87–88).

40. Maximus Planudes, *Praise to Ptolemy* (Μαξίμου τοῦ Πλανούδη ἔπαινος εἰς Πτολεμαῖον), vv. 1–5.

41. Planudes, *Praise to Ptolemy* vv. 16–19.

42. Planudes, *Praise to Ptolemy* v. 10. The image of the green carpet covered with flowers is a recurrent *topos* of beauty in early patristic literature. Basil, for example, extends the image to the vault of heaven, as he compares the stars to flowers (*Hex.* 6.1, 7.1).

43. The reference is made even more explicit in later Western Ptolemaic maps, as wind heads are portrayed as putti (see, for example, the Waldseemüller map, discussed in chapter 4).

44. "In the beginning, Lord, you founded the earth, and the heavens are the work of your hands; they will perish but you remain; they will all wear out like clothing; like a cloak you will roll them up, and like clothing they will be changed. But you are the same, and your years will never fail" (Psalms 102:25–27; also quoted in Heb. 1:10). The theme is also present in the Gospel, for example, when Christ says that heaven and earth will pass, but his word will not (Lk. 21:33).

45. "And the heaven departed as a scroll when it is rolled together; and every mountain and island were moved out of their places" (Rev. 6:14). "They shall perish, but You endure, and they shall wax old like a garment" (Heb. 1:11; Mat. 24:35).

46. "Thus, too, if you survey the earth, loving to clothe herself seasonally [*temporatim uestiri amantem*], you would nearly be ready to deny her identity, when, remembering her green, you behold her yellow, and will before long see her hoary too" (Te., *De pallio* 2.1).

47. Chrysostom then insists on the interrelation of the elements. The sun, he observes, continues to nourish the seeds even when it is veiled by clouds, but only thanks to the assistance of the earth, of the dew, of the rains, of the winds, and the right distribution of the seasons (Chry., *Hom.* 10.8).

48. Chry., *Hom.* 10.8. The same passages are discussed in his *Hom.* 3 (On Hebrews) and 14 (On Romans).

49. *ODB*, "Planudes."

50. Robert Ousterhout, "Temporal Structuring in the Chōra Parekklēsion," *Gesta* 34 (1995): 63–76.

51. The iconography is derived from earlier representations of the Last Judgment, such as the twelfth-century panel icon at the monastery of Saint Catherine on Mount Sinai (Fr. Maximos Constas, "Review of Liliya Berezhnaya and John-Paul Himka's *The World to Come*," *Religion and the Arts* 20 [2016]: 231).

52. The central dark blue circle represents this divine darkness, the totally unapproachable light (1 Tim. 6:16), while the exterior light blue circle symbolizes the light that can be perceived by pure souls (Veronica della Dora, *Landscape, Nature and the Sacred in Byzantium* [Cambridge: Cambridge University Press, 2016], 76). According to Theophanes of Nicaea (775–845 CE), God is hidden, "not by invisibility or darkness, but paradoxically by light itself, that is by the very medium which makes vision possible. . . . It is light that reveals and conceals the presence of God, like a garment covering the body" (Fr. Maximos Constas, *The Art of Seeing: Paradox and Perception in Orthodox Iconography* [Alahambra, CA: Sebastian Press, 2014], 235).

53. In the *mandorla* of light on icons of the Transfiguration, Andreas Andreopoulos envisaged a Christian mandala, or a sacred map of the cosmos based on "the essential representation of the earth around its sacred axis, which in this case is none other than Christ" (Andreas Andreopoulos, *Metamorphosis: The Transfiguration in Byzantine Theology and Iconography* [Crestwood, NY: St. Vladimir's Seminary Press, 2005]). Like the garment, the *mandorla* is a map of the cosmos the Lord disclosed for his disciples to see, a symbol of the uncreated light through which God manifests his energies, while remaining unknowable in his essence.

54. This iconography is found as early as the seventh century, on the Sinai icon of Christ as the Ancient of Days.

55. These two views are effectively summarized by John of Damascus. John notes how some divided the heaven into seven zones, each occupied by a planet, and ascribed it the form of a sphere "equally removed and distant from the earth at all points," while others imagined it as a vault, after evidence from the Old Testament (Ps. 104:2; Isa. 40:22) (Jo. D., *On Heaven* 2.6).

56. See *ODB*, "Cosmos." Saint Paul extends the meaning of the word to the spiritual realm and uses the derivative verb *kosmein* in speaking of human virtues. Origenes explores different uses of the term in Scripture: "For what we call in Latin *mundus*, is termed in Greek κόσμος, and κόσμος signifies not only a world, but also an ornament. Finally, in Isaiah, where the language of reproof is directed to the chief daughters of Sion, and where he says, 'Instead of an ornament of a golden head, you will have baldness on account of your works,' he employs the same term to denote ornament as to denote the world, viz., κόσμος. For the plan of the world is said to be contained in the clothing of the high priest, as we find in the Wisdom of Solomon, where he says, 'For in the long garment was the whole world' " (Or., *De Principiis* 2.16).

57. *Topographia Christiana* 7.1–2, in Maya Kominko, *The World of Kosmas: Illustrated Byzantine Codices of the Christian Topography* (Cambridge: Cambridge University Press, 2013), 214.

58. Constas, *Art of Seeing*, 232–33. The firmament is described in Gen. 1:6–8. Early Christian discussions of the nature of the firmament date back to the second century and often lack consistency. For Tatian, for example, the term "heavens" indicated a finite area beyond which lay the "superior worlds which are not subject to seasons or disease," while for Theophilus of Antioch (d. 185 CE) it was the firmament which was visible to humans and the heavens invisible beyond it, the firmament retaining half the water above it to provide rain and dew. See David Sutherland Wallace-Hadrill, *The Greek Patristic View of Nature* (New York: Barnes & Noble, 1968), 19.

59. Herbert Kessler, "Gazing at the Future: The Parousia Miniature in Vat. Gr. 699," in *Byzantine East, Latin West: Art-Historical Studies in Honor of Kurt Weitzmann*, ed. Doula Mouriki-Charalambous, Christopher Moss, and Katherine Kiefer (Princeton, NJ: Princeton University Press, 1995), 368.

60. For a discussion of alternative explanations on the composition of the *stereoma*, including Gregory of Nyssa and John Chrysostom, see Wallace-Hadrill, *Greek Patristic View of Nature*, 19–20.

61. See, for example, the prayer for the blessing of the waters by Sophronius of Jerusalem (560–638 CE) read at Theophany (*Mēnaion*, January 6).

62. Fr. Maximos Constas, "Beyond the Veil: Imagination and Spiritual Vision in Byzantium," unpublished paper, University of Chicago Divinity School Workshop, March 2015, 1n2.

63. See Jonathan Pennington and Sean McDonough, *Cosmology and New Testament Theology* (London: T&T Clark, 2008), 62–63.

64. Constas, *Art of Seeing*, 237.

65. Eus., *In laudem Const.* 1.2.

66. The narrative of Exodus is itself shrouded in veils: when Moses speaks to the Israelites, he covers his shining face with a veil, as "they were afraid to come nigh him" (Ex. 34:30–35). Compare 2 Cor. 3:13.

67. James Kokkinobaphos, *Hom.* 4, in Constas, *Proclus*, 328. Textile symbology likewise threads through the magnificent illuminations of Vat. Gr. 1162 and Par. Gr. 1208, two twelfth-century codices containing Kokkinobaphos's *Homilies*. The images are discussed in Maria Evangelatou, "Threads of Power: Clothing Symbolism, Human Salvation, and Female Identity in the Illustrated Homilies by Iakobos of Kokkinobaphos," *Dumbarton Oaks Papers* 68 (2014): 241–323.

68. Compare Cyril of Jerusalem: "God has covered his divinity in the heaven like a curtain [*parapetasma*]" (*Catechesis* 9.1, cited in Constas, *Art of Seeing*, 233n28).

69. The identification of the flesh of Christ with the veil of the Temple is made explicit in the New Testament: "We have confidence to enter the Most Holy Place by the blood of Jesus, by a new and living way opened for us through the curtain, that is, his body" (Heb. 10:19–20). However, the early life of the Mother of God and her work on the veil of the Temple is expounded in the *Protoevangelion of John*, upon which Byzantine commentators and icon painters drew (see Constas, *Art of Seeing*, 105–48).

70. Kuryluk, *Veronica and Her Cloth*, 97.

71. These metaphors are described at some length by Onians, *The Origins of European Thought*; and in Scheid and Svenbro, *Craft of Zeus*, 157–63. They are also used in Greek medical treatises, such as those of Hippocrates (*Nat. Mul.* 1.1; *Gland.* 16.8), whereby the membranes wrapping vital organs are called "veils" (see Constas, *Proclus*, 341). Among early Christian writers, Lactantius (240–320 CE), for example, talks about three Parcae, "one who warps the web of life for men; the second, who weaves it; the third, who cuts and finishes it" (*Divinae institutiones* 2.11).

72. Porphyry, *On the Caves of the Nymphs in the Thirteenth Book of the Odyssey*, trans. Thomas Taylor (London: J. M. Watkins, 1917), 6.

73. "And to souls that descend into generation and are occupied in corporeal energies, what symbol can be more appropriate than those instruments pertaining to weaving? Hence, also the poet ventures to say, 'that on these, the nymphs weave purple webs admirable to the view.' For the formation of flesh is on and about the bones, which in the bodies of animals resemble stones. Hence these instruments of weaving consist of stone, and not of any other matter. But the purple webs will evidently be the flesh which is woven from the blood, and the wool is dyed from animal juice. The generation of flesh, also, is through and from blood. Add, too, that the body is a garment with which the soul is invested, a thing wonderful to the sight" (Porph., *Cave of the Nymphs* 6, 19–20). An elemental geographical feature, the cave, observes Porphyry, is the most apt symbol for what the world contains, on account of the matter with which it is connected: "Through matter, therefore, the world is obscure and dark; but through the connecting power, and orderly distribution of form, from which also it is called world, it is beautiful and delightful." He then observes how Pherecydes of Syros also talks about caves to obscurely indicate the generation of souls (11, 36).

74. Constas, *Proclus*, 318, 320n8. A similar metaphor was used by Heraclitus, in which "the soul weaves and reweaves its body like a spider" (Schein and Svenbro, *Craft of Zeus*, 161).

75. Cited in Constas, *Proclus*, 319. The womb, in contrast, was used by Maximos the Confessor as a metaphor to express the condition of terrestrial life: "Both man and the Word of God, the Creator and Master of the universe, exist in a kind of womb, owing to the present condition of our life. In this sense-perceptible world, just as if He were enclosed in a womb, the Word of God appears only obscurely. . . . Human beings, on the other hand, gazing through the womb of the material world, catch but a glimpse of the Word who is concealed within beings" (Max. Conf., *Amb.* 6.3).

76. From the verbs *con-capere* and συν-λαμβάνω.

77. Constas, *Proclus*, 333. Ephraim the Syrian (306–73 CE) defined the difference between the two women in terms of their own clothing: Eve, who "sewed fig leaves together" (Gen. 3:7) into an apron, could not protect her shame; Mary's "garment of glory," by contrast, clothed the nudity of all humans in addition to her own (*Hymn* 1.43, *On the Nativity*). The motif of redemption through the restoration of the garment is expounded by Gregory of Nyssa: "Putting off our sins like some poor and patched garment, we are clothed in the holy and most fair garment of

regeneration" (Gr. Nys., *On the Baptism of Christ* 519). The *apolitykion* ("dismissal hymn") of the feast of Christ's circumcision likewise reminds the faithful that Christ "received the circumcision of the flesh to take away the veil of the passions' (*apolytikion*, January 1, tone 1). Eusebius uses the metaphor to expound Constantine's virtues: "Himself superior to such feelings, he clothes his soul with the knowledge of God, that vesture, the broidery of which is temperance, righteousness, piety, and all other virtues; a vesture such as truly becomes a sovereign" (Eus., *In laudem Const.* 5.6). "Clothed in the robe of piety" (6.10), Constantine is said to contrast both the physical enemies of the empire and the invisible enemies that "with incorporeal assaults besiege the naked soul itself" (7.3). Thanks to his rule, Eusebius concludes, "the souls of men were no longer enveloped in thick darkness" (8.8). Other authors used the "clothing in flesh" metaphor in a more general sense. Cyril of Jerusalem, for example, asks: "Who knitted us with sinews and bones, and clothed us with skin and flesh?" (*Cath. Lec.* 6.15).

78. In the New Testament the metaphor is extended from the body to material goods: "your riches are corrupted and your garments motheaten" (James 5:2).

79. See, for example, Tatian: "Lay not up treasures on earth, where moth and rust corrupt; and he is not ashamed to add to these the words of the prophet: You all shall grow old as a garment" (*Fragm.* 2). Compare John Cassian, bk. 9.

80. Gr. Nys., *Hom.* 40.3, On the Making of Man.

81. Au., *Sermon* 8, On the New Testament.

82. Jo. D., *Hom.* 11, Concerning Paradise.

83. The scene of the vision of Saint Peter, archbishop of Alexandria (d. 311 CE), the last great martyr of Egypt and a firm opponent to the heresy, features Christ appearing to the saint as a twelve-year-old boy in a torn white linen garment. To Peter's question of who has torn his *chitōn*, the child replies that it was Arius, the proponent of the eponymous heresy, thus giving the saint a symbolic premonition of the future danger of Arian teachings. While the scene featured in monumental painting from the thirteenth century, earlier representations are also found in manuscripts dating back to the eleventh century. Both types of representation are discussed in Jelena Bogdanovich, "The Rhetoric and Performativity of Light in the Sacred Space: A Case Study of the Vision of St. Peter of Alexandria," in *Hierotopy of Light and Fire in the Culture of the Byzantine World*, ed. Alexei Lidov (Moscow: Feorija, 2013), 282–304. See also Sašo Cvetkovski, "The Vision of Saint Peter of Alexandria, from the Church of St. Archangels in Prilep," *Zograf*, 36 (2012): 83–88.

84. Photios, *Hom.* 4, discussed in Robert Ousterhout, "The Virgin of the Chōra: An Image and Its Contents," in *The Sacred Image in East and West*, ed. Leslie Brubaker and Robert Ousterhout (Urbana: University of Illinois Press, 1995), 95. The *maphorion* was sheltered in the monastery of Blachernae, one of the most important monasteries of the city.

85. The epithet "*chōra tou achōrētou*" comes from a stanza of the hymn *akathistos*, but it is also used as a word play on the monastery of Chōra. "*Platytera ton ouranōn*" is a common epithet of the Mother of God.

86. Philo Alex., *De somniis*, in *Works*, trans. F. H. Colson and G. H. Whitaker, vol. 5 (London: William Heinemann, 1936), 405–7.

87. Constas, *Art of Seeing*, 238.

88. Archim. Aimilianos Simonopetritēs, Λόγοι εόρτοι μυσταγωγικοί (Athens: Indiktos, 2014), 287.

89. In the economy of signs, symbols, and prefigurations of the Orthodox Church, the Mother of God is associated with symbols such as the burning bush (Ex. 3:1–8), the cloud

(Ex. 19:9, 16–18), and the mountain (Ex. 19:16–20; Dan. 2:334; Hab. 3:3; Isa. 2:2). These symbols, Mary Cunningham notes, reinforce her association with creation. "However, Mary is understood to represent a transfigured creation: the objects that such types evoke are those special places in which God has chosen to reveal himself to humanity" (Cunningham, "The Mother of God," 45).

90. A third space, the narthex, was occupied by not-yet-baptized faithful and was associated with the earth. Symeon also traced a further correspondence between these three divisions of the church and the regions of the visible world: heaven, terrestrial paradise and earth. See Constas, *The Art of Seeing*, 217–58.

91. Mircea Eliade, cited in Sharon Gerstel, "Introduction," in *Thresholds of the Sacred*, ed. Sharon Gerstel (Washington, DC: Dumbarton Oaks Research Library and Collection, 2006), 3.

92. Constas, *The Art of Seeing*, 234.

93. Nicholas Constas, "Symeon of Thessalonica and the Theology of the Icon Screen," in *Thresholds of the Sacred*, ed. Gerstel, 163.

94. Constas, "Beyond the Veil," 1.

95. Constas, "Symeon of Thessalonica," 173.

96. Constas, *Proclus*, 358.

97. Antiphon 15, Vespers of Holy Thursday, *Mēnaion*.

Chapter Three

1. Bernard of Clairvaux, *Life and Works of Saint Bernard Abbot of Clairvaux*, ed. John Mabillon, vol. 2 (London: John Hodges, 1880), 464–65.

2. Bernard of Clairvaux, *Life and Works*, 2:464.

3. Ralph Glaber, "Miracles de Saint Benoit," in George Coulton, *Life in the Middle Ages* (Cambridge: Cambridge University Press, 1967), 1:3.

4. Thomas Aquinas, *Summa Theologica* 1.69.2. Having given substance, shape, and order to the Chthonic creation, God beautifies it by embroidering the garment: lights are created "to adorn the heavens by their movements;" birds and fish to "make beautiful the intermediate element;" land animals to "move upon the earth and adorn it." "For the perfection of the heaven and the earth regards, seemingly, those things that are extrinsic, just as the perfection of a man lies in his proper parts and forms, in clothing or such like" (1.70.5).

5. Glacken, *Traces on the Rhodian Shore*, 253.

6. Bruce Foltz, "Nature Godly and Beautiful: The Iconic Earth," *Phenomenology* 1 (2001): 120.

7. Glacken, *Traces on the Rhodian Shore*, 253.

8. Glacken, *Traces on the Rhodian Shore*, 253.

9. Jo. D., *On Heaven* 2.6.

10. The images of the constellations and their accompanying descriptions are taken from the Aratus texts circulating in contemporary "star catalogues." These were compilations of qualitative descriptions of the constellations, including illustrations of the mythological figure represented by the constellation. See Stephen McCluskey, *Astronomies and Cultures in Early Medieval Europe* (Cambridge: Cambridge University Press, 2000), 135, 143.

11. *O decus Europae, Caesar Enrice, beate / Augeat Imperium tibi rex qui regnat in aevum.*

12. David Ganz, "Pictorial Textiles and Performance: The Star Mantle of Henry II," in *Dressing the Part: Textiles as Propaganda in the Middle Ages*, ed. Kate Dimitrova and Margaret Goehring (Turnout: Brepols, 2014), 26. This idea is given visual expression in the splendid illumination in the Sacramentary of King Henry II (1002–14) (Munich BSB Clm 4456, fol. 33c), in which

Henry II, wearing a golden tunic topped by a blue *chlamys*, stands directly beneath Christ, by whom he is crowned. Pierre Béhar sees in the king wrapped in the cosmic mantle the personification of the divine *axis mundi* around which the universe revolves ("Pour une géopolitique de la papauté," *Géopolitique* 58 [1997]: 12).

13. Eliza Garrison, *Ottonian Imperial Art and Portraiture: The Artistic Patronage of Otto III* (Farnham, UK: Ashgate, 2011), 122. The mantle was most probably presented to Henry II on the occasion of these festivities.

14. Byzantine "silken diplomacy," Muthesius notes, was aimed mainly at the Latin West. Between the eighth and the twelfth centuries, precious silks were used as dowry gifts in imperial intermarriages between Byzantines and Latins (for example, in the wedding between Theophano and Otto II in 972). Up to the thirteenth century, silk trade concessions, granted in exchange for political alliance, dominated Byzantine foreign policy, and, in the same period, Byzantine prisoners could be ransomed from their Muslim captors in exchange for Byzantine silk (Muthesius, *Silk in Byzantium*, 8). Henry's star mantle as we see it today is the result of a fifteenth-century restoration, in which all the embroideries were detached from the old foundation fabric and subsequently applied to the current electric-blue, pomegranate-patterned silk damask; for a discussion of the restoration, see Ganz, "Pictorial Textiles and Performance."

15. In the continuing tension between imperial and papal authority, Henry styled himself "servant of the servants of God," the traditional papal title, thus claiming that his dominion came directly from God and setting, by implication, the autonomy of his power over that of any terrestrial rival (McCluskey, *Astronomies and Cultures in Early Medieval Europe*, 144).

16. Garrison, *Ottonian Imperial Art*, 123.

17. Cosgrove, *Apollo's Eye*, 56.

18. Garrison, *Ottonian Imperial Art*, 123. A comparison can also be drawn with the mantle of Roger II of Sicily, which dates from 1134. The decoration of this mantle is, however, more symbolic than Henry's, and "geographical," rather than cosmographic. It features a central palm tree flanked on either side by a lion attacking a camel. The lion was an emblem of the Norman kings of Sicily, while the camel represented the Arab world.

19. Earlier examples of cosmographic mantles featured celestial deities ranging from the Persian Mithras to the Celtic Epona (see McCluskey, *Astronomies and Cultures in Early Medieval Europe*, 143).

20. Muthesius, *Silk in Byzantium*, 30. Muthesius also suggests that the astronomical theme might have been a Byzantine import, given the strong influence of astronomy at the Byzantine court and the several Byzantine silks featuring astrological motifs (for example, the Brussels charioteer silk featuring the emperor in the form of Helios). McCluskey, by contrast, links the design to the Frankish interest in astronomy that followed the Carolingian revival and sets it beside other, similar courtly artifacts, such as Charlemagne's silver tablet featuring a description of the entire celestial sphere and an astronomical manuscript associated with the court of Louis the Pious (McCluskey, *Astronomies and Cultures in Early Medieval Europe*, 140). Although astrology was fiercely opposed by the early Church Fathers, in twelfth-century Europe it underwent a strong revival, which culminated with Roger Bacon (1214–92). The "Doctor Mirabilis," as the English philosopher, theologian, and alchemist was known by the scholastic accolade, placed further emphasis on the mantle of the heavens, rather than on the mantle of the earth, for celestial things, he argued, "are vaster than anything on earth," and their influence on terrestrial life and on the course of human affairs is therefore greater (*Opus Majus* 2.7). Competence in astrology, he argued, would have helped the Roman Catholic Church foresee war and other

disasters, and act accordingly; see Amanda Powers, "The Cosmographical Imagination of Roger Bacon," in *Mapping Medieval Geographies: Geographical Encounters and Cartographic Cultures in the Latin West and Beyond: 300–1600*, ed. Keith Lilley (Cambridge: Cambridge University Press, 2013).

21. *The Marriage of Mercury and Philology*, in which Capella defined the seven liberal arts, was one of the most influential works in the early Middle Ages. See Jacques Le Goff, *Medieval Civilisation, 400–1500* (London: Folio Society, 2011), 119.

22. Proserpina is first shown embroidering a panorama of the cosmogony and the natural order, but her work is interrupted by Venus; the description can be situated within the ancient Greek tradition of textile *ekphrasis* described in chapter 1. Claudianus's work is said to have also inspired Bernardus Silvestris's *Cosmography* in the twelfth century. See Winthrop Wetherbee, *The Cosmographia of Bernardus Silvestris* (New York: Columbia University Press, 1973), 50. Remigius d'Auxerre is cited in McCluskey, *Astronomies and Cultures in Early Medieval Europe*, 143.

23. See McCluskey, *Astronomies and Cultures in Early Medieval Europe*, 144.

24. Quoted in Alexandra Gajewski and Stefanie Seeberg, "Having Her Hand in It? Elite Women as 'Makers' of Textile Art in the Middle Ages," *Journal of Medieval History* 42 (2016): 37. A connection could be made here between Enide's textile offer to secure her a child and earlier traditions of the Virgin "weaving" Christ's body.

25. Gajewski and Seeberg, "Having Her Hand in It," 33.

26. The two inscriptions translate as: "Peace to Ismahel [Melus of Bari] who commissioned [this mantle]" and "May this gift of the emperor be welcome to the Highest Being."

27. As Ganz notes, while the zone beneath the hem forms the garment's border between the person supposed to wear it and the rest of the world (including his donor), "the zone beneath the Majestas Domini may be defined as a threshold that would open a door to God who was represented in the square above" (Ganz, "Pictorial Textiles and Performance," 28). The inscription, Ganz suggests, was added by the same workshop that crafted the cloak (probably in southern Germany), hence indicating that Henry's donation to the cathedral must have occurred very soon after the king received the gift from Melus. The donation of imperial mantles to cathedrals was not uncommon among Frankish rulers. For example, Charles de Bald presented his own to the treasury of St. Denis, while Otto III donated his own to St. Alessio in Rome (see Garrison, *Ottonian Imperial Art*, 123). Ekkehard IV of St. Gall (d. 1060) noted that a Swabian duchess had donated to his monastery an alb "with the marriage of Philologia embroidered in Gold" (Peter Dronke, *Bernardus Silvestris' Cosmographia* [Leiden: Brill, 1978], 27). Byzantine emperors and queens likewise set their garments as tablecloths for the altars; "under Pulcheria, sister of Emperor Theodosius II (408–450), precious imperial cloths were made to feature large in newly emerging Christian cult ceremonial. Although controversial, it became the practice to place the imperial robe of Pulcheria upon the Christian altar" (Muthesius, *Silk in Byzantium*, 25).

28. See, for example, the personifications in Bernardus Silvestris's *Cosmographia*, which deeply influenced de Lille's work. Intriguingly, Silva, the formless material of created life (the Western medieval equivalent of Chthonia), "cries to be clothed with a finer appearance" (Bernardus Silvestris, *Cosmographia*, trans. Wetherbee, 68).

29. Wetherbee, *The Cosmographia of Bernardus Silvestris*, 14.

30. Quoted in Wetherbee, *The Cosmographia of Bernardus Silvestris*, 14. "The concept of involucrum as an element consistently present in the works of the auctores had been known to the early Middle Ages; implicit in the commentaries of Remigius and Eriugena on Martianus, it is clearly enunciated in a well-known poem of Theodulph of Orleans," while "Abelard, in his theological

writings, speaks freely of involucra as a mode of allusion common to the prophets and the philosophers, employed by Christ in his parables and capable even in pagan writings of intimating things so mysterious as the persons of The Trinity, and he cites Plato as especially favored with the power thus to convey intuitions of truth" (Wetherbee, *The Cosmographia of Bernardus Silvestris*, 14).

31. Alain de Lille, *The Complaint of Nature*, trans. Douglas Moffat, Yale Studies in English 36 (New Haven, CT: Yale University Press, 1908), 11.

32. Alain de Lille, *The Complaint of Nature*, trans. Moffat, 11.

33. De Lille's encyclopedic "catalogue" of animal species was probably inspired by Bernardus Silvestris (see Bernardus Silvestris, *Cosmographia*, trans. Wetherbee, 80).

34. "The virgin, furthermore, on tiles, with the aid of a reed pen, called up and pictured various images of things. Still the pictures would not keep closely but quickly vanish'd to the material beneath them, and died away, leaving no traces. Although she often quickened them and caused them to live, yet they could not endure in the plan of her composition" (de Lille, *The Complaint of Nature*, 19).

35. Nature is nonetheless subordinated to God the Creator: "But after the universal Maker had clothed all things with the forms for their natures, and had wedded them in marriage with portions suitable to them individually, then, wishing that by the round of mutual relation of birth and death there should to perishable things be given stability through instability, infinity through impermanence, eternity through transiency, and that a series of things should be continually woven together in unbroken reciprocation of birth, he decreed that similar things, stamped with the seal of clear conformity, be brought from their like along the lawful path of sure descent. Me, then, he appointed a sort of deputy, a coiner for stamping the orders of things" (de Lille, *The Complaint of Nature*, 43–44).

36. De Lille, *The Complaint of Nature*, 34. Nature also uses textile metaphors to reproach the poet: "What! In thine asking dost thou clothe in the likeness of a doubt a question which is not worthy to take the form of a doubt? . . . Can it be that thou dost not know how poets expose naked falsehood to their hearers with no protecting cloak, that they may intoxicate their ears, and, so to speak, bewitch them with a melody of honeyed delight; or how they cloak that same falsehood with a pretense of credibility, that, by means of images of objective things, they may mold the souls of men on the anvil of dishonorable assent" (de Lille, *The Complaint of Nature*, 39–40; my emphasis).

37. De Lille, *The Complaint of Nature*, 41.

38. De Lille, *The Complaint of Nature*, 58–59. Interestingly, about a century and a half later, in the *Inferno* of his *Divina Commedia*, Dante Alighieri (1265–1321) clothes hypocrites with cloaks and hoods that are dazzling with their glitter but lined with lead (canto 23).

39. Hélinand of Froidmont, "The Worms of Death," quoted in Jacques Le Goff, *Il Meraviglioso e il quotidiano nell'Occidente medievale* (Rome: Laterza, 2010), 48.

40. De Lille, *The Complaint of Nature*, 43.

41. Henry's dedication on the star mantle calls God "SUP(ER)NE USYE," that is, "Superior Being (or Essence)." The word *usya* ("essence") is characteristic of Eriugena (*Periph.* 3.35).

42. Gertrud Schiller, *Iconography of Christian Art* (London: Lund Humphries, 1972), 2:110.

43. The dating of this map oscillates between 1208–50 and c. 1300. For a discussion of its dating, see Armin Wolf, "The Ebstorf Mappamundi and Gervase of Tilbury: The Controversy Revisited," *Imago Mundi*, 64 (2012): 1–27.

44. The adornment of creation, Aquinas argued, had come in the form of a green mantle "decorated" with the movements of its living creatures (*Summa theologica* 1.70–73). On *mappae*

mundi as encyclopedias of creation, see Margaret Hoogvliet, *"Mappae Mundi* and Medieval Encyclopedias: Image versus Text," in *Pre-modern Encyclopedic Texts,* ed. Peter Binkley (Leiden: Brill, 1997), 63–88; and Peter Barber, "Visual Encyclopedias: The Hereford and Other *Mappae Mundi," Map Collector* 48 (1989): 2–8.

45. Evelyn Edson, *The World Map, 1300–1492: The Persistence of Tradition and Transformation* (Baltimore: Johns Hopkins University Press, 2007), 11.

46. Alessandro Scafi, *Mapping Paradise: A History of Paradise on Earth* (London: British Library, 2006), 44.

47. On other *mappae mundi* (for example, the British Library Psalter) the disk of the inhabited world is enshrined in the figure of Christ, a distant echo of the biblical cosmographic mantle and of the patristic conception of God encompassing the totality of creation (see, for example, Gr. Naz., *Hom.* 28.25).

48. Samuel Edgerton, "From Mental Matrix to Mappamundi to Christian Empire: The Heritage of Ptolemaic Cartography in the Renaissance," in *Art and Cartography: Six Historical Essays,* ed. David Woodward (Chicago: University of Chicago Press, 1987), 29.

49. Jacob, *The Sovereign Map,* 117.

50. See Scafi, *Mapping Paradise,* 131; and Le Goff, *Medieval Civilisation,* 133. It remained a common belief among later medieval Western theologians and thinkers that a precise knowledge of geography would lead to more accurate exegesis. Things of the world, argued Roger Bacon, "cannot be known except through a knowledge of the places where they are contained" (cited in Glacken, *Traces on the Rhodian Shore,* 284).

51. Marcia Kupfer, "Reflections in the Ebstorf Map: Cartography, Theology and *Dilectio Speculationis,"* in *Mapping Medieval Geographies: Geographical Encounters and Cartographic Cultures in the Latin West and Beyond, 300–1600,* ed. Keith Lilley (Cambridge: Cambridge University Press, 2013), 103–4.

52. The same motif underpins the *Otia imperialia* by Gervase of Tilbury, whom Armin Wolf has identified as the likely author of the map (Wolf, "The Ebstorf Mappa Mundi"). The theory was widespread in medieval literature influenced by neo-Platonism. Bernardus Silvestris, for example, explicitly divides his *Cosmographia* into two parts titled "Macrocosm" and "Microcosm," tracing the creation of the cosmos and of humans, respectively. The latter were fashioned, he writes, by both Nature and Ourania, and embodied the elements of creation: "a single bond of love links unlike natures, though the flesh be of earth and the mind ethereal. . . . His mind and body, though of diverse natures, will be joined into one, such that a mysterious union will render the work harmonious" (Bernardus Silvestris, *Cosmographia,* trans. Wetherbee, 109, 113).

53. Wolf, "The Ebstorf Mappamundi," 14.

54. Jacob, *The Empire of Maps,* 89.

55. The definition appears in the text on the top right of the map. The word *forma* designates a fine shape, beauty, as well as a stamp (of coinage), thus reinforcing the notion of creation as a mirror of the Creator; see Charlton Thomas Lewis, *An Elementary Latin Dictionary* (Oxford: Oxford University Press, 2010).

56. Jacob, *The Empire of Maps,* 33.

57. Jacob, *The Empire of Maps,* 18.

58. Bernardus Silvestris, for example, saw in the totality of creatures, in the macrocosm, "a face inscribed with the image of the Father" (Bernardus Silvestris, *Cosmographia,* trans. Wetherbee, 87), while Bonaventura (1221–74) spoke of God's traces in the sensible world after Romans 1:20. Similar statements, however, are also expounded by the Thomists (Glacken, *Traces on the*

Rhodian Shore, 237). Other medieval exemplars included the "table of destiny," the "mirror of providence," and the "book of memory" (see Bernardus Silvestris, *Cosmographia,* trans. Wetherbee, 115).

59. Kupfer, "Reflections in the Ebstorf Map," 122.

60. It was in the West that the veil (*mandylion*) took the name "veronica." The origins of the word are ascribed to the name of its owner (a Latinization of the Greek name Beronikē). Veronica was the woman with an issue of blood cured by the touch of the hem of Christ's garment; later, she was also identified with the woman who offered her veil to Christ on his way to Golgotha. In his *Speculum Ecclesiae,* Giraldus Cambrensis (1146–1223) offered an alternative interpretation, rooting the name Veronica in the Latin *vera* (true) *icon* (image). According to this interpretation, which scholars have long discarded as post-factum, the woman would have been named after the relic (Andrea Lorenzo Molinari, "St. Veronica: Evolution of a Sacred Legend," *Priscilla Papers* 28 [2014]: 10).

61. Wolf, "The Ebstorf Mappamundi," 13.

62. The spread of devotional practices connected to the veronica and its burgeoning copies went hand in hand with the newly instituted feast of Corpus Domini and veneration of the Eucharist in the West. See Gerhard Wolf, "From *Mandylion* to Veronica: Picturing the 'Disembodied' Face and Disseminating the True Image of Christ in the Latin West," in *The Holy Face and the Paradox of Representation,* ed. Herbert Kessler and Gerhard Wolf, Villa Spelman Studies 6 (Bologna: Nova Alfa, 1998), 168.

63. Armin Wolf has noted the close resemblance between the veronica on the Ebstorf map and the veronica featuring in Matthew Paris's *Chronica majora* (1250) (Wolf, "The Ebstorf Mappamundi," 9). Intriguingly, this work also included a world map, though more reminiscent of Ptolemy's and Strabo's views than of contemporary circular *mappae mundi.* The reader was explicitly warned that "the world in its truest form resembles an extended *chlamys*" (cited in Suzanne Lewis, *The Art of Matthew Paris in the Chronica Majora* [Aldershot, UK: Scholar Press and Corpus Christi College Cambridge, 1987], 372). By contrast, Gervase of Tilbury states that the circular arrangement of the ocean is indicated in Gen. 1:9 and Ps. 104:6 "Who coveredst the deep as with a garment" (see William Bevan and Henry Phillott, *Medieval Geography: An Essay in Illustration of the Hereford Mappa Mundi* [London: Stanford, 1873], xix).

64. Wolf, "From *Mandylion* to Veronica," 161. Innocent III defined the meeting between image and beholder as an anticipation of the eschatological encounter face to face between the beholder and God (Bevan and Phillott, *Medieval Geography,* 167).

65. Kupfer, "Reflections in the Ebstorf Map," 104; Herbert Kessler, "Pictures Fertile with Truth: How Christians Managed to Make Images of God without Violating the Second Commandment," *Journal of the Walters Art Gallery* 49–50 (1991–92): 53–65.

66. Kuryluk, *Veronica and Her Cloth,* 179.

67. Wolf, "From *Mandylion* to Veronica," 177.

68. Wolf, "From *Mandylion* to Veronica," 177.

69. Kuryluk, *Veronica and Her Cloth,* 115.

70. Neil MacGregor and Erika Langmuir, *Seeing Salvation: Images of Christ in Art* (London: BBC, 2000), 92.

71. The relationship between the *Otia imperialia* and the map and its Otto-related topography is discussed by Wolf, "The Ebstorf Mappamundi," 1, 10. The center of *mappae mundi* was not usually stressed until the thirteenth century, with the European involvement in the Crusades (Edson, *The World Map,* 20).

72. Le Goff, *Medieval Civilisation*, 137.

73. On "the great age of forest clearance," see Glacken, *Traces on the Rhodian Shore*, 330–41.

74. Glacken, *Traces on the Rhodian Shore*, 213. This perception of landscape and nature is discussed by Thomas in the context of premodern and early modern England (Thomas, *Man and the Natural World*).

75. Bernard of Clairvaux, *Life and Works*, 460–61.

76. In Bernard's description, Clairvaux is a self-enclosed and self-sufficient microcosm providing its inhabitants with both physical and spiritual sustainment. This approach to landscape as a "*summa* of creation" is also found in secular literature. The courtly world in which the heroes of Chrétien de Troyes and of chivalric literature move is in effect a cosmos in miniature. "Its order and hierarchy are a source of inner stability for the chivalric hero, and to stray from the standard it provides or to seek to rise above this standard is to risk the perils of the dark forest, the Silva where man's vulnerability to passion and confusion is exposed and menaced with death" (Wetherbee, *The Cosmographia of Bernardus Silvestris*, 59). It is interesting that Bernard thinks of his description of the landscape surrounding his abbey as "a mirror" enabling the reader to see the site with his or her mind's eye: "If you wish to know the site of Clairvaux, these lines will describe it for you as if in a mirror." "Nature as *speculum*" is a widespread motif in medieval literature, especially in the writings of Bonaventura (see Glacken, *Traces on the Rhodian Shore*, 237–38).

77. Am., *Ep.* 8, quoted in Glacken, *Traces on the Rhodian Shore*, 299.

78. Richard Jenkyns, *Virgil's Experience: Nature and History: Times, Names, and Places* (Oxford: Oxford University Press, 1998); Glacken, *Traces on the Rhodian Shore*, 143. Cosgrove likened the ages of this universal history to "a mythical map in which the domed structure of an ordered cosmos is focused on the microcosm of domestic and community space (*domus*)." As Apollo gazes across the earth, so the poetic eye looks outward from domestic space over the cultivated realms, into the wilderness, to the ends of the earth, where the chaos of the world's origins reigns (Cosgrove, *Apollo's Eye*, 34).

79. Verg., *G.* 1.120.

80. Verg., *G.* 2.238–40, my emphasis. Vergil makes recurrent use of the verb *vestire* in connection with the land.

81. Bernardus Silvestris, *Cosmographia*, trans. Wetherbee, 76.

82. Compare Verg., *G.* 2.333–34, 2.404–7, 3.287–310, 3.384–91.

83. "In the Middle Ages, Virgil's reputation was such that it inspired legends associating him with magic and prophecy. From at least the third century, Christian thinkers interpreted *Eclogues* 4, which describes the birth of a boy ushering in a golden age, as a prediction of Jesus' birth. As a result, Virgil came to be seen on a similar level to the Hebrew prophets of the Bible as one who had heralded Christianity" (Jan Ziolkowski and Michael Putnam, *The Virgilian Tradition: The First Fifteen Hundred Years* [New Haven, CT: Yale University Press, 2008], xxxiv–xxxv).

84. The body occupies a central place in the medieval system of thought: "The Church defined itself as the mystical body of Christ, and the nascent state saw itself as a living body with the monarch as its head. . . . Bodily imagery played an important role in defining the three orders of the tripartite society. Priests formed the corpus (body) consecrated by ordination. . . . The corps of warriors was ennobled by its prowess in battle and tourney. And the corps of laborers, stooped beneath their heavy burden, was nevertheless exalted in the sculpture of cathedral portals, which depicted peasants engaged in the labors of the months and artisans engaged in the various occupations of the active life" (Jacques Le Goff, *The Medieval Imagination* [Chicago: University of Chicago Press, 2001], 14).

85. Compare John 15:1.

86. Francis Petrarch, *Familiar Letters*, ed. and trans. James Robinson (New York: G. P. Putnam, 1898), 261–75, especially 261–62, my emphasis. The two Latin inscriptions make the connection clearer: "Italy, benevolent country, nourishes famous poets. Thus this one [Vergil] enables you to achieve Grecian genius," and "This is Servius, who recovers the mysteries of the eloquent Vergil, so they are revealed to leaders, shepherds and farmers."

87. Le Goff, *Medieval Civilisation*, 159.

88. Bruce Foltz, "Nature Godly and Beautiful," 124.

89. Kuryluk, *Veronica and Her Cloth*, 197.

90. Petrarch's ascent of Mount Ventoux, which he undertook for the sole purpose of "discovering" the landscape from such an elevated perspective, gained him the fame of "first modern man;" see for example, Kenneth Clark, *Landscape into Art* (London: Folio Society, 2013). In reality, Petrarch's account is deeply allegorical, betraying a typically medieval religious sensibility. Some commentators have identified the episode as a threshold moment between two eras and attitudes, that is, modern empirical knowledge and Renaissance humanism, on the one hand, and medieval religious devotion, on the other; see, for example, Simon Schama, *Landscape and Memory* (New York: Verso, 1995), 421.

Chapter Four

1. Bachelard, *The Poetics of Reverie*, 1.

2. Oster cited in Werner Nell, "Romantic Folk Culture and the Souls of Black Folk: Framing the Beginnings of African-American Culture Studies in Cross-Atlantic Traveling Concepts," in *Traveling Traditions: Nineteenth-Century Cultural Concepts and Transatlantic Intellectual Networks*, ed. Erik Redling (Berlin: De Gruyter, 2016), 152.

3. Jacob Burckhardt, *The Civilization of the Period of the Renaissance in Italy*, trans. Samuel Middlemore (Cambridge: Cambridge University Press, 2015).

4. After the title of David Divine's book *The Opening of the World: The Great Age of Maritime Exploration* (New York: Putnam, 1973). The systematic rediscovery of the classical past and its creative reuse (as opposed to medieval blind acceptance of the *auctores*) opened up new temporal horizons. Pioneered by figures such as Criaco d'Ancona (1391–1452) and Cristoforo Buondelmonti (1386–1430), antiquarianism emerged in the fifteenth century as the forerunner of archaeology. The collection and study of antiquities became a common practice over the next two centuries in different European countries. Propelled by empirical observation and rational inquiry, by the late seventeenth century, scholars and clergymen, such as Thomas Burnet, interrogated themselves on the origins of the earth, paving the way to the terrifying abyss of "deep time" opened by geologists one century thereafter. See Glacken, *Traces on the Rhodian Shore*, 407–8; Veronica della Dora, *Mountain: Nature and Culture* (London: Reaktion, 2016). On the development of antiquarianism, see Alain Schnapp, *The Discovery of the Past* (New York: Harry Abrams, 1997).

5. Dan O' Sullivan, *The Age of Discovery, 1400–1550* (New York: Longman, 1984), 77.

6. Livingstone, *The Geographical Tradition*, 34.

7. Catherine Smith, "The Winged Eye: Leon Battista Alberti and the Visualization of the Past, Present and Future," in *Renaissance from Brunelleschi to Michelangelo: The Representation of Architecture*, ed. Henry Millon and Vittorio Magnago Lampagnani (New York: Rizzoli, 1997), 452–55.

8. Lenoble, cited in Hadot, *Veil of Isis*, 123.

9. Brotton, *History of the World in Twelve Maps*, 155–56.

10. Amerigo Vespucci, quoted in Glacken, *Traces on the Rhodian Shore*, 358.

11. Cosgrove, *Apollo's Eye*, 98. On Fra' Mauro's *mappa mundi*, see Pietro Falchetta, *Fra' Mauro's World Map* (Turnhout: Brepols, 2006).

12. Of course, the transition occurred gradually. The *Geography* became initially popular in Italy for the lists of ancient place names it provided to poets, antiquarians, and storytellers, rather than as a scientific treatise. Flavio Biondo, for instance, was said to have used a Ptolemaic table to compose his *Italia illustrata* (1453), a topographic description of the peninsula aimed at the exploration of the Roman roots of the Renaissance world and its antiquities. On the reception of Ptolemy in early Renaissance Italy, see Patrick Gautier Dalché, "The Reception of Ptolemy's Geography (End of the Fourteenth to Beginning of the Sixteenth Century)," in *The History of Cartography*, vol. 3, ed. David Woodward (Chicago: University of Chicago Press, 2007), 285–364; and Sean Roberts, *Printing a Mediterranean World: Florence, Constantinople, and the Renaissance of Geography* (Cambridge, MA: Harvard University Press, 2013).

13. Jacob, *The Sovereign Map*, 121.

14. The word, for example, is used by Girolamo Benivieni in a fictional dialogue with the humanist Antonio Manetti (the biographer of Filippo Brunelleschi) appended to the 1506 Florentine edition of Dante's *Divina Commedia* and including a series of maps of hell. Here "knowledge of the *mantellino*" is listed as a prerequisite to map hell (along with geometry, astronomy, and portolan charts); see Girolamo Benivieni, *Dialogo di Antonio Manetti: Cittadino fiorentino circa al sito, forma, & misure del lo infero di Dante Alighieri poeta excellentissimo* (Florence: F. di Giunta, 1506), 5, 28. An example of Ptolemaic map commonly referred to as "*planisfero a mantellino*" is Giovanni Matteo Contarini's, which was printed in Francesco Rosselli's workshop also in 1506. The inventory of the workshop compiled by Rosselli's son Alessandro in 1527 features a "*mappamondo a mantellino, in 2 fogli reali*" ("a world map drawn in the *mantellino* fashion, on two royal sheets"). Another variant of the inventory includes "*5 mantellini di Tolomeo in tela dipinti*" ("five Ptolemaic little mantles painted on canvas"); see Vladimiro Valerio, "Per una nuova ecdotica dei testi scientifici figurati. Tradizioni grafiche delle proiezioni tolemaiche dell'ecumene nel primo libro della *Geografia*," *Humanistica* 7 (2012): 64n12. The word *mantellino* has also been linked to Henricus Martellus, a German miniator active in Florence in the late fifteenth century, who adopted Ptolemy's second projection in his work; see Sebastiano Gentile, *Firenze e la scoperta dell'America* (Florence: Olschki, 1992), 247.

15. Brotton, *History of the World in Twelve Maps*, 175.

16. Jacob, *The Sovereign Map*, 113. A similar iconography is found on the Ptolemaic map in Hartmann Schedel's *Nuremberg Chronicle* (1493) and in Ortelius's *Typus Orbis Terrarum* (1570), among other maps.

17. Best known from Piero della Francesca's 1445 painting, this iconography was especially popular in Italy from the late Middle Ages, but was banned by the Council of Trent for its suspicious pagan influence (the same idea as the Marian mantle is found, for example, in a legend from the Krishna myth of India, in which the goddess Govardhana lifts a hill from the ground, creating a gigantic shelter for the village people and their animals). Venice alone numbers more than thirty paintings and bas-reliefs of the Madonna della Misericordia (a list and map can be found in "Madonna della Misericordia, a Venetian Icon," http://www.aloverofvenice.com/MisericordiaHome/Madonna5.htm [accessed 15 March 2017]). Western medieval Madonnas were often depicted wearing blue mantles embroidered with a universe of golden stars, as opposed to

Byzantine iconography featuring only three stars (usually on red fabric). In Renaissance paint-ing the mantle of the Virgin became a hybrid liminal space with its hem sometimes representing a point of contact between different worlds (following the tradition of the woman with an issue of blood healed by touching the hem of Christ's garment, Mt. 9:20–22; Mk. 5:25–34; Lk. 8:43–48). Filippo Lippi, for example, embroidered the hem of the blue mantle of the Virgin in the Pala Barbadori (1437) with golden pseudo-Kufic inscriptions, perhaps as a reminder of Mary's Middle Eastern origins.

18. Piani and Baratono, "Teofanie cosmografiche." A contemporary version of the protective Marian mantle was the subject of the *drappellone* (Palio banner) of the Palio di Siena in 2015 painted by the artist Francesco Mori. I am grateful to the authors for bringing this *drappellone* to my attention.

19. Linda Biesele Hall and Teresa Eckmann, *Mary, Mother and Warrior* (Austin: University of Texas Press, 2004), 46.

20. Piani and Baratono, "Teofanie cosmografiche." The map, which has been dated to the last part of the sixteenth century, was copied from Caspar Vopel's map (1570), probably the most popular cartographic representation of the time celebrating exploration, after the Waldseemül-ler map. It features an organic representation of the newly formed earth (including the Ameri-cas) framed by a golden ribbon. The continental landmasses are painted in green and set against a troubled blue ocean. The Red Sea is painted in red, according to the medieval convention of the *mappa mundi*.

21. See Denis Cosgrove, "Prospect, Perspective and the Evolution of the Landscape Idea," *Transactions of the Institute of British Geographers* 10 (1985): 45–62; see also his *Social Formation and Symbolic Landscape* (Madison: University of Wisconsin Press, 1998).

22. Berger, quoted in John Pickles, *A History of Spaces: Cartographic Reason, Mapping and the Geo-coded World* (London: Routledge, 2003), 85. The technique had "worked out experi-mentally for constructing a visual triangle which allowed the painter to determine the shape and measurement of a gridded square placed on the ground when viewed along the horizontal axis, and to reproduce in pictorial form its appearance to the eye. The *costruzione legittima* gave the realist illusion of three-dimensional space on a two-dimensional surface. This construction, the foundation of linear perspective, depended upon concepts of the vanishing point, distance point, and intersecting plane. Alberti describes it as a triangle of rays extending outwards from the eye and striking the object of vision" (Cosgrove, "Prospect," 48).

23. Cosgrove, "Prospect," 48.

24. "There is nothing more powerful, swift, or worthy than the eye. In short, it is the fore-most of the body's members, a sort of king or god. Didn't the ancients regard God as similar to the eye, since he surveys all things and reclaims them singly? On the one hand, we are enjoined to give glory for all things to God, to rejoice in him, to embrace him with all our mind and vig-orous virtue, and to consider him as an ever-present witness to all our thoughts and deeds. On the other hand, we are enjoined to be as vigilant and circumspect as we can, seeking everything which leads to the glory of virtue, and rejoicing whenever by our labor and industry we achieve something noble or divine" (Alberti, quoted in Smith, "The Winged Eye," 454).

25. Martin Jay, "Scopic Regimes of Modernity," in *Vision and Visuality*, ed. Hal Foster (Se-attle: Bay Press, 1988), 7.

26. Bryson, cited in Jay, "Scopic Regimes."

27. Edgerton, "From Mental Matrix to *Mappamundi*," 36.

28. Alpers, "The Mapping Impulse," 70.

29. Alpers, "The Mapping Impulse," 59.

30. Jacob, *The Sovereign Map*, 113.

31. See Pickles, *A History of Spaces*, 80–85.

32. "Nulla si può trovare, quanto io estimo, più acommodata cosa altra che quel velo, quale io tra i miei amici soglio appellare intersegazione. Quello sta così. *Egli è uno velo sottilissimo, tessuto raro, tinto di quale a te piace colore, distinto con fili più grossi in quanti a te piace paraleli*, qual velo pongo tra l'occhio e la cosa veduta, tale che la pirramide visiva penetra per la rarità del velo. Porgeti questo velo certo non picciola commodità: primo, che sempre ti ripresenta medesima non mossa superficie, dove tu, posti certi termini, subito ritruovi la vera cuspide della pirramide, qual cosa certo senza intercisione sarebbe difficile; e sai quanto sia impossibile bene contraffare cosa quale non continovo servi una medesima presenza. Di qui pertanto sono più facili a ritrarre le cose dipinte che le scolpite" (Alberti, *De pict.* 2.31; my emphasis). Alberti was also the first to devise a system of coordinates for locating and representing the major monuments of Rome (see Smith, "The Winged Eye," 452).

33. Edgerton, "From Mental Matrix to *Mappamundi*," 39.

34. Alberti, *De pict.* 2.36, my translation. This concept was popular among humanists and artists. Art manuals, including *On Technique* (1550), stated that a solid knowledge of anatomy is necessary for the artist to re-create natural movement and muscular structure. In his *Idea del tempio della pittura* (1590) Lomazzo takes the idea further, arguing that knowledge of anatomy is necessary in order to re-create the movements and emotions of the soul, which manifest themselves on the exterior of the body, that is, on the "garment of skin." In revealing what lies behind it, the fleshy mantle becomes in a way analogous to Alberti's *velum*. See Giorgio Vasari, *On Technique*, trans. Louisa Maclehose (New York: Dover Publications, 1960); Giovanni Paolo Lomazzo, *Idea del tempio della pittura* (Milan: Paolo Gottardo da Ponte, 1590).

35. "Più linee, quasi come nella tela più fili accostati fanno superficie" (Alberti, *De pict.* 1.2).

36. Alberti, *De pict.* 1.5.

37. On the gendered aspect of this image, see, for example, Barbara Freedman, *Staging the Gaze: Postmodernism, Psychoanalysis, and Shakespearean Comedy* (Ithaca, NY: Cornell University Press, 1991), 1–2; and Lynda Nead, *Female Nude: Art, Obscenity, and Sexuality* (New York: Routledge, 1992), 11–18.

38. As Rose notes, here lies another parallel between the female body and landscape: "the techniques of perspective used to record landscapes were also used to map female nudes, and the art genre of naked women emerged in the same periods as landscape painting" (Gillian Rose, *Feminism and Geography: The Limits of Geographical Knowledge* [Oxford: Polity Press, 1993], 96–97).

39. Carter, quoted in Peta Mitchell, *Cartographic Strategies of Postmodernity: The Figure of the Map in Contemporary Theory and Fiction* (London: Routledge, 2012), 45–46.

40. Denis Cosgrove, *The Palladian Landscape: Geographical Change and Its Cultural Representations in Sixteenth-century Italy* (Leicester: Leicester University Press, 1994).

41. In his introduction to the *Masque of Blackness*, Jonson writes: "First, for the scene, was drawn a Landschap consisting of small woods . . . [and then] an artificial sea was seen to shoot forth, as if it flowed to the land, raised with waves which seemed to move, and in some places the billow to break, as imitating that orderly disorder which is common in nature. . . . [T]he scene behind seemed a vast sea . . . that flowed forth, from the termination, or horizon of which (being the level of the state [the monarchial seat on a rectangular platform], which was placed in the upper end of the hall) was drawn, by the lines of perspective, the whole work shooting

downwards from the eye. . . . So much for the bodily part, which was of Master Inigo Jones's design and act." Quoted in Kenneth Olwig, "Performance, Etherial Space and the Practice of Landscape/Architecture: The Case of the Missing Mask," *Social and Cultural Geography* 12 (2011): 308.

42. Kenneth Olwig, *Landscape, Nature and the Body Politic: From Britain's Renaissance to America's New World* (Madison: University of Wisconsin Press, 2002), 141.

43. "Non senza gran ragione (Splendidissimi Ascoltatori) uscì quell'accorto conseglio dalla bocca di colui, c'havea sale in zucca, che l'arte deve esser à tutto potere imitatrice della generosa natura, poscia che essendosi ella cotanto dilettato nella produttion delle cose sotto, e sopra lunari, anzi preso trastullo, e gioco in *pinger il manto della terra di vezzosi, e vaghi fiori*, le cavernose tane del liquido mare di guizzosi pesci, gli antri e secrete spelonche di gemme, perle, topati & amatisti, l'aria di canori augelli, e tutt'il globo sopra il Ciel della bella Febe di scintillanti lumi, hà voluto come sagace maestra istruir anco noi, che à suo esempio fossimo de' doni suoi larghi, e splendidi dispensieri. Quindi e,' che gli Astrologi con mirabil artificio han composti sfere imitatrici di quelle della natura, e Cieli posticci, come un'Archimede, i Matematici fatto volar per aria colombe di legno, come Archita. I Pittori e Statuarii con ritratti quasi al vivo ingannar la vista e i sensi à chi gli guarda, come Zeusi; altri formar un cristallino fonte, e far che con l'arte si contrasti come in duello con la natura" (Prologue to *Nuova Comedia cavata dall'opra della Contralesina del Pastor Monopolitano intitolata le nozze d'Antilesina* [Venice: n.p., 1603], my italics). In his *Iconologia* (1593), Cesare Ripa represented the allegory of "The Terrestrial Machine" as a woman dressed in a tripartite garment featuring the sky with the clouds, the waves of the sea, and a green cloth embroidered with mountains, cities, and castles.

44. Olwig, *Landscape, Nature and the Body Politic*, xxx.

45. The ruler's gaze on the land symbolizes political power and its extension over the world, a motif well established in sixteenth-century portraiture, from Queen Elizabeth I's "Ditchley portrait" (1592) to the Beauvais tapestry series illustrating the "Fabulous History of the King of Gaul" (ca. 1530), where the monarch commands extensive bird's-eye views of the kingdom under his feet. Turned into actual fabric, the fabric of the French landscape becomes a stage for regal performance, a stage that continues beyond the horizon. Image and medium together express "a desire for the absolute mastery of the space and ground of France through a panoptic fantasy and the king's gigantic body, which reduces the kingdom to the dimensions of a little lovely garden." As Jacob observes, however, "the truly synoptic gaze is from outside the tapestry and belongs to the spectator," who can frontally view the overall form of both the land and its ruler (Jacob, *The Sovereign Map*, 320–21). Other cartographic tapestries include the impressive "Sheldon Tapestry Map of Warwickshire," the only complete survivor from a set of four tapestry maps commissioned in the 1590s by Ralph Sheldon for his new home (Weston House) in South Warwickshire. As with the Beauvais tapestry, here Shakespeare's county is captured from a bird's-eye view, though from a higher angle and at a much smaller scale. In different ways and at a different scales, these earlier large tapestries share the scenic quality of Jonson's masques.

46. Knowledge of the human body was rooted in the medicine of third-century BCE Alexandria, amplified by Galen (130–210 CE), and subsequently transmitted by his followers. With the capture of Alexandria by the Arabs in 642, medical knowledge became the preserve of Arabic commentators. As in the case of Ptolemy's *Geography*, with the fall of Constantinople and the influx of Greek scholars into western Europe, the Greco-Arabic medical tradition was extended by Renaissance humanists. See Jonathan Sawday, *The Body Emblazoned: Dissection and the Human Body in Renaissance Culture* (New York: Routledge, 1996), 39.

47. Sawday, *The Body Emblazoned*, 26.

48. For example, William Harvey (1578–1657), the first known physician to accurately describe the circulatory system, leaned on a system of beliefs inherited from Aristotle, which held that microcosm and macrocosm are united in the common bond of correspondence: "body, world, and heavens are ordered in a pattern of replication" (Sawday, *The Body Emblazoned*, 22).

49. Raphael Cuir, *The Development of the Study of Anatomy from the Renaissance to Cartesianism: Da Carpi, Vesalius, Estienne Bidloo* (Lewiston, NY: Edwin Mellen Press, 2009), 1.

50. Sawday, *The Body Emblazoned*, 23. The connection between the Ptolemaic table and human anatomy is clearly illustrated in Bartolomeo Eustachio's *Tabulae anatomicae* (1552) featuring details of human anatomy framed by ruled strips, exactly as on Ptolemaic maps. The author could therefore discuss in his text any pictorial detail simply by referring to its "coordinates," or map locations. As Samuel Edgerton observed, "Cinquecento readers were able to see meridians and parallels in any visual situation. Alberti's *velum* was impressed in their very optical process" (Edgerton, "From Mental Matrix to *Mappamundi*," 43).

51. Kuryluk, *Veronica and Her Cloth*, 200.

52. On the moral dimension of human anatomy, see Sawday, *The Body Emblazoned*, 85–140; and Cuir, *The Development of the Study of Anatomy*, 83–104.

53. Titles of anatomical and geographical (or cosmographic) works alike often contained the word "*fabrica*," as if to convey the manufactured structure of microcosm and macrocosm; see, for example, Blancanus's *De mundi fabrica* (1635) and Mercator's *Atlas sive cosmographicae meditationes de fabrica mundi* (1595).

54. Sawday, *The Body Emblazoned*, 70.

55. Interestingly, the same gesture recurs in the frontispieces of Renaissance maps and atlases as a sign of demonstration (see Jacob, *The Sovereign Map*, 115–16; Pickles, *A History of Spaces*, 3).

56. Here one can trace interesting parallels between the two horizontal, "dissected" female bodies. On the relationship between art and anatomy, see Cuir, *The Development of the Study of Anatomy*.

57. Sawday, *The Emblazoned Body*, 116. Likewise, as on Dürer's woodcut, on Vesalius' title page a third dimension is opened up: the eye is pushed deep into the theatrical space of the page and, imaginatively, beyond it.

58. Yves Hersant, "Foreword," in Cuir, *The Development of the Study of Anatomy*, vii.

59. Sawday, *The Emblazoned Body*, 116.

60. Tom Koch, "Review of Matthew Smallman-Raynor and Andrew Cliff's *Atlas of Epidemic Britain*," *Society and Space*, https://www.societyandspace.org/articles/atlas-of-epidemic-britain-by-matthew-smallman-raynor-and-andrew-cliff (accessed 15 April 2017).

61. Cuir, *The Development of the Study of Anatomy*, 29.

62. Cosgrove, "Globalism and Tolerance." Examples of "cartographic" theaters range from Abraham Ortelius's *Theatrum orbis terrarum* ("Theater of the Earthly Globe"), the first printed atlas (1570), to John Speed's *Theatre of the Empire of Great Britaine* (1611), and Vincenzo Coronelli's *Teatro delle città* ("Theater of Cities") and *Teatro della guerra* ("Theater of War"), published respectively in 1697 and 1706.

63. These atlases were conceived of as memory theaters enabling geographical learning through the spatialized memorization of *loci memoriae*. In the introduction to his *Theatrum*, Ortelius explicitly stated that the maps of his atlas "being placed as if were certaine glasses before our eyes, will the longer be kept in memory and make *the deeper impression* in us" (quoted in Alpers, "The Mapping Impulse," 88, my italics). Elaborated cartouches, exotic people and animals, mythical creatures, colorful vignettes served as privileged *loci* for the effective imprinting of geographical information in memory (see Veronica della Dora, "Mapping 'Melancholy-Pleasing

Remains': Morea as a Renaissance Memory Theater," in *Viewing the Morea: Land and People in the Late Medieval Peloponnese*, ed. Sharon Gerstel [Washington, DC: Dumbarton Oaks Research Library and Collection, 2013], 455–75).

64. In the 1587 version of the *Typus*, the Ciceronian quotation was complemented with four medallions inscribed with other Stoic quotations taken from Cicero and Seneca. The map and the passage are discussed in Cosgrove, "Globalism and Tolerance."

65. These motifs are made explicit in a number of sixteenth-century cordiform projections, and especially in the anonymous Fool's Head map (printed probably between 1580 and 1590, possibly in Antwerp). This map features the opening mantle of the world, drawn in Ptolemy's second projection, in place of the fool's face. The rest of the fool's head is embellished with the grotesque attributes of his calling: ass ears, little bells, and staff. The ancient Delphic motto "*Nosce te ipsum*" features at the top of the map in place of a title. Other inscriptions also fit the theme of *contemptus mundi*. See Giorgio Mangani, "Abraham Ortelius and the Hermetic Meaning of the Cordiform Projection," *Imago Mundi* 50 (1998): 59–83.

66. Tomaso Porcacchi, *L'isole più famose del mondo* (Venice: Simon Galignani, 1572), 109–10.

67. Porcacchi's *tovaglia* calls to mind the tablecloth on the frontispiece of the *Commentationes in Psalmos Euthymii Zigaboni* (1530). Surrounded and surveyed by the twelve apostles dressed up in mantles and wearing Roman Catholic tiaras, Jerusalem (the visible world) is set upon a table covered by a *tovaglia*, underneath which we are not given to see.

68. See Edward Casey, *Representing Place: Landscape Painting and Maps* (Minneapolis: University of Minnesota Press, 2002), 238; Mangani, *Cartografia morale*, 114–15.

69. Alpers, "The Mapping Impulse," 55. Incidentally, Vermeer came from the same town as Leeuwenhoek, and the two knew each other.

70. Alpers, "The Mapping Impulse," 58.

71. Hersant, "Foreword," vii.

72. Koyré, *From the Closed World to the Infinite Universe* (Baltimore: Johns Hopkins University Press, 1957), 1.

73. Koyré, *From the Closed World to the Infinite Universe*, 2.

74. Blaise Pascal, *Thoughts* 25.18.

Chapter Five

1. Hadot, *The Veil of Isis*, 238.

2. Francis Bacon, *Works*, ed. James Spedding, vol. 8 (London: Longman & Co., 1874), 257.

3. Hadot, *Veil of Isis*, 123.

4. Marjorie Hope Nicholson, *Mountain Gloom and Mountain Glory: The Development of the Aesthetics of the Infinite* (Seattle: University of Washington Press, 1997), 132–33.

5. Koyré, *From the Closed World to the Infinite Universe*, 2.

6. Peter Hulme and Ludmilla Jordanova, *The Enlightenment and Its Shadows* (London: Routledge, 1990), 4.

7. Light and sight, however, were more than metaphors for knowledge, or instruments for achieving it, as they had been in the previous two centuries. They were themselves objects subjected to the systematic scrutiny of the relentless Promethean gaze. If Copernicus's and Galileo's cosmological models set the sun at the center of the universe, its light became itself the subject of the greatest scientific investigations of the century. In works like Newton's *Opticks* (1703), "the laws of light seemed to have been laid bare" (Hulme and Jordanova, *Enlightenment*, 3).

8. Denis Cosgrove, "Global Illumination and Enlightenment in the Geographies of Vincenzo Coronelli and Athanasius Kircher," in *Geography and Enlightenment*, ed. David Livingstone and Charles Withers (Chicago: University of Chicago Press, 1999), 48.

9. Francis Bacon, *The New Organon*, bk. 1.84, ed. Lisa Jardine and Michael Silverthorne (Cambridge: Cambridge University Press, 2000), 69.

10. See John Henry, "National Styles in Science: A Possible Factor in the Scientific Revolution?" in *Geography and Revolution*, ed. David Livingstone and Charles Withers (Chicago: University of Chicago Press, 2005), 52.

11. Francis Bacon, *The Great Instauration and New Atlantis* (Arlington Heights, IL: Harlan Davidson, 1980), 24.

12. Livingstone, *Geographical Tradition*, 104. See also Charles Withers, "Geography, Science and the Scientific Revolution," in *Geography and Revolution*, ed. David Livingstone and Charles Withers (Chicago: University of Chicago Press, 2005), 77. Bacon was himself the author of an essay on travel, first published in 1597. See Francis Bacon, "XVIII. Of Travel," in *Essays, Civil and Moral*, vol. 3 (New York: P. F. Collier & Son, 1909).

13. John Ray, *Wisdom of God Manifested in the Works of Creation* (London: William Innys & Richard Manby, 1735), 199–200.

14. By which he meant "science."

15. Dianne Harris and Dede Fairchild-Ruggles, "Landscape and Vision," in *Sites Unseen: Landscape and Vision* (Philadelphia: University of Pennsylvania Press, 2007), 10–11.

16. Harris and Fairchild-Ruggles, "Landscape and Vision," 11.

17. "All men naturally desire knowledge. An indication of this is our esteem for the senses; for apart from their use we esteem them for their own sake, and most of all the sense of sight. Not only with a view to action, but even when no action is contemplated, we prefer sight, generally speaking, to all the other senses. The reason of this is that of all the senses sight best helps us to know things, and reveals many distinctions' (Arist., *Metaph.* 980a).

18. Henry Power, *Experimental Philosophy, in Three Books Containing New Experiments Microscopical, Mercurial, Magnetical: With Some Deductions, and Probable Hypotheses, Raised from Them, in Avouchment and Illustration of the Now Famous Atomical Hypothesis* (New York: Johnson Reprint Corp., 1966), xlviii.

19. Bacon, *New Organon*, bk. II.39, 171.

20. Bacon, *New Organon*, bk. II.39, 171.

21. Adam Dodd, "Size Is in the Eye of the Beholder: On the Cultural History of Microfaunae in Seventeenth-century Europe," in *Assigning Cultural Values*, ed. Kjerstin Aukurst (New York: Peter Lang, 2013), 18–19.

22. Dodd, "Size Is in the Eye of the Beholder," 18.

23. See Hadot, *The Veil of Isis*. Robert Boyle goes as far as to compare the lens of the telescope to the framed glass of the picture of a beloved one (a metaphor for God), which he calls "solid veil": "To make the rightest use of fading beauties, you must consider God and them, as you were wont to do your mistress's picture and its crystal cover: where, though that native glass were pure and lovely and very richly edg'd, yet to gaze on it was not the chiefest business of your eye; nor did you in it terminate your sight, but greedily look through and beyond it, upon the adored image, that solid veil betrayed. Methinks, Seraphic and our common lovers behold exterior beauties with a difference resembling that, wherewith children and astronomers consider Galileo's optic glasses (with one of which telescopes, that I remember I saw at Florence, he merrily boasted, that he had *trovato la corte a Giove*) which the one prizes most for what they appear;

the other for what they discover. For children contenting themselves to wonder at the length, and fall in love with the workmanship and gildings of the tube, do thus but gaze upon them; whereas astronomers look through them, and, scarce taking notice of the unusual ornaments, or the shape, employ them to find out unknown lights in the sky, and to descry in heaven bright stars, unseen before, and other celestial novelties and beauties" ("Seraphic Love," in *The Works of the Honourable Robert Boyle*, vol. 1 [London: Printed for J. & F. Rivington et al., 1772], 264).

24. Hadot, *The Veil of Isis*, 239–40.

25. Van Leeuwenhoek has been credited with the pioneering of microbiology, thanks to the use of the microscope, which he considerably perfected, achieving up to 300 times magnification. Then he rediscovered red blood cells and spermatozoa, and helped popularize the use of microscopes to view biological ultrastructure. In 1676, he reported the discovery of microorganisms. The *Natura arcana* (or "Mysteries of Nature Revealed") is a collection of Van Leeuwenhoek's numerous letters and reports in which he announced his discoveries to the academies of London and Paris.

26. Seventeenth-century studies in insectology often contain the word "theater" in their title, like the great geographical atlases of the time. Examples include Thomas Moffet's *Theatrum insectorum* (1634) and Samuel Purchas's *A Theatre of Politicall Flying-Insects* (1657).

27. Stuart Clark, *Vanities of the Eye: Vision in Early Modern European Culture* (Oxford: Oxford University Press, 2013), 11.

28. The engraving is signed by Francesco Curti (Bologna, 1603–70). Another example is found on the frontispiece to A. Robert's *Geographische Kleinod aus zweyen sehr ungemeinen Edelgesteinen bestehend* (1689), where Europe, unveiled amid the allegories of other continents, features a luminous eye over her breast.

29. The inscriptions read as follow: "*Dies diei eructat [v]erbum . . . Et Nox nocti indicat scientiam*" ("one day speaks to another . . . and night to night imparts his knowledge" [Ps. 19:2], upper part left and right); "*Numerus, mensura, pondus*" ("number, measure, weight" [Wis. 11:20], upper part center); "*Videbo Caelos tuos, opera digitor[um] tuor[um] . . . Non inclinabitur in saeculum saeculi*" ("I see Thy heavens work of Thy hand . . . Never be brought down," middle left and right); "*Ponderibus librata suis*" ("The earth is balanced by its own weight," middle).

30. Maurice Finocchiaro, *Defending Copernicus and Galileo: Critical Reasoning in the Two Affairs* (New York: Springer, 2012), 170.

31. The system is epitomized by the putti in the top right corner holding Jupiter, Saturn, and the moon, which rotate around the earth together with the sun.

32. Giovanni Battista Riccioli, *Almagestum novum astronomiam veterem novamque complectens observationibus aliorum et propriis novisque theorematibus, problematibus ac tabulis promotam in tres tomos distributam* (Bologna: Ex typographia eredi Victorii Benatii, 1651–65).

33. Joseph Addison, "An Oration in Defence of the New Philosophy, Spoken in the Theatre at Oxford, July 7, 1693," in D. M. Fontanelle, *A Conversation on the Plurality of Worlds: Translated from the French of M. De Fontanelle, to Which Is Added Mr Addison's Defence of the Newtonian Philosophy* (London: E. Curll, 1728), 197. Similar celebratory tones were used by Power in his description of the advances enabled by telescopes and microscopes: "Dioptrical Glasses (which are now wrought up to that height and curiosity we see) are but a Modern Invention: Antiquity gives us not the least hint thereof, neither do their Records furnish us with anything that does Antedate our late discoveries of the Telescope, or Microscope. The want of which incomparable Artifice made them not only erre in their fond Coelestial Hypothesis, and Crystalline wheel-work of the Heavens above us, but also in their nearer Observations of the minute Bodies and

smallest sort of Creatures about us, which have been by them but sleightly and perfunctorily described, as being the disregarded pieces and huslement of the creation; when (alas!) those sons of Sense were not able to see how curiously the minutest things of the world are wrought, and with what eminent signatures of Divine Providence they were inrich'd and embellished without our Dioptrical assistance. Neither do I think that the Aged world stands now in need of Spectacles, more than it did in its primitive Strength and Lustre: for howsoever though the faculties of the soul of our Primitive father Adam might be more quick & perspicacious in Apprehension, than those of our lapsed selves; yet certainly the Constitution of Adam's Organs was not divers from ours, nor different from those of his Fallen Self, so that he could never discern those distant, or minute objects by Natural Vision, as we do by the Artificial advantages of the Telescope and Microscope" (Power, *Experimental Philosophy*, n.p.). By the end of the seventeenth century, the microscope had started to reveal millions of microorganisms in a world of beauty and intricacy previously unknown to the human eye. "Leeuwenhoek found 8,280,000 living creatures in a single drop of water and declared in 1683 that there were more animals in his own mouth than there were people in the United Provinces" (Thomas, *Man and the Natural World*, 167–68).

34. This iconography is discussed in Cosgrove, "Global Illumination."

35. The work covered the phenomena of light and shadow by way of "their physical and metaphysical qualities and sources, as well as astronomy and astrology of the heavenly bodies, the phenomena of luminescence, phosphorescence, and fluorescence, the properties and transmission of colour and radiation, and the laws of optics and perspectival projection." As Clark comments, the last book is devoted to "*Magia lucis et umbrae*" and "presented in the traditional manner of the natural magicians as a natural and wholly undemonic account of the rarities, prodigies and *occulta* of its subject matter—mid seventeenth-century visual experience" (Clark, *Vanities of the Eye*, 101).

36. Jocelyn Godwin, *Athanasius Kircher: A Renaissance Man and the Quest for Lost Knowledge* (New York: Thames & Hudson, 1979); Daniel Stolzenberg, *Egyptian Oedipus: Athanasius Kircher and the Secrets of Antiquity* (Chicago: University of Chicago Press, 2015).

37. Clark, *Vanities of the Eye*, 101–7.

38. Cosgrove, *Apollo's Eye*, 160.

39. Cosgrove, *Apollo's Eye*, 141.

40. See Athanasius Kircher, *Mundus subterraneus* (Amsterdam: Joannes Jansson & Elizeum Weyerstraten, 1665).

41. An antecedent to Kircher's cartographic veil is found in a now lost painting described in Jean-François Niceron's *Thaumaturgus opticus* (1646), where the robes of Saint John are said to feature a landscape "containing several trees, shrubs, flowers, etc." (cited in Clark, *Vanities of the Eye*, 94).

42. "Since the sixteenth century it had become common to inscribe the Hebrew Tetragrammaton (script and number of creation) onto cosmographic images and world maps, signifying divine illumination parting the clouds of material unknowing in order to illuminate the true nature of the cosmos. The Tetragrammaton could readily be supplemented, or even replaced, on geographic maps by the Jesuit JHS, signifying the Son's task of bringing light to a world fallen into darkness" (Cosgrove, "Global illumination," 53–54).

43. Cosgrove, "Global illumination," 33.

44. See Paula Findlen, "The Economy of Scientific Exchange in Early Modern Europe," in *Patronage and Institutions: Science, Technology, and Medicine at the European Courts, 1500–1750*, ed. Bruce Moran (Woodbridge, UK: Boydell Press, 1990), 5–24.

45. Cosgrove, "Global Illumination," 50.

46. Daniel Stolzenberg, "Kircher among the Ruins: Esoteric Knowledge and Universal History," in *The Great Art of Knowing: The Baroque Encyclopedia of Athanasius Kircher*, ed. Daniel Stolzenberg (Stanford: Stanford University Libraries, 2001), 127–39.

47. See Haun Saussy, "China Illustrata: The Universe in a Cup of Tea," in *The Great Art of Knowing: The Baroque Encyclopedia of Athanasius Kircher*, ed. Daniel Stolzenberg (Stanford: Stanford University Libraries, 2001), 105–13.

48. Cosgrove, *Apollo's Eye*, 139.

49. Clark, *Vanities of the Eye*, 83.

50. Clark, *Vanities of the Eye*, 109.

51. Cited in Clark, *Vanities of the Eye*, 101.

52. See, for example, Chardin's *Journal du voyage en Perse et aux Indes Orientales* (1686). A similar strategy is creatively adopted in some seventeenth-century atlases. For example, the frontispiece of Jan van Keulen's *Zee-Atlas* (1681) features two putti lifting the curtains over a trafficked sea. The textile motif is repeated in the cartouches of a number of maps in the atlas: sometimes the title of the map is inscribed in drapes held by the inhabitants of the region depicted, other times it features on a sheet tied on trees, at once concealing and revealing the landscape behind it. Likewise, John Seller (1630–97), hydrographer to James II, uses airborne drapery to complement cartographic and coastal views with strategic topographical details, such as forts. See Charles Bricker, *Landmarks of Cartography* (Ware, Hertfordshire, UK: Wordsworth, 1989), 170.

53. Cosgrove, "Global Illumination," 37.

54. Cosgrove, "Global Illumination," 41.

55. Vincenzo Coronelli, *Atlante Veneto. Tomo I nel quale si contiene la descrittione geografica, storica, sacra, profana, e politica, degl'imperii, regni, provincie, e stati dell'Universo, loro divisione, e confini, coll'aggiunta di tutti li paesi nuovamente scoperti, accresciuto di molte tavole geografiche, non più publicate* (Venice: Domenico Padovani, 1690).

56. Cosgrove, "Global Illumination," 39.

57. Vincenzo Coronelli, *Isolario, descrittione geografico-historica, sacro-profana, antico-moderna, politica, naturale, e poetica. Mari, golfi, seni, piagge, porti, barche, pesche, promontori, monti, boschi, fiumi . . . ed ogni più esatta notitia di tutte l'isole coll'osservationi degli scogli sirti, scagni, e secche del globo terracqueo. Aggiuntivi anche i ritratti de' dominatori di esse. Ornato di trecento-dieci tavole geografiche, topografiche, corografiche, iconografiche, scenografiche . . . a' maggiore dilucidatione, ed uso della navigatione, et in supplimento dei 14 volumi del Bleau, tomo dell'Atlante veneto*, pt. 1 (Venice: By the author, 1696), 69.

58. Coronelli, *Isolario*, 163.

59. See Leonora Navari, "Vincenzo Coronelli and the Iconography of the Venetian Conquest of the Morea: A Study in Illustrative Methods," *Annual of the British School at Athens* 90 (1995): 505–19.

60. della Dora, "Mapping Melancholy-Pleasing Remains." Transposed on fabrics, islands and cities in these works become trophies akin to war banners, like the Ottoman one the Venetians captured at Coroni (to which Coronelli dedicated two separate engravings).

61. Charles Dempsey, *Inventing the Renaissance Putto* (Chapel Hill: University of North Carolina Press, 2001), 20.

62. In the context of human anatomy, Cuir and Sawday talk about a slow and gradual shift from the Vesalian body toward the Cartesian body, which culminates with the disappearance of self-demonstrating figures in the landscape in favor of the neutral background of the white page.

While the Vesalian body was a subject with a life of its own and responsible for its own attitudes, the Cartesian body was a mere anatomical object governed, like the universe, by mechanical laws. Reduced to machines, cosmos and microcosm no longer mirrored each other.

63. Bode cited in Dempsey, *Inventing the Renaissance Putto*, 20.

64. Other examples of Father Time lifting his veil over terrestrial globes are found, for example, on the frontispiece of Van Leeuwen's *Illustrious Holland, or a Treatise on the Origin, Progress, Traditions, State and Religion of Old Batavia* (1685) and in François van Bleyswijck's frontispiece to Nicolas Gueudeville's *Le nouveau theatre du monde, ou La géographie royale* (1713) in which Chronos (this time featuring an hourglass on the top of his bald head) unveils Asia, while Chorographia measures distances on the globe with her compass. Both illustrations are discussed in Veronica della Dora, "Lifting the Veil of Time: Maps, Metaphor and Antiquarianism, 17–18th c.," in *Time in Space: Representing the Past in Maps*, ed. Kären Wiegen and Caroline Winterer (Chicago: University of Chicago Press, 2020), 111–12.

65. The same motif is used in Francesco Primaticcio's *Astronomia* (1552). This time a star-crowned allegory of Astronomy has taken the place of Time and the heavenly sphere has taken the place of the earthly globe. The resemblance to Stradanus's frontispiece here is even more striking, with a veiled figure lying under the weight of the globe in a fashion similar to Stradanus's Neptune.

66. Hadot, *The Veil of Isis*, 169.

67. See Fritz Saxl, "Veritas filia temporis," in *Philosophy and History: Essays Presented to Ernst Cassirer*, ed. Raymond Klibansky (Oxford: Clarendon, 1936), 197–222; and Erwin Panofsky, *Studies in Iconology* (San Francisco: Harper & Row, 1972), 69–91.

68. The popularization and wide spread of the theme across Europe is owing to texts such as Andreas Alciatus's *Emblemata* (1542, 1602) and Cesare Ripa's *Iconologia* (1593, 1603) and their various editions and translations, including the influential Dutch translation by Dirck Pietersz Pers (1644).

69. As, for example, in Bernini's *The Triumph of Truth* (1652) in the funerary monument of Pope Alexander VII Chigi (1678), and, even more explicitly, in George Withers's poem *For the Day Present, or the Last Day* (1622):

> And when time's veil is rent away,
> Whereby eternity is hid,
> When though shalt all things open lay,
> Which ere we thought, or said, or did,
> Among time's ruins bury so
> Our failings through our tract of time
> That from these dungeons here below
> We to celestial thrones may clime,
> And there to our eternal king,
> For ever Hallelujah sing.

70. Francis Bacon, *Novum Organum*, bk. 1, in Basil Montagu, *The Works of Francis Bacon, Lord Chancellor of England*, vol. 3 (Philadelphia: Carey and Hart, 1844), 158.

71. Hadot, *The Veil of Isis*, 242.

72. The *Encyclopédie, ou Dictionnaire raisonné des sciences, des arts et des métiers* ("Encyclopedia, or a Systematic Dictionary of the Sciences, Arts, and Crafts") was a general encyclopedia published in eleven volumes between 1751 and 1772. It included 2,569 plates depicting every possible facet of contemporary life: from nature and medicine to commerce, the arts, and

manufacturing. These images complemented philosophical discussions of religion, statehood, history, morals, and literature. Overall, the *Encyclopédie* advocated a move toward a secularized "scientific" learning. Ultimately, Diderot's aim was to create a work that would incorporate all of the world's knowledge and disseminate all this information to the public. Chaplin connects the idea of global knowledge epitomized by the modern *Encyclopédie* to the concept of "world blanketing," or the totalizing impression produced by eighteenth-century circumnavigations. In her words, "the fantasy of having a reference work that contained everything in the world matched the goals of mapping all the parts of the world" (*Round about the Earth*, 134).

73. Jean d'Alembert, "Preliminary Discourse," in Denis Diderot, *The Encyclopedia*, ed. and trans. Stephen Gendzier (New York: Harper Torchbooks, 1967), 19–20.

74. Descartes, for example, "taught good minds how to shake off the yoke of scholasticism;" Newton "made light known to mankind by decomposing it;" Pascal was "a prodigy of wisdom and penetration;" and so on (d'Alembert, "Preliminary Discourse," 19–20).

75. D'Alembert, "Preliminary Discourse," 19.

76. Diderot, quoted in "Introduction," in Denis Diderot, *The Encyclopedia*, ed. and trans. Stephen Gendzier (New York: Harper Torchbooks, 1967), vii; and d'Alembert, "Preliminary Discourse," 34. This point is emphasized in various parts of d'Alembert's text; for example, further on he states that the goal of the *Encyclopédie* is to "contribute to the certitude and to the progress of human knowledge that in multiplying the number of scholars, distinguished artists and enlightened amateurs will spread throughout society" ("Preliminary Discourse," 41).

77. "We now declare that we owe our tree of knowledge to Chancellor Bacon" (d'Alembert, "Preliminary Discourse," 20).

78. Note on the frontispiece, in Diderot, *Encyclopedia*, viii.

79. Diderot, *Encyclopedia*, 175.

80. Diderot, *Encyclopedia*, 176; my emphasis.

81. Hadot, *The Veil of Isis*, 129.

82. Jay, "Scopic Regimes," 17. In this sense, these mantles perpetrated the illusionistic "unrolled" heavens painted by Giotto in the Scrovegni chapel, beneath which is the eternal presence of God.

Chapter Six

1. Johann Wolfgang Goethe, "Dedication," *NOH London Magazine* 9 (1824): 187.

2. Dorothea von Muecke, *Virtue and the Veil of Illusion: Generic Innovation and the Pedagogical Project in Eighteenth-Century Literature* (Stanford, CA: Stanford University Press, 1991), 6.

3. Quoted in Hadot, *Veil of Isis*, 249.

4. Hadot, *Veil of Isis*, 251.

5. Chenxi Tang, *The Geographic Imagination of Modernity: Geography, Literature, and Philosophy in German Romanticism* (Stanford, CA: Stanford University Press, 2008), 13.

6. Hadot, *Veil of Isis*, 213.

7. Muecke, *Virtue and the Veil*, 13.

8. Hadot, *Veil of Isis*, 282.

9. Tang, *Geographic Imagination*, 3.

10. See Denis Cosgrove, "Modernity, Community and the Landscape Idea," *Journal of Material Culture* 11 (2006): 49–66.

11. Tang, *Geographic Imagination*, 59.

12. Nicholas Halmi, *The Genealogy of the Romantic Symbol* (Oxford: Oxford University Press, 2011), 24.

13. Tang, *Geographic Imagination*, 14.

14. See, for example, Thomas Burnet, *The Sacred Theory of the Earth: Containing an Account of the Original Creation of the Earth and all the General Changes which It Hath Already Undergone, or Is to Undergo till the Consummation of All Things* (London: John Hooke, 1719 [1684]), 193–94.

15. See Johann Jacob Scheuchzer, *Ouresiphoites Helveticus, sive itinera per Helvetiae alpinas regiones* (Bologna: Libreria alpina degli Esposti, 1970 [1702–11]). Scheuchzer was a Swiss physician, and even though he is usually cited for having produced a catalogue of Alpine dragons, in reality his scientific work, which blended natural history and natural theology, was crucial in reorienting attitudes toward the Alps. During his wanderings, he explored every corner of the Alps and examined plants, as well as fossils, glaciers, and avalanches. His influence is discussed in Peter Hansen, *The Summits of Modern Man: Mountaineering after the Enlightenment* (Cambridge, MA: Harvard University Press, 2013), 40–43.

16. The journey took place between 1728 and 1732. Haller was accompanied by his friend Johannes Gessner. The two started out in Biel and subsequently traveled a route through the Swiss Alps that took them also to the cities of Lausanne, Geneva, Lucerne, Zurich, and Bern. See Dorothy Roller Wiswall, *A Comparison of Selected Poetic and Scientific Works of Albrecht von Haller* (Bern: Peter Lang, 1981), 14.

17. Tang, *Geographic Imagination*, 61. More specifically, see Wiswall, *A Comparison*. On the reception of the poem, see Alison Martin, "Natural Effusions: Mrs J. Howorth's English Translation of Albrecht von Haller's 'Die Alpen,'" *Translation Studies* 5 (2012): 17–32.

18. The translation in prose is by Mrs J. Howorth, *Poems by Baron von Haller* (London: J. Bell, 1794), 32, 25, 26.

19. Roller Wiswall, *A Comparison of Selected Poems*, 45; Tang, *Geographic Imagination*, 63. "A mind enlightened by philosophy, and exercised in study, a sublime genius, who from this vast theatre of creation can recur to the Creator himself, must in these grand scenes find the most delightful occupation for his thoughts. Carry the light of knowledge into those subterraneous mines where metals vegetate, where the gold is prepared which our rivers draw along in their rapid course; look upon the fields enriched with various herbage, and shining with liquid pearls: here shall you find exhaustless subjects for warm devotion and scientific research; here shall you find pleasures equally animated and lasting" (Howorth, *The Poems of Baron Haller*, 30–31).

20. Howorth, *Poems of Baron Haller*, 29. Dorothy Roller Wiswall interprets the rays of light dispelling the fog as a symbol for the light of knowledge dispelling the fog of ignorance, but the light of knowledge being too bright for human eyes (Roller Wiswall, *A Comparison*, 82).

21. Tang, *Geographic Imagination*, 62. Haller, who following his studies in Leiden had traveled to England, was indeed a passionate admirer of Newton, whom he took as the prototype of the great scientist. See Roller Wiswall, *A Comparison*, 11–12.

22. Tang, *Geographic Imagination*, 62.

23. The translation of this passage is from Tang, *The Geographic Imagination*, 63.

24. Howorth, *Poems of Baron Haller*, 31. The zooming-in movement structures the poem and characterizes Haller's "scientific way of seeing." As in his eight-volume compendium on physiology, *Elementa physiologiae corporis humani* (1757–66) (a vast system organizing knowledge of the physiology of the human body from the tiniest detail of blood vessels to general systems, such as the nervous system), ideas and phenomena are presented in a highly organized structure. Nature is revealed through sequential vignettes, or "pictures." "Each picture is

displayed in a stanza of ten lines of alexandrine verse with the first two lines introducing the subject, the next six lines developing the idea, and the final two lines reaching a succinct conclusion" (Roller Wiswall, *A Comparison*, 55). Aesthetic descriptions of plants and flowers similar to those in his poem on the Alps are also found in Haller's botanical works; in a way, scientific compiling and naming is akin to description in poetry (see Roller Wiswall, *A Comparison*, 93–110).

25. In his *Reflections on the Study of Nature* Linnaeus characterized nature as a "splendid theatre" that dazzles our eyes with a magnificent display of beauties: "For the gratification of our eyes, the earth is everywhere covered with verdure" (quoted in Tang, *Geographic Imagination*, 65).

26. The verses are from Uz's "Spring Songs" and Drummond's "Seventh Sonnet." These poems and the genre are discussed in Biese, *Development for the Feeling of Nature*, 224–34.

27. Tang, *Geographic Imagination*, 66. This idea is elaborated in the later part of the poem, through the figure of the wise man who researches the natural world—a sort of self-made botanist, geologist, and meteorologist: "Behold that man who enjoys so vigorous an old age, and who is surrounded by a circle of auditors; he discovers to them their secret recedes where Nature hides her troves. From the exercise of his own sagacity he has learned the virtues of plants, the varieties of their species, and the name of the humble moss. His penetrating glance darts into the deep bosom of the mine, or, parting through the fields of air, recognizes those sulphury vapors which bear the latent thunder in their humid volumes" (Howorth, *The Poems of Baron Haller*, 28). The stanza is commented upon by Roller Wiswall, *A Comparison*, 75–77.

28. Albrecht von Haller, "Die Falscheit menschlicher Tugenden," in *Versuch schweizerischer Gedichte* (1732), cited in Hadot, *Veil of Isis*, 253.

29. This effect of transparency was achieved through naturalistic description: his biographer and friend Johann Georg Zimmermann, for example, referred to the poem as exhibiting an accuracy in natural descriptions which only the eye of a trained scientist could achieve. He thus went to great lengths to unsettle the boundary between poetic text and the factual account of Haller's expedition, so as "to highlight its unsurpassable faithfulness in imitating nature" (Tang, *Geographic Imagination*, 61). More characteristically, Diderot achieved the illusion of transparency and the reader's absorption into the represented worlds through antitheatrical techniques that veiled the materiality of the signifier, the structures of semiotic mediation, and the labor of artistic production. See Muecke, *Virtue and the Veil*, 10; and Malcom Howard Dewey, "Herder's Relation to the Aesthetic Theory of His Time: A Contribution Based on the Fourth Critical Waeldchen" (PhD diss., University of Chicago, 1920), 38–40. While representing the pietistic facet of the Enlightenment and opposing Voltaire for his libertine views, Haller was himself involved in the *Encyclopédie* project, to which he contributed with entries on physiology and the history of physiology and anatomy (see Roller Wiswall, *A Comparison*, 13–14).

30. Halmi, *Genealogy of Romantic Symbol*, 11.

31. Goethe, quoted in Hadot, *Veil of Isis*, 253.

32. In *Faust* (pt. 1), for example, he mocked scientific instruments: "to nature's portals ye should be the key; / Cunning your wards, and yet the bolts ye fail to stir," arguing of nature that "Inscrutable in broadest light, / To be unveil'd by force she doth refuse, / What she reveals not to thy mental sight, / Thou wilt not wrest me from her with levers and with screws." (Johann Wolfgang Goethe, *Faust*, vv. 325–30, trans. A. S. Kline, www.poetryintranslation.com [accessed 29 January 2016]).

33. Hadot, *Veil of Isis*, 254.

34. Martin Swales, *Goethe: Selected Poems* (New York: Oxford University Press, 1975), 13.

35. Johann Wolfgang Goethe, "Welcome and Departure," in Nathan Haskell Dole, *The Works of J. W. von Goethe*, vol. 9 (Boston: Francis A. Nicolls & Co., 1902), 40.

36. Johann Wolfgang Goethe, "Ganymede," in Dole, *The Works of Goethe*, 209.

37. The passage refers to his visit to Chamonix and is discussed in Biese, *Development of the Feeling for Nature*, 317.

38. Biese, *Development of the Feeling for Nature*, 290.

39. Cited in Biese, *Development of the Feeling for Nature*, 294.

40. Tang, *Geographic Imagination*, 57.

41. Muecke, *Virtue and the Veil*, 13. A theorization of these ideas is found in Herder's *Fourth Critical Waeldchen* (1769): "The sense of sight is the most artificial and the most philosophical of the senses; it is attained, as those who have been blind tell us, only with the greatest effort and practice" (Dewey, *Herder's Relation to the Aesthetic Theory*, 114). "The world of the one who feels is purely a world of immediate presence; he has no eye, hence no distance as such: hence no surface, no colors, no imagination, no sentiment of the imagination; everything present, in our nerves, unmediated in us" (cited in Tang, *Geographic Imagination*, 71). It is interesting that Herder recurs to the veil metaphor: "There are always marvels that shine through the veil of language, shine with two-fold charms, tear away the veil and they disappear" (Ernest Menze and Karl Menges, *Johann Gottfried Herder: Selected Early Works, 1764-1767* [University Park, PA: Pennsylvania State University, 1992], 32).

42. Johann Wolfgang von Goethe, *The Sorrows of the Young Werther*, bk. 1 (Cambridge, MA: Harvard Classics Shelf of Fiction, 1917), 5-10.

43. Biese, *Development of the Feeling for Nature*, 323.

44. Remo Bodei, *Paesaggi sublimi. Gli uomini davanti alla natura selvaggia* (Milan: Bompiani, 2008), 106.

45. Robert Rehder, *Wordsworth and the Beginnings of Modern Poetry* (London: Croom Helm, 1981), 23. Novalis referred to Nature as "a mirror of the spirit," or the spirit unaware of itself. In his unfinished work "The Disciples at Saïs," he uses the "unveiling Nature" metaphor: "one of them succeeded—he raised the veil of the goddess of Saïs. Yet what did he see? He saw— wonder of wonders!—himself" (the passage is discussed in Hadot, *Veil of Isis*, 273).

46. Percy Bysshe Shelley, "Mont Blanc: Lines Written in the Vale of Chamouni," in *The Poetical Works of Coleridge, Shelley, and Keats*, ed. Cyrus Redding (Washington, DC: Woodstock Books, 2002), 218.

47. In his "Defence of Poetry," Shelley repeatedly uses veil metaphors. For example, he writes, poetry "lifts the veil from the hidden beauty of the world, and makes familiar objects be as if they were not familiar; it reproduces all that it represents, and the personifications clothed in its Elysian light stand forward in the minds of those who have once contemplated them." "All high poetry is infinite. . . . Veil after veil may be undrawn, and the inmost naked beauty of the meaning exposed." Or again, poetry "arrests the vanishing apparition which haunt the interlunations of life, and veiling them, or in language or in form, sends them forth among mankind," yet it also "strips the veil of familiarity from the world, and lays bare the naked and sleeping beauty, which is the spirit of its forms." In other words, "whether it spreads its own figured curtain, or withdraws life's dark veil from before the scene of things, [poetry] equally creates for us a being within our being . . . it purges from our inward sight the film of familiarity which obscures from us the wonder of our being" (*Selected Prose Works of Shelley* [London: Watts & Co., 1915], 87, 105, 113). In "Do Not Lift the Painted Veil," life is a painted veil suspended between twin destinies—Fear (of death) and Hope (in an afterlife)—"ever weave/ Their shadows, o'er the chasm, sightless and drear." The poet who dared to lift the veil did not find anything, but a dark void.

48. Shelley, "Mont Blanc."

49. For example, in Coleridge's "Song of the Pixies" (1793), a description of a summer vacation and his childhood home in Devon, textile metaphors are used repeatedly throughout the nine stanzas, from the "wildest texture" of the earth to the "light drapery" of the clouds floating in the sky. In the "Sonnet to the Autumnal Moon" (1788) the moon's watery light glimmers "through a fleecy veil." In "Hymn before Sun-rise, in the Vale of Chamouni" (1802), the poet uses the *topos* of the clouds "veiling" the mountain's breast; he then talks about his eyes being "veiled by tears" as he raises his gaze to the mountain and a vapory cloud rises "like a cloud of incense from the earth" in a hymn of cosmic praise reminiscent of Psalm 141, which is found again, and more explicitly, in his sonnet "To Nature" (1820). Richard Holmes calls Coleridge a "master of imaginary or dream topographies." However, these vistas were usually inspired by physical places; "the exterior landscapes of his travels are gradually converted into interior heartlands of his poetry. . . . Beneath the material surface always lies a spiritual mystery waiting to be recognized" (Richard Holmes, "Introduction," in *Samuel Taylor Coleridge: Selected Poetry* [New York: Penguin Classics, 1996], xii–xix).

50. Coleridge, "Over My Cottage (Fragments from a Notebook, 1796–1798)," in Ernest Hartley Coleridge, *The Complete Poetical Works of Samuel Taylor Coleridge*, vol. 2 (Oxford: Clarendon Press, 1975), 997.

51. "Anacreon" (1786), vv. 37–46. A similar motif is found in "Septimi Gades" (1794), vv. 59–66: "The silvery morning vapours glide / And half the landscape veil. / . . . So shall our still lives, half betrayed, / Show charms more touching from their shade, / Though veiled not unseen." What is veiled is not necessarily unknown. In "The Prelude," bk. 8 (1805–6), the poet describes the shepherd's dog moving freely in and out of the silvery vapors of the mist, once again, making the reader aware that "beneath the mist there is a complete and definite landscape waiting to be explored" (Robert Rehder, *Wordsworth and the Beginnings of Modern Poetry* [London: Croom Helm, 1981], 174).

52. William Wordsworth, "Michael," in *Lyrical Ballads* (1798), vv. 58–60.

53. William Wordsworth, *The Excursion: A Poem* (London: Edward Moxon, 1836), 72.

54. Farinelli, *Geografia*, 44.

55. Alexander von Humboldt, *Cosmos: A Sketch of the Physical Description of the Universe*, trans. Elise Otté, vol. 1. (Baltimore: Johns Hopkins University Press, 1997), 55.

56. Cosgrove, "Modernity;" Tang, *Geographic Imagination*, 59.

57. Humboldt, *Cosmos*, 1:79 (my emphasis).

58. Humboldt, *Cosmos*, 1:33.

59. Humboldt, *Cosmos*, 1:33.

60. See Michael Dettelbach, "Introduction," in Humboldt, *Cosmos*, vol. 2.

61. Humboldt, *Cosmos*, 1:41.

62. Humboldt, *Cosmos*, 2:96. Interestingly, the mantle metaphor is also found in descriptions of other tropical landscapes, such as Nathaniel Ogle's description of Western Australia: "Throughout the wide field of animate and inanimate creation, we see the combination of perfect adaptation with beauty, comprising every hue and form in the mantle of the earth, and in its living myriads that creep, or prowl, or graze upon it" (Nathaniel Ogle, *The Colony of Western Australia: A Manual for Emigrants to that Settlement or Its Dependencies* [London: James Fraser, 1839], 226).

63. Alexander von Humboldt and Aimé Bonpland, *Essay on the Geography of Plants*, ed. Stephen Jackson and Sylvie Romanovski (Chicago: University of Chicago Press, 2013), 76, 115–16.

64. The *Tableau* was meant to illustrate the *Essay on the Geography of Plants*. The word *tableau* is ambivalent, as "it refers most of the times to the picture, but sometimes to the text.

The ambiguity is even greater in French, where *tableau* can mean what is covered in English by the words *board, picture, table,* and *list*." As Sylvie Romanowski observes, "this suggests that the words and the picture are inseparable—the two together form one work, and only combined into one work can they convey Humboldt's knowledge. The illustration is an adjunct to the text, but the text one long caption to the illustration. What we have here is a plate accompanying a text, but also a text accompanying a picture" (Romanowski, "Humboldt's Pictorial Science: An Analysis of the *Tableau Physique des Andes et Pays Voisins*," in Humboldt and Bonpland, *Essay on the Geography of Plants*, 160–61).

65. According to Jorge Cañizares-Esguerra, Humboldt borrowed the idea of geographical plant distribution from local Creole informants. Already in 1801, when he encountered Humboldt, the self-taught Colombian naturalist Francisco José de Caldas was already charting the geographical distribution of plants in the northern Andes, followed by various publications (Jorge Cañizares-Esguerra, *Nature, Empire, and Nation: Explorations of the History of Science in the Iberian World* [Stanford, CA: Stanford University Press, 2006], 112–28).

66. Cotopaxi is a still active volcano and, at 5,872 meters of altitude, is the second-highest peak in Ecuador.

67. Indeed, on this very mountain Humboldt claimed to have found "all the phenomena that the surface of our planet and the surrounding atmosphere present to the observer;" cited in Michael Dettelbach, "Global Physics and Aesthetic Empire: Humboldt's Physical Portrait of the Tropics," in *Visions of Empire: Voyages, Botany and Representations of Nature*, ed. Daniel Miller and Peter Reill (Cambridge: Cambridge University Press, 1996), 268.

68. Romanowski, "Humboldt's Pictorial Science," 162.

69. Humboldt and Bonpland, *Essay on the Geography of Plants*, 17.

70. He thus criticized Linnaean botanists for exhausting their energies with "details," ignoring "the big picture," occupied as they were with the discovery of new species and their classification, and claimed that he himself "would much rather know the exact elevational limits of an already known species than discover fifteen new ones" (Malcolm Nicolson, "Humboldtian Plant Geography after Humboldt: The Link to Ecology," *British Journal for the History of Science* 29 [1996]: 297).

71. Humboldt and Bonpland, *Essay on the Geography of Plants*, 22.

72. Humboldt, *Cosmos*, 1:41.

73. Hadot, *Veil of Isis*, x.

74. Hadot, *Veil of Isis*, 213. "The earnest investigator delights in the simplicity of numerical relations, indicating the dimensions of the celestial regions, the magnitudes and periodical disturbances of the heavenly bodies, the triple elements of terrestrial magnetism, the mean of pressure of the atmosphere, and the quantity of heat which the sun imparts in each year, and in every season of the year, to all points of the solid and liquid surface of our planet. These sources of enjoyment do not, however, satisfy the poet of Nature, or the mind of the inquiring many. To both of these the present state of science appears as a blank, now that she answers doubtingly, or wholly rejects as unanswerable, questions to which former ages deemed they could furnish satisfactory replies. In her severer aspect, and clothed with less luxuriance, she shows herself deprived of that seductive charm with which a dogmatizing and symbolizing physical philosophy knew how to deceive the understanding and give the rein to imagination" (Humboldt, *Cosmos*, 1:81).

75. See Franco Farinelli, *I Segni del mondo* (Florence: Scandicci, 1992). Humboldt was a promoter of panoramas and other devices for the education of the general public (see *Cosmos*, 2:98).

76. Humboldt, *Cosmos*, 1:40.

77. Humboldt, *Cosmos*, 1:26.

78. Humboldt, *Cosmos*, 1:37. Direct sensory contact with nature was for Humboldt essential to accurate scientific analysis. As he wrote to Goethe, "those who only observe and reach abstractions can spend a lifetime classifying plants and animals in the hot tropics and believe that they can describe nature, but they will never get close to it." In a letter to Caroline von Wolzogen (Berlin, May 14, 1805) he wrote: "In the Amazon forest, as on the peaks of the Andes, I had the feeling that the same life infiltrates stones, plants and animals, as well as the swelling breast of humankind, as if animated by a single spirit from pole to pole. Everywhere I felt strongly how powerfully those relationships forged at Jena influence me now, and—thanks to Goethe's perspectives on Nature—I have acquired virtually new organs of perception;" cited in Anne Buttimer, "Alexander von Humboldt and Planet Earth's Green Mantle," *Cybergeo: European Journal of Geography* (2012): 16, 27.

79. Paul Smethurst, *Travel Writing and the Natural World, 1768–1840* (New York: Palgrave Macmillan, 2013), 102.

80. Humboldt and Bonpland, *Essay on the Geography of Plants*, 180.

81. Weitsch's tropical landscapes include those in Humboldt's portrait and in a painting of Humboldt and Aimé Bonpland at the foot of the Chimborazo volcano (1810). In both cases, these are imaginary re-creations of landscape based on Humboldt's accounts. By contrast, Church's numerous views rested on actual observations conducted by the artist himself.

82. Edmund Bunkse, "Humboldt and Aesthetic Tradition in Geography," *Geographical Review* 71 (1981): 14.

83. *The Heart of the Andes* was first exhibited in 1859 in New York. The event attracted more than twelve thousand people. The painting was displayed in a casement window–like frame and measured about 13 feet high and 14 feet wide. Drawn curtains were fitted, creating the sense of a view out a window. A skylight directed at the canvas heightened the perception that the painting was illuminated from within, as did the dark fabrics draped on the studio walls to absorb light (Kevin Avery, "The Heart of the Andes Exhibited: Frederic E. Church's Window on the Equatorial World," *American Art Journal* 18 [1986]: 52–72).

84. Humboldt, *Cosmos*, 2:100.

85. Edward Casey, *Representing Place: Landscape Painting and Maps* (Minneapolis: University of Minnesota Press, 2002), 268.

86. "Landscape painting, and fresh and vivid descriptions of nature alike," Humboldt argues, "conduce to heighten the charm emanating from a study of the external world, which is shown to us in all its diversity of form by both, while both are alike capable, in a greater or lesser degree, according to the success of the attempt, to combine the visible and the invisible in our contemplation of nature" (Humboldt, *Cosmos*, 2:82).

87. Humboldt, *Cosmos*, 2:25.

88. Humboldt, *Cosmos*, 2:22.

89. Humboldt, *Cosmos*, 2:31.

90. Humboldt, *Cosmos*, 2:85.

91. Humboldt, *Cosmos*, 2:90.

92. Humboldt, *Cosmos*, 2:80.

93. Humboldt, *Cosmos*, 2:81–82. On Humboldt and humans' continuing urge to see beyond the horizon, see Geoffrey Martin, *All Possible Worlds: A History of Geographical Ideas*, 14th ed. (New York: Oxford University Press, 2005), 107–21.

94. Thomas, *Man and the Natural World*, 89.

95. Textile metaphors also feature posthumously, in a statue of the Prussian naturalist executed by Reinhold Begas and "unveiled" at the Humboldt University in Berlin in 1883. In his right hand Humboldt holds an exotic vegetal specimen, while with his other hand he pulls aside a heavy drape, partly revealing a globe. The monument effectively captures the scalar tension between detail and cosmos at the heart of Humboldt's vision.

96. Humboldt, *Cosmos*, 2:82.

97. Marshall Berman, *All That Is Solid Melts into Air: The Experience of Modernity* (New York: Penguin Books, 1988).

98. della Dora, *Mountain*, 174–77.

99. Tang, *Geographic Imagination*, 198; Farinelli, *Geografia*, 44.

Chapter Seven

1. Robert Dickinson, *The Makers of Modern Geography* (New York: Praeger, 1969), 78.

2. "Of late, it has been a commonplace to speak of geographical exploration as nearly over," Sir Halford Mackinder wrote. "In Europe, North America, South America, Africa, and Australasia there is scarcely a region left for the pegging out of a claim of ownership" (Halford Mackinder, "The Geographical Pivot of History" [1904], *Geographical Journal* 170 [2004]: 298).

3. "From the present time forth, in the post-Columbian age, we shall again have to deal with a closed political system, and none the less that it will be one of world-wide scope" (Mackinder, "Geographical Pivot," 299). Indeed, the world map Mackinder uses to illustrate his argument bears a somewhat uncanny resemblance to the closed order of medieval *mappae mundi* (the map is discussed in Cosgrove, *Apollo's Eye*, 222).

4. With the exception, perhaps, of Humboldt, who was criticized by contemporary commentators for his lack of direct references to God in his work (see Dettelbach, "Introduction," in Humboldt, *Cosmos*, 1:xxiii–xxix).

5. Kapp cited in Tang, *Geographical Imagination*, 225. On Ritter, see Dickinson, *Makers of Modern Geography*, 34–50.

6. Cited in Tang, *Geographical Imagination*, 225. As Geoffrey Martin comments, "Ritter saw in all his geographical studies the evidence of God's plan. A Supreme Being, an all-wise Creator, was identified as the author of a plan for building the earth as the home for man, and all through Ritter's writings and lectures are words of praise for the divine creation. Even the arrangement of the continents Ritter saw as evidence of God's purpose" (Martin, *All Possible Worlds*, 123). Asia, said Ritter, was "the land of morning and the source of civilization"; Europe was "the land of the evening and seat of advanced civilization"; Africa was a nondescript and undistinguished midday; and the polar regions were a dark night. Looming on Europe's horizon was the New World, whose discovery created a bright "new Orient" (Glacken, *Traces*, 277). Like Humboldt, Ritter believed the earth to be an organism: "as the body is made for the soul, so is the physical globe made for mankind" (cited in Dickinson, *Makers of Modern Geography*, 37).

7. Mackinder, "Geographical Pivot," 298.

8. Clements Markham, "The Field of Geography," *Geographical Journal* 11 (1898): 1.

9. Carl Troll, "Die geographische Landschaft und ihre Erforschung," *Studium Generale* 2 (1950): 163–81; cited in Tang, *Geographical Imagination*, 84.

10. Farinelli, *I segni del mondo*.

11. Higman, *Flatness*, 8.

12. Halford Mackinder, "The Human Habitat," *Scottish Geographical Magazine* 47 (1931): 335.

13. Cited in Elena dell'Agnese, "Halford John Mackinder (1861–1947)," in *Cos'è il mondo? È un globo di cartone. Insegnare geografia fra Otto e Novecento*, ed. Marcella Schmidt (Milan: Unicopli, 2010), 260–61.

14. Markham cites another evocative metaphor: "Sir Richard Strachey has poetically compared geographical knowledge to a setting in which are gloriously held together the bright gems of science, to form an intellectual diadem for man" ("The Field of Geography," 1).

15. Halford Mackinder, "On the Scope and Methods of Geography," *Proceedings of the Royal Geographical Society and Monthly Record of Geography* 9 (1887): 143.

16. Francis Younghusband, "Address at the Anniversary Meeting, 31 May 1920," *Geographical Journal* 56 (1920): 3.

17. Younghusband, "Address," 3. See also Christina Kennedy, James Sell, and Ervin Zube, "Landscape Aesthetics and Geography," *Environmental Review* 12 (1988): 37.

18. Cornish, "On Kumatology."

19. Vaughan Cornish, "Harmonies of Scenery: An Outline of Aesthetic Geography," *Geography* 14 (1928): 277.

20. Mackinder, "Scope and Methods," 149. As M. J. Wise comments, "Mackinder was still not quite 26 years of age. His lecture 'On the Scope and Methods of Geography' advocated the unity of the subject and the study of the reciprocity between human action and the physical environment. It included the vivid representation of the 'wrinkled tablecloth of south-east England' and the patterns of human life which had grown up upon it. J. N. L. Baker has remarked that the basic ideas were not original: most, if not all, had been employed by William Hughes, of King's College London, before his death in 1876. 'But what mattered in 1887 was the perfection of timing coupled with the brilliance of exposition. Mackinder, with his courage, power of presentation and force of personality, produced a classic document in the history of the development of British geography'" (M. J. Wise, "The Scott Keltie Report 1885 and the Teaching of Geography in Great Britain," *Geographical Journal* 152 [1986]: 375).

21. Mackinder, "Scope and Methods," 156.

22. Thus, Mackinder rejects the early paradigm according to which phenomena were determined by catastrophic events imprinting the landscape. In a later paper he explicitly expanded geography's domain to the hydrosphere. This encompassed not only the seas and oceans clothing the largest part of the terrestrial surface, but also the cycle of rain and erosion that constantly molded the rock and the land. The German geographer Albrecht Penck (1888–1923) talked about the "wearing back" of slopes to indicate the opposite process (Higman, *Flatness*, 78). Thomas Pickles later used the tablecloth metaphor to describe the formation of fold mountains. He invited the reader to imagine "placing a few books on a tablecloth and pushing them towards each other so that the cloth wrinkles around and between them. The books represent stable blocks of the earth's crust; the cloth represents layers of rock which may be many miles thick" (Thomas Pickles, *Physical Geography* [London: Dent, 1962], 33). Johannes Frederick Umbgrove, on the other hand, writes: "A napkin or a tablecloth is pushed into folds over the smooth surface of the table on which it was spread. For most types of mountain-chains (as for example that of the Alps) this apparently instructive experiment is not to the point. But it is often held to be relevant with regard to the Jura Mountains. However, when examining the available evidence it will appear that Nature's way of rumpling the tablecloth of the Juras was essentially different from our usual attempt at imitation" (Johannes Frederik Umbgrove, *Symphony of the Earth* [Dordrecht: Springer, 1950], 55).

23. Mackinder, "Scope and Methods," 154.

24. Leo Waibel, "Place Names as an Aid in the Reconstruction of the Original Vegetation of Cuba," *Geographical Review* 3 (1943): 379.

25. Halford Mackinder, "The Round World and the Winning of the Peace," *Foreign Affairs* 21 (1943): 604–5. The passage is commented upon by Stephen Jones, "Views of the Political World," *Geographical Review* 45 (1955): 317–18. Later examples of the tablecloth metaphor span geology, as well as spatial modeling. A late use of the tablecloth metaphor is found in the work of Peter Gould, who likened spatial diffusion to "spilled wine spreading on a tablecloth" (cited in Richard Symanski, "Good as Gould," *Geographical Review* 90 [2000]: 248). More specifically, he and his team produced a series of maps showing the spread of AIDS in the western United States between 1981 and 1988. The maps were presented sequentially, in order to illustrate the "spatial-geographic logic unfolding over time," as the epidemic extended from urban centers into suburbs and the surrounding countryside (the maps are discussed in Denis Wood, *The Power of Maps* [London: Routledge, 1992], 238n5). Comparing epidemic diffusion to a "wine stain on the tablecloth" implied an outward horizontal movement on a flat surface—the surface of the map. Territory was thus reduced to a smooth Euclidean plane observed from above. The map's relation to the planar surface is embedded in the very etymology of the word (Lat. *mappa*, "cloth"). As a geography textbook reminded students in the 1960s, "just as a cloth may cover, or give coverage to a table, so does a map provide coverage for the earth or any of its parts. By definition, a map is a representation of all or a portion of the earth, drawn to scale and *usually on a plane or flat surface*" (Henry Kendall et al., *Introduction to Geography* [New York: Harcourt, Brace & World, 1962], 40, my emphasis).

26. Friedrich Ratzel, *History of Mankind* (London: Macmillan, 1896), 29.

27. Ellen Churchill Semple, *Influences of Geographic Environment* (New York: Holt & Co., 1911), 1.

28. Semple, *Influences*, 507–8.

29. Paul Vidal de la Blache, *Principles of Human Geography* (London: Christopher, 1926), 9–10.

30. Vidal de la Blache, *Principles of Human Geography*, 10.

31. Livingstone, *Geographical Tradition*, 267.

32. Paul Vidal de la Blache, *The Personality of France (Tableau de la géographie de la France)*, trans. H. C. Brentnall (London: Alfred A. Knopf, 1928), 14. Similar metaphors are found in Semple: "[Ancient colonists] put their stamp on the [Mediterranean] basin" (Ellen Churchill Semple, *The Geography of the Mediterranean Region: Its Relation to Ancient History* [London: Constable, 1932], 66). Otto Schülter's conception of *Kulturlandschaft* likewise placed emphasis on humanity's cultural "imprint" on the land: the material expression of the immaterial throughout the ages (Livingstone, *Geographical Tradition*, 264).

33. Vidal de la Blache, *The Personality of France*, 62, 49.

34. Carl Sauer, *Land and Life: A Selection from the Writings of Carl Sauer* (Berkeley: University of California Press, 1963), 393. The word "surface" is repeatedly used throughout *Land and Life*.

35. Sauer, *Land and Life*, 392.

36. Mapping the distribution of traditional housing types and barns, for example, allowed the geographer to define "cultural landscapes," or "cultural areas." Landscape, Sauer argued, "is the field of geography because it is a naively given, important section of reality, not a sophisticated thesis" (Carl Sauer, "The Morphology of Landscape," in *Human Geography: An Essential Anthology*, ed. John Agnew, David Livingstone, and Alasdair Rodgers [Oxford: Blackwell, 1996], 298). His understanding was akin to the German *Landschaft*, which he imported to North America, reacting against the claims of an increasingly unsophisticated environmental determinism.

37. "Whatever opinion one may hold about natural law, or nomothetic, genetic, or causal relation," Carl Sauer wrote, "a definition of landscape as singular, unorganized, or unrelated has no scientific value" (Sauer, "Morphology of Landscape," 301).

38. Carl Sauer, "The Education of a Geographer," Presidential Address Given by the Honorary President of the Association of American Geographers at its 52nd Annual Meeting, Montreal, Canada, 4 April 1956, http://www.colorado.edu/geography/giw/sauer-co/1941_fhg/1941_fhg _body.html (accessed 1 January 2018). Fieldwork was to be complemented by archival research; indeed, argued Sauer, "the reconstruction of past culture areas is a slow task of detective work, as to collecting of evidence and weaving it together . . . the historical geographer must therefore be a regional specialist, for he must not only know the region as it appears today; he must know its lineaments so well that he can find in it traces of the past" (Carl Sauer, "Foreword to Historical Geography," *Annals of the Association of American Geographers* 31 [1941]: 1–24).

39. Sauer, *Land and Life*, 403–4. In his seminal paper, "The Morphology of Landscape," Sauer pointed out that "the aesthetic qualities of landscape had never been disregarded in the best geography." However, he also suggested that the only known approach to aesthetics was subjective and that such concepts of landscape as Humboldt's "physiognomy" and Banse's "soul" lay beyond science (Kennedy, Sell, and Zube, "Landscape Aesthetics," 36).

40. Cited in Denis Cosgrove, *Geography and Vision* (London: I. B. Tauris, 2008), 121.

41. Mackinder, "The Human Habitat," 329.

42. Richard Hartshorne, *The Nature of Geography: A Critical Survey of Current Thought in the Light of the Past* (Lancaster, PA: Association of American Geographers, 1961), 164. At a philosophical level, the American geographer argues, landscape is a surface for "it includes only that which we can see or feel from the outside" (169). Yet, if the study of the landscape is restricted only to the "designs" of the carpet, Hartshorne observes, "the point of view is [solely] aesthetic. If on the other hand we are interested in the landscape as a manifestation of something else—the complex of related factors in the area—then we are merely using it as a means of studying a different object, whether defined as that total complex, or as the area itself." Landscape, Hartshorne concludes, is therefore "only the outward manifestation of things that are fundamental—the interrelated factors of area" (169). Hartshorne criticizes Sauer's use of the concept of landscape as an unnecessary synonym for "area."

43. Higman, *Flatness*, 107.

44. Wassily Kandinsky, *Point and Line to Plane*, trans. Howard Dearstyne and Hilla Rebay (Bloomfield Hills, MI: Cranbrook Press, 1947), 53.

45. Cited in Higman, *Flatness*, 132.

46. Higman, *Flatness*, 133; Walter Christaller, *Central Places in Southern Germany*, trans. Carlisle Baskin (Englewood Cliffs, NJ: Prentice-Hall, 1966).

47. As well as consumers with the same income level and same shopping behavior (that is, they would visit the nearest central place to meet their demands).

48. Higman, *Flatness*, 133; see also Mechtild Rössler, "Applied Geography and Area Research in Nazi Society: Central Place Theory and Planning, 1933–1945," *Environment and Planning D: Society and Space* 7 (1989): 419–31.

49. Cited in Higman, *Flatness*, 131.

50. Aerial photography had been used in urban planning and documentation ever since the first balloons were launched into the atmosphere. In particular, aerial perspective and modern urbanism became closely linked in those countries that were being colonized by Europeans during the nineteenth century. See Denis Cosgrove and William Fox, *Photography and Flight* (London: Reaktion Books, 2010).

51. Cosgrove and Fox, *Photography and Flight*, 58. After the conflict, the airman's view endured as a trope for imagining that the clarity of the pilot's perspective captured in air photography allowed for a better world to be created from the disaster of war.

52. Cited in Roland Courtot, "Les Paysages et les hommes des Alpes du sud dans les carnets de Paul Vidal de La Blache," *Méditerranée* 109 (2007): 13.

53. Beaver's research focused on the mining and manufacturing industries, and on the geography of transports. It belonged to the new field of applied geography. The overarching theme of his work was the spatial impact of technical change, which he illuminated through his grasp of geology and historical events and processes. Between 1941 and 1943 he worked at the University of Cambridge. His contributions, especially to railroad issues in western and southwestern Europe, proved of great value to the geographic department of the British Naval Intelligence Department. His work on sand and gravel resources was important for construction projects during the war (for example, for airfields), as well as for reconstruction work and new urban development in its aftermath. See Hugh Clout, "Stanley Henry Beaver (1907–84)," in *Geographers: Biobibliographical Studies*, ed. Elizabeth Baigent and André Reyes Novaes, vol. 36 (London: Bloomsbury, 2017), 150; "Stanley Henry Beaver, 1907–1984," *Transactions of the Institute of British Geographers* 10 (1985): 504–6; Michael John Wise and Elizabeth Baigent, "Beaver, Stanley Henry (1907–1984)," in *Oxford Dictionary of National Biography* (Oxford: Oxford University Press, 2011).

54. Stanley Beaver, "Geography from a Railway Train," *Geography* 21 (1936): 265.

55. As Beaver explains, a large vertical exaggeration was necessary in drawing the gradient profiles, because the slopes involved were so small.

56. Clout, "Beaver," 150.

57. Derwent Whittlesey, "The Horizon of Geography," *Annals of the Association of American Geographers* 35 (1945): 1.

58. Whittlesey, "The Horizon of Geography," 8.

59. Whittlesey, "The Horizon of Geography," 16.

60. Higman, *Flatness*, 134.

61. See, for example, Richard Edes Harrison, *Look at the World: The Fortune Atlas for World Strategy* (New York: Alfred A. Knopf, 1944).

62. The fireside chat was broadcasted on February 23, 1942. Its impact on popular World War II mapping is discussed in Susan Schulten, *The Geographical Imagination in America, 1880–1950* (Chicago: University of Chicago Press, 2001).

63. Susan Schulten, "Richard Edes Harrison and the Challenge to American Cartography," *Imago Mundi* 50 (1998): 180.

64. Ellsworth Huntington, "Geography and Aviation," in *Geography in the Twentieth Century*, ed. Griffith Taylor (New York: Methuen, 1951), 528–40.

65. Cosgrove and Fox, *Photography and Flight*, 36.

66. Cosgrove and Fox, *Photography and Flight*, 41.

67. Barbara Bender, "Time and Landscape," *Current Anthropology* 43 (2002): 103.

68. The cross-sectional method is best exemplified in the landscape reconstructions in Darby's edited collection *An Historical Geography of England before A.D. 1800* (Cambridge: Cambridge University Press, 1936); and, more notably, in his seven-volume work *Domesday Geography of England* (Cambridge: Cambridge University Press, 1952–67), a regional reconstruction of the human geography of medieval England based on the Domesday Book of 1086.

69. H. C. Darby, "The Changing English Landscape," *Geographical Journal* 117 (1951): 378.

70. H. C. Darby, "On the Relations of Geography and History," *Transactions and Papers (Institute of British Geographers)* 19 (1953): 9.

71. Cited in Alan Baker, *Geography and History: Bridging the Divide* (Cambridge: Cambridge University Press, 2003), 114.

72. W. G. Hoskins, *The Making of the English Landscape* (Dorset: Little Toller, 2013), 14.

73. Hoskins, *English Landscape*, 13–14.

74. Hoskins, *English Landscape*, 21.

75. Hoskins, *English Landscape*, 9.

76. Hoskins, *English Landscape*, 39.

77. Hoskins, *English Landscape*, 25.

78. Hoskins, *English Landscape*, 27.

79. Hoskins, *English Landscape*, 19.

80. Hoskins, *English Landscape*, 271–72.

81. David Matless, *Landscape and Englishness* (London: Reaktion Books, 2016), 371–72.

82. On the other end of the spectrum is architect E. A. Gutkind's *Our World from the Air* (1952). The book contained four hundred air photographs organized to demonstrate the need for a rational planning of human impacts on and transformations of the earth's surface. Altogether, the images present an aestheticized vision of postwar planning. Another example of celebration of modernist planning is found in William Garnett's photographs documenting the instant city that was being erected near Long Beach, California (1950). Reflecting Garnett's modernist geometrical aesthetic, the views were taken from a low oblique angle sharply raking light with long shadows and cutting out any horizon. In a sense, these images are a celebration of flatness and horizontality (see Cosgrove and Fox, *Photography and Flight*, 60).

83. Hoskins, *English Landscape*, 270.

84. Darby, "English Landscape," 392–93.

85. See, for example, David Matless, "One Man's England: W. G. Hoskins and the English Culture of Landscape," *Rural History* 4 (1993): 187–207; John Wylie, *Landscape* (London: Routledge, 2007).

86. Hoskins, *English Landscape*, 210–11.

87. James Gibson, *The Perception of the Visual World* (Boston: Houghton Mifflin, 1950).

88. James Gibson, *The Ecological Approach to Visual Perception* (New York: Psychology Press, 2015), 22.

89. Tim Ingold, "Earth, Sky, Wind, and Weather," *Journal of the Anthropological Institute* 13 (2007): S26.

Chapter Eight

1. *New York Times*, 5 October 1957.

2. Fraser MacDonald, "Perpendicular Sublime: Regarding Rocketry and the Cold War," in *Observant States*, ed. Fraser MacDonald, Rachel Hughes, and Klaus Dodds (London: I. B. Tauris, 2010), 267.

3. Pioneered by geologist Eduard Suess in 1875, the concept of "biosphere" was originally defined as the place on the earth's surface where life dwells. Life, Suess argued, was at the interface between the atmosphere and the lithosphere (see Oldroyd, *Thinking about the Earth*, 287).

4. Whittlesey, "Horizon of Geography," 17. "What is it that really gives a unique interest to the surface of this earth? Surely not its dead features; there are mountains also on the moon,

ruins from a live past. Is it not the fluid envelopes, the water and the air, which by their circulations, their physical and chemical reactions, and their relation to life, impart to the earth's surface an activity almost akin to life itself? Which is the fundamental—the living, palpitating being, or the dead skeleton which it shapes and leaves behind as a monument? . . . It seems to me today that it is in the water rather than in the rocks that we must look for our salvation" (Mackinder, "The Human Habitat," 323).

5. Carl Troll, "Die geographische Landschaft und ihre Erforschung," *Studium Generale* 2 (1950): 163–81; cited in Tang, *Geographical Imagination*, 84.

6. Gibson, *The Perception of the Visual World*, 59.

7. Gibson, *The Perception of the Visual World*, 60.

8. Fraser MacDonald, "High Empire: Rocketry and the Popular Geopolitics of Space Exploration, 1944–62," in *New Spaces of Exploration: Geographies of Discovery in the Twentieth Century*, ed. Simon Naylor and James Ryan (London: I. B. Tauris, 2010), 196.

9. Whittlesey, "Horizon of Geography."

10. In the speech Churchill termed Russia's occupation line "the iron curtain which at present divides Europe in twain" (cited in John Phillips, "The Iron Curtain: Behind It Russia Controls Destiny of All Nations of Eastern Europe," *Life*, 29 April 1946, 27). See also Ignace Feuerlicht, "A New Look at the Iron Curtain," *American Speech* 30 (1955): 186–89; Patrick Wright, *Iron Curtain: From Stage to Cold War* (Oxford: Oxford University Press, 2007).

11. Franklin Rogers, "Iron Curtain Again," *American Speech* 27 (1952): 141.

12. Wright, *From Stage to Cold War*, 16.

13. Audra Wolfe, *Competing with the Soviets: Science, Technology, and the State in Cold War America* (Baltimore: Johns Hopkins University Press, 2013), 6.

14. Victor Zorza, "Minuteman, Polaris, Bombers, Carriers, Submarines, Mobile Troops Comprise the Flexible U.S. Strategy Which Caused Khrushchev's Losing Fight with His Marshals," *Life*, 6 November 1964, 76.

15. Matthew Godwin, "Britnik: How America Made and Destroyed Britain's First Satellite," in *New Spaces of Exploration: Geographies of Discovery in the Twentieth Century*, ed. Simon Naylor and James Ryan (London: I. B. Tauris, 2009), 173.

16. Hugh Dryden, "The International Geophysical Year: Man's Most Ambitious Study of His Environment," *National Geographic*, February 1956, 285; Walter Sullivan, *Assault on the Unknown: The International Geophysical Year* (New York: Hodder & Stoughton, 1962), 4.

17. Christy Collis and Klaus Dodds, "Assault on the Unknown: The Historical and Political Geographies of the International Geophysical Year (1957–8)," *Journal of Historical Geography* 34 (2008): 555.

18. Sullivan, *Assault on the Unknown*, 4.

19. Dryden, "The International Geophysical Year."

20. Dryden, "The International Geophysical Year," 285. *Life* magazine also presented the IGY as "the single most significant peacetime activity of mankind since the Renaissance" ("Portrait of Our Planet: Great Discoveries of the IGY Are Revealed," *Life*, 7 November 1960, 79). And recuperating neo-Stoic motifs, Sullivan went as far as commenting that "many saw a great hope for mankind setting a pattern of cooperation that could lead to permanent peace. . . . By gazing more intently beyond his narrow environment, man might be persuaded that his conflicts are petty" (Sullivan, *Assault on the Unknown*, 48).

21. MacDonald, "High Empire," 220.

22. "The World Studies the World," *Life*, 15 July 1957, 19.

23. Dryden, "The International Geophysical Year," 298. The global scale is often conveyed through "girdling" metaphors: from the geodetic survey ("a girdle for the earth") to observation outposts "girdling" the planet and representations of the globe "girdled" by the flags of different nations (Robert Conly, "Men Who Measure the Earth: Surveyors from 18 New World Nations Invade Trackless Jungles and Climb Snow Peaks to Map Latin America," *National Geographic*, March 1956, 335–62; "Portrait of Our Planet," 79).

24. For example, the presence of two American astronomers, as well as top astronomers from other countries, at the dedication of a Russian observatory in 1954 was believed to indicate a break in the Iron Curtain: "[Observers] interpret this break in the scientific Iron Curtain as a further indication that Russian policy has changed, that the curtain is gradually softening, that Soviet leaders now believe the two major powers can coexist on one planet. . . . During their two-week stay behind the Iron Curtain, Drs. Brouwer and Nassau will be guests of Dr A. N. Nesmeyanov, president of the Academy of Sciences of the USSR. . . . Another indication of the break in the Iron Curtain, at least scientifically, is the reported invitation to British physiologists to visit a similar sort of meeting this spring in the USSR" (*Science News Letter*, 29 May 1954, 351). International scientific conferences were also envisaged as fulfilling a similar role. For example, see Simone Turchetti and Peder Roberts, "Introduction: Knowing the Enemy, Knowing the Earth," in *The Surveillance Imperative: Geosciences during the Cold War and Beyond*, ed. Simone Turchetti and Peder Roberts (New York: Palgrave Macmillan, 2014), 4. Appropriately, the IGY emblem featured a world divided in light and shadow (see Collis and Dodds, "Assault on the Unknown," 556).

25. IGY data were not always fully shared, and the most notable and controversial example is Sputnik itself. As a result, Sullivan commented, "the feeling of wonder and excitement at man's escape from the earth was largely lost in fear" (Sullivan, *Assault on the Unknown*, 3). Scientific communities were also not always attuned to each other. For example, Soviet scientists were excluded from Western geological debates and worked with their own tectonic plate theory (Jacob Hamblin, "Science in Isolation: American Marine Geophysics Research, 1950–1968," *Physics in Perspective* 2 [2000]: 301). "Many scientists accepted the civilian-military nexus and even opted for a new epic Cold War competition (because of increased funding opportunities via institutions such as the Office of Naval Research); some even saw it as a historic opportunity . . . for the first time to triangulate the whole earth" (Collis and Dodds, "Assault on the Unknown," 558). On government sponsorship and the significance of scientific activity during the Cold War, see Naomi Oreskes, "Science in the Origins of the Cold War," in *Science and Technology in the Global Cold War*, ed. Naomi Oreskes and John Krige (Cambridge, MA: MIT Press, 2016), 11–29. On Big Science during the Cold War see, for example, Wolfe, *Competing with the Soviets*, 41–54; and Asif Siddiqi, "Fighting Each Other: The N-1, Soviet Big Science, and Cold War at Home," in *Science and Technology in the Global Cold War*, ed. Oreskes and Krige, 189–250.

26. Cited in Oreskes, "Science in the Origins of the Cold War," 13.

27. George Reisch, "When Structure Met Sputnik: On the Cold War Origins of the Structure of Scientific Paradigms," in *Science and Technology in the Global Cold War*, ed. Oreskes and Krige, 376.

28. Simon Naylor, Katerina Dean, and Martin Siegert, "The IGY and the Ice Sheet: Surveying Antarctica," *Journal of Historical Geography* 34 (2008): 582.

29. Hugh Robert Mill, "Antarctic Research," *Nature* 57 (1898): 415.

30. "Waiting to Be Won," *Punch*, 5 June 1875, 242–43.

31. "A Cold Reception (Arctic Regions 1875)," *Punch*, 11 November 1876, 203–4.

32. Heidi Hansson, "*Punch, Fun, Judy* and the Polar Hero: Comedy, Gender and the British Arctic Expedition 1875–1876," in *North and South: Essays on Gender, Race and Region*, ed. Christine DeVine and Mary Ann Wilson (Cambridge: Cambridge Scholars Publishing, 2012), 68–70.

33. Hansson, "*Punch*," 63. A similar scene captioned "Waiting to Be Won" appeared in the 5 June 1875 issue of the magazine, only instead of the expedition party it featured two polar bears, as if to stress the region's desolate inaccessibility and lack of human presence.

34. The metaphor is repeatedly used throughout the book with reference to explorers' feats. The following words (from the introduction) refer to his party's reaching the South Pole on December 14, 1911: "Thus the veil was torn aside for all time and one of the greatest of our earth's secrets had ceased to exist. Since I was one of the five who, on that December afternoon, took part in this unveiling, it has fallen to my lot to write—the history of the South Pole" (Roald Amundsen, *The South Pole: An Account of the Norwegian Antarctic Expedition in the Fram, 1910–1912* [New York: Cooper Square Press, 2001], 2).

35. Roland Huntford, *Race for the South Pole: The Expedition Diaries of Scott and Amundsen* (New York: Continuum, 2011), 125.

36. David Boyer, "Year of Discovery Opens in Antarctica," *National Geographic*, September 1957, 339–80. "Tearing the veil of Antarctica" was an expression used by Amundsen.

37. Klaus Dodds, "Assault on the Unknown: Geopolitics, Antarctic Science and the International Geophysical Year (1957–8)," in *New Spaces of Exploration*, ed. Naylor and Ryan, 159.

38. Richard Byrd, "All-Out Assault on Antarctica: Operation Deep Freeze Carves Out United States Bases for a Concerted International Attack on Secrets of the Frozen Continent," *National Geographic*, August 1956, 141–80. See also Paul Frazier, "Across Frozen Desert to Byrd Station: Tracker-Borne Explorers Conquer Yawning Chasms of Ice and Snow to Set Up an IGY Outpost in Uncharted Marie Byrd Land," *National Geographic*, September 1957, 383–98. Elsewhere, IGY activities in Antarctica were described as "man's most ambitious assault on the southern continent"; "the biggest assault ever made on the secrets of the white continent"; a "giant international assault on the cold, silently resisting emptiness of Antarctica"; a conquest of an "invincible" enemy; or an "air and sea invasion" enacted by an army of intrepid scientists in collaboration with US Navy men (David Boyer, "Year of Discovery Opens in Antarctica: Daring Scientists of Dozens of Nations, Pooling Knowledge and Resources Launch Man's Most Ambitious Assault on the Whole Continent," *National Geographic*, September 1957, 339, 342; Richard Byrd, "To the Men at South Pole Station," *National Geographic*, July 1957, 4; Paul Siple, "Man's First Winter at the South Pole: Cold Fiercer than Men Had Ever Faced, Ceaseless Winds, a Sixth-month Night— Yet 18 Pioneers of Science Survived and Thrived," *National Geographic*, February 1958, 439–78).

39. Byrd, "All-out Assault," 160, 173; Byrd cited in Meeville Grosvenor, "Admiral at the Ends of the Earth," *National Geographic*, July 1957, 48.

40. Walter Sullivan, for example, later commented that "the assault on Antarctica" was "the first really thorough world effort to uncover the geophysical secrets of the continent at the bottom of the world" (Sullivan, *Assault on the Unknown*, 30–31).

41. Byrd, "All-out Assault," 147.

42. The treaty, which froze territorial claims, set aside the continent as a scientific preserve and banned military activity from Antarctica. It was opened for signature in 1959 and entered into force in 1961. The original signatories were the twelve countries active in Antarctica during the IGY (Argentina, Australia, Belgium, Chile, France, Japan, New Zealand, Norway, South Africa, the Soviet Union, the United Kingdom, and the United States). The treaty was a diplomatic expression of the operational and scientific cooperation that had been achieved thus far. On

the geopolitics of Antarctic science during the IGY, see Dodds, "Assault on the Unknown;" and Naylor, Dean, and Siegert, "The IGY and the Ice Sheet."

43. Dodds, "Assault on the Unknown," 149.

44. Naylor, Dean, and Siegert, "The IGY and the Ice Sheet," 578. Much of the impetus came from the possibility of finding uranium and other minerals under the ice. Indeed, at some point there was even some discussion of the use of atomic power to "melt" the mantle to allow mineral exploitation (Dodds, "Assault on the Unknown," 161). Details on "plumbing the Antarctic sheet" are provided in Dryden, "The International Geophysical Year," 289–97; and Boyer, "Year of Discovery Opens in Antarctica," 358–81. Geophones are instruments that convert ground movements into voltage, which may be recorded at a recording station; they are used to detect seismic activity.

45. Paul-Emile Victor, "Wringing the Secrets from Greenland Icecap," *National Geographic*, January 1956, 121, 131.

46. Naylor, Dean, and Siegert, "The IGY and the Ice Sheet," 589.

47. As a result of the erratic working of altimeters, between 1955 and 1961, US missions in Antarctica experienced no fewer than nineteen deaths (Simone Turchetti, Katrina Dean, Simon Naylor, and Martin Siegert, "Accidents and Opportunities: A History of the Radio Echo-Sounding of Antarctica, 1958–79," *British Journal for the British History of Science* 41 [2008]: 418–23). The transparency of ice to radio waves is described by Byrd: "During Deepfreeze I, Bud Waite made valuable studies of radio propagation in snow and ice. He confirmed that snow, unlike water, won't short-circuit a copper antenna wire laid across its surface. In the snow, Waite and his men dug two pits 20 feet deep and a mile apart. Even with very low-power equipment, they were able to talk easily by radio through the intervening barrier of hard-packed snow" (Byrd, "All-out Assault," 157).

48. Turchetti et al., "Accidents and Opportunities," 418.

49. Dodds, "Assault on the Unknown," 169; Naylor, Dean, and Siegert, "The IGY and the Ice Sheet," 579.

50. Emil Schulthess, "Fantasia of the Antarctic," *Life*, 21 November 1960, 72.

51. Sullivan, *Assault on the Unknown*, 15–17.

52. Lincoln Barnett, "The Canopy of Air," *Life*, 8 June 1953, 80. Siple used similar metaphors and poetic tropes in his description of the Scott-Amundsen base at the end of the Antarctic winter: "As I looked out at our vapor-shrouded camp, I became increasingly aware of the beauty of the sky. Directly behind the camp, away from the sun, rose a shallow, slate-blue arc of sky darker than the rest. I recognized it as the last remnant of the earth's shadow that had subtly slid back over our heads like a dark canopy during the past month" (Siple, "Man's First Winter," 443).

53. Barnett, "The Canopy of Air," 82. Russian scientists likewise described the effects of auroras as "multicolored curtains and draperies," or vast violet "canopies" (Anatolii Malakhov, *The Mystery of the Earth's Mantle* [Moscow: Mir Publishers, 1970], 145–46).

54. Sullivan, *Assault on the Unknown*, 108, 189–95.

55. Byrd, "To the Men at South Pole Station," 311.

56. MacDonald, "High Empire," 202.

57. Jacques Arnould, "Space Conquest and Ritual Practices: Lighting Candles for Ariane," *Theology and Science* 11 (2013): 45.

58. Arnould, "Space Conquest," 46.

59. MacDonald, "High Empire;" Sullivan, *Assault on the Unknown*, 22.

60. MacDonald, "High Empire," 197.

61. Barnett, "The Canopy of Air," 98.

62. Newman Bumstead, "Rockets Explore the Air above Us: Scientists, Firing Missiles Equipped with Electronic Eyes and Ears, Probe Mysteries on the Borders of Outer Space," *National Geographic*, February 1957, 562–80.

63. Bumstead, "Rockets," 570.

64. Wolfe, *Competing with the Soviets*, 40.

65. Sullivan, *Assault on the Unknown*, 247.

66. Godwin, "Britnik," 178.

67. Sullivan, *Assault on the Unknown*, 23.

68. Malcolm Ross and Lee Lewis, "To 76,000 Feet by Stratolab Balloon," *National Geographic*, February 1957, 272.

69. Joseph Kaplan, "How Man-made Satellites Can Affect Our Lives," *National Geographic*, December 1957, 807. Frazier, "Across Frozen Desert," 383. Like rockets, satellites set their own imaginative patterns and textile metaphors. At the start of the IGY, the globe was imagined to be one day "wrapped" by a web of satellite paths, "like strings on a ball" (Heinz Haber, "Space Satellites, Tools of Earth Research," *National Geographic*, April 1956, 505–7).

70. Ira Bowen, "Sky Survey Charts Universe," *National Geographic*, December 1956, 780. The Palomar telescope was said to have "penetrated space" to an approximate depth of 6 sextillion miles—1 billion light years.

71. Sullivan, *Assault on the Unknown*, 137. Tests took eleven days from start to finish, with the first launch on August 27 and the final launch on September 6. They were performed by the Defense Nuclear Agency, in conjunction with the Explorer 4 space mission.

72. Defense Nuclear Agency, *Operation ARGUS, 1958* (Washington, DC: Department of Defense Documents, 1958), 1–143.

73. During the duration of the operations, the newly launched American satellite Explorer 4 was said to have "woven in and out" of these shells some 250 times to observe the growth and decay of the radiation (Sullivan, *Assault on the Unknown*, 150).

74. Cited in Robert Poole, "What Was Whole about the Whole Earth? Cold War and Scientific Revolution," in *The Surveillance Imperative: Geosciences during the Cold War and Beyond*, ed. Simone Turchetti and Peder Roberts (New York: Palgrave Macmillan, 2014), 216. See also Sullivan, *Assault on the Unknown*, 163.

75. Cited in Godwin, "Britnik," 182. Commenting on the 1962 super-high-altitude tests, Malakhov wrote: "What a mighty force is this in the hands of mankind; nature herself is giving us warning of the great hazard of radioactive radiation. Cosmic particles have been established to be fatal to animals and plants. If the earth had no magnetic field and thick atmosphere, life could hardly exist on the surface of our planet. Radiations are destructive to the human organism. Nature set up a shield against the menacing force of outer space, and it is intolerable that the same hazard should be created here on earth. That is why the peoples of the world so gladly received the news of the conclusion of the Moscow Treaty prohibiting nuclear tests" (Malakhov, *The Mystery*, 150–51).

76. Kaplan, "How Manmade Satellites Can Affect Our Lives," 804.

77. Klaus Dodds and Lisa Funnell, "Going Atmospheric and Elemental: Roger Moore's and Timothy Dalton's James Bond and Cold War Geopolitics," in *Media and the Cold War in the 1980s: Between Star Wars and Glasnost*, ed. Henrik Bastiansen, Martin Klimke, and Rolf Werenskjold (London: Palgrave Macmillan, 2019), 63–86.

78. Poole, "Whole Earth," 216. Malakhov, for example, provides a detailed account of the diversion of a river near the town of Porkovs-Uralsky (Malakhov, *The Mystery*, 81–82). The

explosion was detected by several seismic stations around the world (see *Congressional Record*, Senate, 14 April 1958, 6306–7, https://www.gpo.gov/fdsys/pkg/GPO-CRECB-1958-pt5/pdf/GPO -CRECB-1958-pt5-7.pdf [accessed 23 March 2018]). Descriptions of a wide range of underground nuclear explosions for industrial purposes can be found in Bruce Bolt, *Nuclear Explosions and Earthquakes: The Parted Veil* (San Francisco: W. H. Freeman, 1976). In spite of the "veil of silence" hanging over much of Soviet nuclear research, "peaceful uses of nuclear explosives" were said to have been "aggressively pursued in the Soviet Union" (Bolt, *Nuclear Explosions*, 123, 187). While condemning the military employment of nuclear power, Bolt, like many other scientists of his time, was an enthusiastic supporter of its "peaceful use." He employs the "parted veil" metaphor to describe the need to provide an "objective" assessment of nuclear power that would not prevent "good" applications of it: "My own conclusion is that, at this stage, an extension of the Limited Test Ban Treaty of 1968 and the Threshold Treaty of 1974 to include all underground tests would indeed be beneficial to the forces of disarmament and peace. I would want, however, to exclude the stringently controlled nuclear explosives for peaceful purposes. . . . A CTBT, if negotiated with care, should contribute towards reduction of suspicions, more open dealings between nations, and more realistic assessment of the role of nuclear power in future ages. It should increase the obligation of other members of the nuclear club not to continue testing nuclear weapons and of potential members not to start. Unfortunately, unlike the distinction between natural and unnatural earthquakes, the veil across the future history of civilization has not been parted" (Bolt, *Nuclear Explosions*, 248).

79. Sullivan, *Assault on the Unknown*, 410–16.

80. Malakhov, *The Mystery*, 96. In 1906, Oldham evocatively wrote how "just as the spectroscope opened up a new astronomy by enabling the astronomer to determine some of the constituents of which distant stars are composed, so the seismograph, recording the unfelt motion of distant earthquakes, enables us to see into the earth and determine its nature with as great a certainty, up to a certain point, as if we could drive a tunnel through it and take samples of the matter passed through" (Oldham, "Constitution of the Interior of the Earth," 456).

81. Damon Teagle and Benoit Ildefonse, "Journey to the Mantle of the Earth," *Nature* 471 (2011): 437–39.

82. Bolt, *Nuclear Explosions*, 18.

83. Bolt, *Nuclear Explosions*, 16, 41.

84. Malakhov, *The Mystery*, 83–85. Such was the enthusiasm for the new technique and for nuclear power during the Cold War that, after the Apollo 11 mission successfully landed on the moon in 1969, it was proposed that a small nuclear bomb be detonated on its surface to provide a source of seismic waves (Bolt, *Nuclear Explosions*, 233).

85. Teagle and Ildefonse, "Journey to the Mantle," 437. "Drilling for scientific purposes was much less common. It had begun with efforts to determine the structure, composition, and history of coral islands. In 1877 the Royal Society of London sponsored a borehole that went down 350 m (1,140 ft) on Funafuti in the South Pacific. In 1947, pre-bomb-test drilling of Bikini reached 780 m (2,556 ft). In 1952 drilling on Eniwetak finally reached basaltic crust beneath coralline rock at a depth of over 1,200 m (4,000 ft)" (John Heilbron, "Mohole Project and Mohorovich Discontinuity," in *The Oxford Companion to the History of Modern Science* [Oxford: Oxford University Press, 2003], 541–42). For a fuller account of drilling in the Pacific, see Willard Bascom, *A Hole on the Bottom of the Sea: The Story of the Mohole Project* (Garden City, NY: Doubleday, 1961), 35–40.

86. William Tonking, "Project Mohole: Exploring the Earth's Crust," *Journal of the Royal Society of Arts* 114 (1966): 981; Daniel Greenberg, "Mohole: Geopolitical Fiasco," in *Understanding*

the Earth: A Reader in the Earth Sciences, ed. Ian Gass, Peter Smith, and Richard Wilson (Sussex, UK: Artemis Press, 1971), 343; Teagle and Ildelfonse, "Journey to the Mantle," 438; Harry Hess, "The AMSOC Hole to the Earth's Mantle," *EOS* 40 (1959): 340–45. The brainchild of AMSOC, the American Miscellaneous Society, the project was funded by NSF.

87. Daniel Greenberg, *The Politics of Pure Science* (Chicago: University of Chicago Press, 1999), 174; Greenberg, "Mohole," 345. On the necessity to distinguish between natural and artificial earthquakes, see Bolt, *Nuclear Explosions*.

88. Cited in Samuel Matthews, "Scientists Drill Sea to Pierce Earth's Crust," *National Geographic*, November 1961, 694.

89. Bascom, *A Hole*, 47; Greenberg, *The Politics*, 175.

90. Greenberg, *The Politics*, 177.

91. "On December 6, 1957, the next meeting of the AMSOC Committee was held at Dr William Rubey's house in Washington. One government agency, jittery in the uproar over the recent Russian success in launching the first satellite and sensitive to the remarks at Toronto about their deep-drilling abilities, actually stationed a security guard around the house, presumably to protect whatever advantage the United States might have in a drilling race" (Bascom, *A Hole*, 50).

92. Greenberg, "Mohole," 346.

93. John Steinbeck, "High Drama of Bold Thrust through Ocean Floor," *Life*, 14 April 1961, 111. In a similar vein, Bascom complained that "we have spent millions on observations to study the craters of the sun and moon, but we have only studied a small number of craters on the earth" (Bascom, *A Hole*, 42). The workers in the geological sciences and other fields, he then explained, "had adjusted to satellites—and gladly accepted the boost the Russian Sputnik gave to public interest in all forms of American science. The Mohole project might achieve a similar result" (55). Finally, "touching the edge of the mantle," Bascom claimed, would bring mutual benefits: "anything learned about the Earth's interior is helping those who wish to know about the moon and Mars" (25).

94. Steinbeck, "High Drama," 111, 118.

95. As with the race for space, both American and Soviet commentators mobilized Baconian metaphors. Special emphasis was placed on vision and direct evidence: scientists were to "see for themselves what the Earth's interior is like;" "man looks deep inside the Earth" (Bascom, *A Hole*, 19). The hole was "a geological key" that would enable scientists to unlock many important secrets of our earth (Bascom, *A Hole*, 18). On the other side of the Iron Curtain, Malakhov wrote about cores and stones put "under interrogation," in order to reveal their name and age. Scientific practice and instruments here are described as "instruments of torture" (Malakhov, *The Mystery*, 169–74).

96. Steinbeck, "High Drama," 111.

97. Steinbeck, "High Drama," 118.

98. Bascom, *A Hole*, 66–67.

99. According to Malakhov, the basaltic core from the second layer of the crust was estimated to be up to 212 million years old (Malakhov, *The Mystery*, 164).

100. On the disputes, see Greenberg, "Mohole." Namely, the Deep Sea Drilling Project, begun in 1968 (Heilbron, "Mohole Project").

101. Malakhov, *The Mystery*, 161.

102. Malakhov, *The Mystery*, 225–26.

103. Malakhov, *The Mystery*, 226–27.

104. Malakhov, *The Mystery*, 231–43.

105. Malakhov, *The Mystery*, 243–44. Science fiction accounts had long predated the Mohole. Bascom devotes a whole chapter of his book to the subject, ranging from Jules Verne's *Journey to the Centre of the Earth* (1864) to pseudo-scientific theories formulated in the 1930s. One of the most interesting science fiction novels is Arthur Conan Doyle's *When the World Screamed* (1922). Here the earth is likened to a giant creature; moors and heaths are said to resemble "the hairy side of a giant animal," and earthquakes "fidgetings and scratchings." The protagonist, Professor Challenger, conceives a project akin to Mohole: "I propose to let the earth know that there is at least one person, George Edward Challenger, who calls for attention. Like the mosquito who explores the surface of the human body, we are unaware of its presence until it sinks its proboscis through the skin, which is our crust, then we are reminded that we are not altogether alone. . . . Your artesian drill, a hundred feet in length and as sharp as possible will be my stinging proboscis" (cited in Bascom, *A Hole*, 60).

106. Alicia Ault, "What's the Deepest Hole Ever Dug?" *Smithsonian*, 19 February 2015, https://www.smithsonianmag.com/smithsonian-institution/ask-smithsonian-whats-deepest-hole-ever-dug-180954349/ (accessed 27 March 2018).

107. "Introducing a New Life Series: The New Portrait of Our Planet. Great Discoveries of the IGY Are Revealed," *Life*, 7 November 1960, 79.

Chapter Nine

1. Rachel Carson, *Silent Spring* (New York: Mariner Books, 2002), 2.

2. Peter Dreier, "How Rachel Carson and Michael Harrington Changed the World," *Contexts* 11 (2012): 42. DDT was first employed on a large scale in the Naples typhus epidemic of 1943–44. It continued to be used during the rest of World War II to protect millions of soldiers and civilians against insect-borne diseases. Thanks to the pesticide, World War II is thought to be "the first major war in which more people died from enemy action than from disease" (Cheryll Glotfelty, "Cold War, Silent Spring: The Trope of War in Modern Environmentalism," in *And No Birds Sing: Rhetorical Analyses of Rachel Carson's Silent Spring*, ed. Craig Waddell [Carbondale: Southern Illinois University Press, 2000], 158). The enthusiasm and high level of publicity DDT received in postwar America is therefore not surprising. The role of the pesticide during the war was indeed exploited by pesticide producers as a key argument in the massive campaign they set up to discredit Carson (Mark Hamilton Lytle, *The Gentle Subversive: Rachel Carson, Silent Spring, and the Rise of the Environmental Movement* [New York: Oxford University Press, 2007], 133–95).

3. The phrase is from Lytle's book *The Gentle Subversive*.

4. Alison Steinbach, "Metaphor and Visions of Home in Environmental Writing," https://green.harvard.edu/news/metaphor-and-visions-home-environmental-writing (accessed 20 May 2018).

5. Emory Jerry Jessee, "Radiation Ecologies: Bombs, Bodies, and the Environment during the Atmospheric Nuclear Weapons Testing Period, 1942–1965," PhD diss., Montana State University, 2013.

6. Turchetti and Roberts, "Knowing the Enemy," 8.

7. *King Lear* 3.4.127.

8. "What time the jocund rosie-bosom'd hours / Led forth the train of Phoebus and the Spring, / And Zephyr mild profusely scatter'd flowers, / On Earth's green mantle from his musky wing, / The Morn unbarr'd th' ambrosial gates of light, / Westward the raven-pinion'd Darkness flew, / The Landscape smil'd in vernal beauty bright, / And to their graves the sullen Ghosts withdrew" ("The Tomb of Shakespeare" [1755], in John Gilbert Cooper and Nathaniel Cotton,

The British Poets, vol. 72 [Whittingham, UK: Cooper & Cotton, 1822], 90). Cooper seems to have been inspired by Thomas Gray's "Elegy Written in a Country Churchyard," "an elegiac rhapsody or gothic fantasy in which the dreaming poet is led through a series of tableaux based on Shakespeare's plays, before the rising sun terminates the vision and inspires several stanzas of moral reflection." http://spenserians.cath.vt.edu/TextRecord.php?action=GET&textsid=37833 (accessed 1 May 2018).

9. Paolo Aresi, *Della tribolatione e suoi rimedi: Lettioni di monsignor Paolo Aresi*, vol. 2 (Tortona: Niccolò Viola, 1624), 757; Ferdinando Zucconi, *Lezioni sacre sopra la divina Scrittura*, vol. 1 (Venice: Stamperia Baglioni, 1741), 101; Domenicantonio Capalbo, *Panegirici e sermoni sacri*, vol. 1 (Naples: Andrea Migliaccio, 1775), 96. "Pigliamo à mostrar questa bellezza il chiaro testimonio della primavera, la sensibil pruova dei Giardini; e certi segni de colli e monti, gli aperti inditii delle selve, de gli antri e spechi, che certo non fora dubbio mai più agli mortali di simirabil beltà stassi gloriosa la bella primavera per lo verde manto della terra" (Tommaso Buoni, *Discorsi accademici de' mondi: Parte prima, nella quale con stile oratorio si parla dell'Archetipo de Mondi: del Mondo Angelico; del Mondo inferiore reato; della nobiltà, eccellenza, bellezza, meraviglie, forze e differenze de Mondi; del Cielo; della Luce: de gli Elementi: de Misti; delle piante; e de gli animali* [Venice: Battista Colosini, 1605], 108). "Apri gli occhi, o sconsigliata, e mira nel Teatro di questo Mondo, com'è bella la serenità del Cielo; com'è vago l'aureo lampo del Sole; com'è tranquilla l'azzura calma del mare; com'è dilettevole il verde manto della Terra" (Luca Assarino, *Raguagli del Regno di Cipro* [Venice: Turrini, 1654], 34). "Quanto all'astro, che c'illumina, ci arde ancora e ci consuma, quando i più vasti spazi campestri non offrono che nudità e privazione, quando gli ardori tengono in incendio la Terra, che sembra la febbre della natura, questa pianta privilegiata, abbellita del più vago e piacevole colore sorge di mezzo agli altri campi di sterilità, e par che porti il risorgimento ed il ravvivamento della Vegetazione. Tali sono i campi de' Risi che sembrano quasi tante preziose gemme sparse qua e là sul variegato manto della Terra, che se dilettano il semplice spettatore, molto più allettano l'industre colono o il felice Proprietario, che sperano ampiamente di raccoglierli" (Melchiorre Delfico, *Memoria sulla coltivazione del riso nella provincia di Teramo* [Naples: G. M. Porcelli, 1783], 51–52). The earth's green mantle as "the couch where life with joy reposes" is from John Pierpont, "For a Lady's Album," in Samuel Kettell, *Specimens of American Poetry with Critical and Autobiographical Notices* (Boston: Goodrich & Co., 1829), 272.

10. Steinbach, "Metaphor."

11. M. H. Abrams, cited in Steinbach, "Metaphor."

12. Henry David Thoreau, *Walden, and Other Writings* (New York: Modern Library, 2004), 436. Parallels can be drawn with Alexander Pope's belief that "a tree was a nobler object than a prince in his coronation robes," or Ruskin's detailed descriptions of trees "as if statues or horses." As Thomas shows, while forests in England had traditionally been deemed wilderness and clearing was equated with progress and civilization, by the late eighteenth century they became "Romantic," and by the beginning of the following century they started to be cherished as escapes from cities and factories (see Thomas, *Man and the Natural World*, chap. 4).

13. John Muir, *John of the Mountains: The Unpublished Journeys of John Muir* (Temecula, CA: CRSC, 1991), 25.

14. Basil, *Hexaem.* 6.6; Gr. Naz., *Hom.* 2.26; Giorgio di Pisidia, *Esaemerone*, trans. Fabrizio Gonnelli (Pisa: Edizioni ETS, 1998), 189.

15. Muir, *John of the Mountains*, 38. At the same time, as in German-speaking and English Romantic literature, Muir's diaries feature a vast repertoire of "mantles" and "veils" of ice, snow, spray, light, and mist that have the power to transfigure the landscape and bewilder the viewer.

16. Thoreau, *Maine Woods*, cited in Christine Avery and Michael Colebrook, *The Green Mantle of Romanticism* (London: Green Spirit Press, 2008), 74.

17. Avery and Colebrook, *The Green Mantle*, 76.

18. Muir, *John of the Mountains*, 137–38.

19. Avery and Colebrock, *The Green Mantle*, 100.

20. Muir, *John of the Mountains*, 215.

21. Indeed, in 1903 President Theodore Roosevelt visited Muir and declared that he wanted trees preserved "because they are monuments in themselves;" cited in Avery and Colebrook, *The Green Mantle*, 85. See Schama, *Landscape and Memory*, 184–85; Denis Cosgrove, "Images and Imagination in 20th-Century Environmentalism: From the Sierras to the Poles," *Environmental Planning A* 40 (2008): 1862–80.

22. Robert Marshall, "The Problem of the Wilderness," *Scientific Monthly* 30 (1930): 141.

23. Marshall, "The Problem of the Wilderness," 144–45.

24. Sydney Mangham, *Earth's Green Mantle: Plant Science for the General Reader* (London: English Universities Press, 1939), 19.

25. Arthur Hill, "Foreword," in Mangham, *Earth's Green Mantle*, 6.

26. Mangham, *Earth's Green Mantle*, 15.

27. Mangham, *Earth's Green Mantle*, 23.

28. Mangham, *Earth's Green Mantle*, 42, 83.

29. Ecology, argues Mangham, is a branch of botany "which may be expected to make very important contributions in connection with the exploitation of plants of economic value, since it aims at finding as much as possible of the relations of plants to the conditions of growth in their natural environments" (Mangham, *Earth's Green Mantle*, 29).

30. Mangham, *Earth's Green Mantle*, 49. Mangham states: "Left to themselves plants tend to compete with one another and to form in time relatively stable communities of the types best fitted for survival in the different kinds of situations presented by the earth's changing surface. . . . Usually in the absence of human interference, a mantle of vegetation is put on and is worn for varying periods before being exchanged for one of a different pattern better suited to some local alteration which has occurred in the conditions of life" (279).

31. Mangham, *Earth's Green Mantle*, 52.

32. Heuristic usages of the "green mantle" metaphor are found, for example, in Richard Pohl, *How to Know the Grasses* (Dubuque, IA: Wm. C. Brown, 1954); Lincoln Barnett, "The Rain Forest," *Life* 37 (1954): 76–102; Garret Hardin, *Biology: Its Principles and Implications* (San Francisco: Freeman, 1961), 259; and many other botany and biology texts.

33. See, for example, George Jobberns, "The Maintenance of New Zealand's Green Mantle," *New Zealand Geographer* 8 (1952): 72–73; Robert Mackechnie, "Scotland's Green Mantle," *Nature* 205 (1965): 429–31; Richard Carrington, "The Green Mantle," in *The Mediterranean: Cradle of Western Culture* (New York: Viking Press, 1977); Mary Gillham, *Swansea Bay's Green Mantle* (Cowbridge, UK: D. Brown, 1982); Commission of the European Communities, *Europe's Green Mantle: The Heritage and Future of Our Forests* (Luxembourg: EUR-OP, 1984); Michael Davis, "Hertfordshire's Green Mantle," *Arboricultural Journal* 11 (1987): 53–71.

34. Arthur Tansley, *Britain's Green Mantle: Past, Present and Future* (London: Allen & Unwin, 1949), v. In concluding the book, Tansley discusses more and less desirable transformations of land use and the extent to which the loss of certain "interesting" botanical species is legitimated by land productivity and economic gain. He then makes recommendations for specific types of trees and plants to be protected. "In these ways," Tansley concludes, "we can maintain

the beauty and interest of our Green Mantle, modifying its character here and there in accordance with modern needs, but preserving most of its essential historic character. It would be unreasonable and ridiculous to ignore legitimate economic demands, but it is nonetheless essential to take other values—of beauty, sentiment, and scientific interest—into full consideration. If we do not, we shall bequeath to our descendants a defaced and ugly land. The modern movement to establish National Parks and Nature Reserves can contribute in an important degree to the preservation of the Green Mantle" (266).

35. Conservationists went as far as calling *Silent Spring* "an *Uncle Tom's Cabin* for nature." Supreme Court justice William O. Douglas praised the book as "the most important chronicle of this century for the human race." Sierra Club director David Brower credited the book for changing his perspective; Carson, he claimed, "removed a veil that had concealed from me what the life force consists of and how interrelated are all of us who share in it" (quoted in Linda Sargent Wood, *A More Perfect Union: Holistic World Views and the Transformation of American Culture after World War II* [Oxford Scholarship Online, 2010], https://www.oxfordscholarship .com/view/10.1093/acprof:oso/9780195377743.001.0001/acprof-9780195377743 [accessed 20 May 2017]). Patrick defined the book as a "rhetorical archetype of modern environmental literature" (Amy Patrick, "Apocalyptic or Precautionary? Revisioning Texts in Environmental Literature," in *Coming into Contact: Explorations in Ecocritical Theory and Practice*, ed. Annie Merrill Ingram, Ian Marshall, Daniel Philippon, and Adam Sweeting [Athens: University of Georgia Press, 2007], 144). On the posthumous reception and legacy of *Silent Spring*, see Perry Parks, "Silent Spring, Loud Legacy: How Elite Media Helped Establish an Environmental Icon," *Journal of Mass Communication Quarterly* 94 (2017): 1215–38.

36. Kennedy's Science Advisory Committee issued a legal report in 1963 largely agreeing with Carson's findings, thus marking a first step toward the DDT ban of 1972 and other legislation, including the Clean Air Act (1970), Clean Water Act (1972), Safe Drinking Water Act (1974), Endangered Species Act (1973), and Toxic Substances Control Act (1976) (see Wood, *A More Perfect Union*).

37. Ralph Lutts, "Chemical Fallout: *Silent Spring*, Radioactive Fallout, and the Environmental Movement," in *And No Birds Sing*, ed. Waddell, 17.

38. Lutts, "Chemical Fallout," 18.

39. Lutts, "Chemical Fallout," 22; Ralph Lapp, *The Voyage of the Lucky Dragon* (New York: Harper & Brothers, 1958), 178.

40. Jessee, *Radiation Ecologies*, 2.

41. Jessee, *Radiation Ecologies*, 26. Ironically, radiation was understood by contemporary professional ecologists as a revolutionary tool for scientific discovery. Among other things, it allowed, for example, the study of patterns of oceanic and atmospheric motions. It could be also used to study the effects of stress on living systems. In the decades after World War II, the Atomic Energy Commission thus became a vital source of funding for ecological research. In 1954 the Odum brothers, for example, were granted funding to study the flow of energy in the coral reef ecosystem at Eniwetok Atoll, the site of the Pacific hydrogen bomb tests. Their vision of an "ecosystem" (a term coined by Tansley to equate nature to an energy system) was mechanistic, and dictated less by concerns for human health than by pure research. Nuclear power, they believed, had the potential to solve the problems it created. More than a fabric threatened to be undone by humans, the earth's mantle was a veil hiding nature's secrets, and atomic energy was the instrument to breach the veil and discover them. As Joel Hagen commented, "the delight of making new discoveries seemed impossible to resist. There was an almost universal

optimism that the potential benefits of atomic energy outweighed its destructive power, that nuclear swords could eventually be beaten into plowshares" (Joel Hagen, *An Entangled Bank: The Origins of Ecosystem Ecology* [New Brunswick, NJ: Rutgers University Press, 1992], 120). See also Peter Taylor, "Technocratic Optimism, H. T. Odum, and the Partial Transformation of Ecological Metaphor after World War II," *Journal of the History of Biology* 21 (1988): 213–44.

42. Carson, *Silent Spring*, 2.

43. Carson, *Silent Spring*, 3.

44. In an earlier draft, Carson had explicitly written that the powder reminded the villagers of the dust that fell upon the Japanese fishing boat and made them wonder if perhaps the wind had carried fallout from a bomb test. See Lutts, "Chemical Fallout," 35.

45. That year the United States reestablished its own testing program. By the end of it, nearly ninety devices were detonated by the Americans, as compared to about forty by the Soviet Union (Lutts, "Chemical Fallout," 31).

46. Lutts, "Chemical Fallout," 27–30. "What Dr. No began other evil geniuses such as Karl Stromberg (*The Spy Who Loved Me*) and Hugo Drax (*Moonraker*, 1979, dir. Gilbert), continued, and they were deadly serious in their globalizing ambition. They sought to destroy the world and/or remake/reshape it with their distinct geopolitical and biopolitical imprimatur . . . in *On Her Majesty's Secret Service* (1969, dir. Hunt), as Bond discovers a sinister plot to poison the world's food supply via bacterial agents" (Dodds and Funnell, "Going Atmospheric and Elemental," 69–70).

47. Paul Boyer, *By the Bomb's Early Light: American Thought and Culture at the Dawn of the Atomic Age* (New York: Pantheon, 1985).

48. Carson, *Silent Spring*, 53; Rachel Carson, *The Edge of the Sea* (Boston: Houghton Mifflin, 1955), 2. "The shore is an ancient world, for as long as there has been an earth and sea there has been this place of the meeting of land and water. Yet it is a world that keeps alive the sense of continuing creation and of the relentless drive of life. Each time that I enter it, I gain some new awareness of its beauty and its deeper meanings, sensing that intricate fabric of life by which one creature is linked with another, and each with its surroundings" (Carson, *The Edge of the Sea*, 14).

49. "The web of our life is a mingled yarn, good and ill together" (Shakespeare, *All's Well That Ends Well* 4.3). In an elegy titled "The Funeral Bell" Thoreau recorded his feelings about the sounds of the distant bell. "In this poem the lives of people are threads and the poet is the weaver. . . . The poem compares the fancy sable hangings of the expensive funeral and finds them inferior to the blue sky which is a 'cope' or hood" (Norwick, *The History of Metaphors*, 423). See also James McKusick, *Green Writing: Romanticism and Ecology* (New York: Palgrave Macmillan, 2010), 152–53.

50. "I am tempted to give one more instance showing how plants and animals, most remote in the scale of nature, are bound together by a web of complex relations. . . . We can clearly see how it is that all living and extinct forms can be grouped together in one great system; and how the several members of each class are connected together by the most complex and radiating lines of affinities. We shall never, probably, disentangle the inextricable web of affinities between the members of any class; but when we have a distinct object in view, and do not look at some unknown plan of creation, we may hope to make sure but slow progress" (Charles Darwin, *On the Origin of Species by Means of Natural Selection or, The Preservation of Favored Races in the Struggle for Life* [New York: Cosimo Classics, 2007], 47, 272). These and other metaphors in *On the Origins of Species* are discussed in Hagen, *An Entangled Bank*; Wood, *A More Perfect Union*, 13–14; Norwick, *The History of Metaphors*, 427–31.

51. Darwin, *Origin of Species*, 307.

52. Arthur Thomson, *The System of Animate Nature* (London: William & Norgate, 1920), 58–59. Web of life metaphors were also common in early twentieth-century poetry and popular writing. Various examples can be found in Norwick, *The History of Metaphors*, 343–438. The metaphor was also used by the American philosopher Alfred North Whitehead, whose approach paralleled Carson's (see Lytle, *The Gentle Subversive*, 88–89).

53. Wood, *A More Perfect Union*.

54. The holistic view of the environment comes from Frederic Clements. His theory that plant communities were an organism was subsequently challenged by Tansley's concept of ecosystem. See Fred Bosselman, "The Influence of Ecological Science on American Law: An Introduction in the Symposium Ecology and the Law," *Chicago-Kent College of Law Review* 69 (1994): 847–73.

55. Cited in Patrick, "Apocalyptic or Precautionary," 147.

56. Carson, *Silent Spring*, 42. The Cappadocian Fathers similarly insisted on the complexity and interconnection of the planet's hydrological system by comparing, for example, rivers to the circulatory system of living beings. Rivers and subterranean waters, they believed, sustained the diversity in the unity of creation. Just as blood nourished the different organs of the human body, so did water flows nourish plants according to their individual needs and characteristics (see, for example, Gregory of Nyssa's *Homilies* 13 and 29).

57. "This death-by-indirection," the biologist explains, "now finds its counterpart in what are known as systemic insecticides. These are chemicals with extraordinary properties which are used to convert plants or animals into a sort of Medea's robe by making them actually poisonous" (Carson, *Silent Spring*, 32).

58. Steinbach, "Metaphor." Carson did not have an academic affiliation, and could not rely on a scientific consensus to defend her arguments. She thus turned to a series of narrative strategies to secure public consensus. These comprised vivid prose and literary devices, such as metaphors. See Mollie Murphy, "Scientific Argument without Scientific Consensus: Rachel Carson's Rhetorical Strategies in the *Silent Spring* Debates," *Argumentation and Advocacy* 54 (2018): 1–17.

59. Glotfelty, "Cold War, Silent Spring," 164.

60. "As crude a weapon as the cave man's club, the chemical barrage has been hurled against the fabric of life—a fabric on the one hand delicate and destructible, on the other miraculously tough and resilient, and capable of striking back in unexpected ways. These extraordinary capacities of life have been ignored by the practitioners of chemical control who have brought to their task no 'high-minded orientation,' no humility before the vast forces with which they tamper. The 'control of nature' is a phrase conceived in arrogance, born of the Neanderthal age of biology and philosophy, when it was supposed that nature exists for the convenience of man. The concepts and practices of applied entomology for the most part date from that Stone Age of science. It is our alarming misfortune that so primitive a science has armed itself with the most modern and terrible weapons, and that in turning them against the insects it has also turned them against the earth" (Carson, *Silent Spring*, 297).

61. Like "The Earth's Green Mantle," "Elixirs of Death" is the title of one of the chapters of *Silent Spring*.

62. Fulvio Apollonio, "Tormento e speranza nelle monete di Vivarelli," http://en.numista.com /catalogue/pieces5983.html (accessed 1 May 2018).

63. Michael Soulé and Bruce Wilcox, "Conservation Biology: Its Scope and Its Challenge," in *Conservation Biology: An Evolutionary-Ecological Perspective*, ed. Michael Soulé and Bruce Wilcox (Sunderland, MA: Sinauer Associates, 1980), 7–8. Similar metaphors were appropriated

by other biologists, including E. O. Wilson. In his afterword to the 2002 edition of Carson's *Silent Spring*, for example, he claimed that "Nature is dying the torture-death of a thousand cuts" (Carson, *Silent Spring*, 362).

64. Turchetti and Roberts, "Introduction," 8.

65. Cited in Robert Poole, "What Was Whole about the Whole Earth? Cold War and Scientific Revolution," in *The Surveillance Imperative: Geosciences during the Cold War and Beyond*, ed. Simone Turchetti and Peder Roberts (New York: Palgrave Macmillan, 2014), 213.

66. Cosgrove, *Apollo's Eye*, 260–62.

67. John Foster, *The Sustainability Mirage: Illusion and Reality in the Coming War on Climate* (London: Routledge, 2012), 17.

68. "The earth reminded us of a Christmas tree bauble hung in the darkness of space. With increasing distance it became smaller and smaller. Finally it shrunk to the size of a marble, the most beautiful marble you can imagine" (Irwin, cited in Kevin Kelley, *The Home Planet* [Reading, MA: Addison-Wesley, 1988], 38).

69. Foster, *Sustainability Mirage*, 17.

70. James Lovelock, *Gaia: A New Look at Life on Earth* (Oxford: Oxford University Press, 2009), xvi.

71. Lovelock, *Gaia*, xii.

72. Bruno Latour, *Facing Gaia: Eight Lectures on the New Climatic Regime* (Cambridge: Polity Press, 2017), 75.

73. Latour, *Facing Gaia*, 79.

74. Lovelock, *Gaia*, 22–23.

75. Lovelock, *Gaia*, 78, 118.

76. James Lovelock, *The Revenge of Gaia: Earth's Climate Crisis and the Fate of Humanity* (London: Penguin Books, 2006), 17.

77. See Patrick, "Apocalyptic or Precautionary," 146–47.

78. Cosgrove, "Images and Imagination."

79. Turchetti and Roberts, "Introduction."

80. Cosgrove, "Images and Imagination," 1873.

Chapter Ten

1. Thomas Pynchon, *The Crying of Lot 49* (New York: Harper Perennial, 2006), 11.

2. Pynchon, *The Crying*, 11.

3. David Cowart, "Pynchon's *The Crying of Lot 49* and the Paintings of Remedios Varo," *Critique* 18 (1977): 25.

4. Lewis Carroll, *Sylvie and Bruno Concluded* (London: Macmillan, 1893), 168–69.

5. Jorge Luis Borges, "Of Exactitude in Science," reprinted in *A Universal History of Infamy* (Harmondsworth, UK: Penguin, 1975). Both Carroll's and Borges's passages are discussed in James Corner, "The Agency of Mapping: Speculation, Critique and Invention," in *Mappings*, ed. Denis Cosgrove (London: Reaktion, 1999), 221–22.

6. The term was first used in 1995 by Nancy Peluso and refers to counterhegemonic mappings, usually proclaiming the spatial interests of oppressed groups of people, or simply challenging dominant spatial visions. See Robert Rundstrom, "Counter-mapping," in *International Encyclopaedia of Human Geography*, ed. Rob Kitchin and Nigel Thrift (Amsterdam: Elsevier, 2009), 314–18; Denis Wood, *Rethinking the Power of Maps* (New York: Guilford Press, 2010), 189–231.

7. Judith Tyner, *Stitching the World: Embroidered Maps and Women's Geographical Education* (Aldershot, UK: Ashgate, 2015).

8. The novel, *Donna Scalotta*, is dated conjecturally before 1321. See Christopher Ricks, *The Poems of Tennyson* (Berkeley: University of California Press, 1987), 387.

9. Verses 2–3, 16.

10. Verses 64–72. On the mirroring effect, see Joseph Chadwick, "A Blessing and a Curse: The Poetics of Privacy in Tennyson's 'The Lady of Shalott,'" in *Critical Essays on Alfred Tennyson*, ed. Herbert Tucker (New York: Macmillan, 1993), 83–99.

11. Verses 109–17.

12. Verses 118–21.

13. Chadwick, "A Blessing and a Curse," 90.

14. Alison Smith, "The Pre-Raphaelites and the Arnolfini Portrait: A New Visual World," in *Reflections: Van Eyck and the Pre-Raphaelites*, ed. Alison Smith et al. (London: National Gallery, 2017), 58.

15. Smith, "The Pre-Raphaelites," 57.

16. Alan Sinfield, "The Mortal Limits of the Self: Language and Subjectivity," in *Critical Essays on Alfred Lord Tennyson*, ed. Herbert Tucker (New York: G. K. Hall, 1993), 242.

17. Chadwick, "A Blessing and a Curse," 93.

18. "Remedios Varo: Nota biográfica," http://remedios-varo.com/biografia/ (accessed 5 July 2018).

19. Janet Kaplan, *Unexpected Journeys: The Art and Life of Remedios Varo* (New York: Abbeville, 2000).

20. Janet Kaplan, "Art Essay: Remedios Varo," *Feminist Studies* 13 (1987): 39.

21. Ricki O'Rawe and Roberta Ann Quance, "Crossing the Threshold: Mysticism, Liminality, and Remedios Varo's *Bordando el manto terrestre* (1961–2)," *Modern Languages Open* (2016): 1.

22. Kaplan, *Unexpected Journeys*, 18.

23. Kaplan, *Unexpected Journeys*, 18.

24. Cited in Kaplan, *Unexpected Journeys*, 8. Varo's mantle of the earth resembles the blanket in *Dream of Malinche* (1939) by Mexican Surrealist painter Antonio Ruiz. Here a blanket covering a sleeping woman becomes a landscape of winding roads and nestled villages in a fashion similar to the tapestry produced by the girls in the tower in Varo's painting (the painting is discussed in Kaplan, *Unexpected Journeys*, 255). While in Ruiz's painting the female body is objectified and passive, however, Varo's counters this typically Surrealist tendency by making women active protagonists of the scene; see Inés Ferrero Cándenas, "Reconfiguring the Surrealist Gaze: Remedios Varo's Images of Women," *BHS* 88 (2011): 455–67.

25. Kaplan, *Unexpected Journeys*, 21.

26. One can draw parallels with other paintings by Varo. In *Tejido espacio-tiempo* (1954), for example, the viewer contemplates a medieval lady and her suitor through the normal warp and weft of a magically woven cosmic egg. See O'Rawe and Quance, "Crossing the Threshold," 8.

27. Kaplan, *Unexpected Journeys*, 223.

28. Kaplan, *Unexpected Journeys*, 223–24.

29. Octavio Paz, cited in Kaplan, *Unexpected Journeys*, 231.

30. Luca Cerizza, *Alighiero e Boetti: Mappa* (London: Afterall, 2008).

31. Jacob, The Empire of Maps, 50.

32. "Ladies from the Women's Voluntary Service Make Maps Out of Cloth for the Army," http://www.britishpathe.com/video/map-making-by-w-v-s-issue-title-quite-so/query/cloth +map (accessed 3 July 2018).

33. Slow routines spent between the hotel and the bazaar, between local houses, mosques, and textile workshops, were occasionally punctuated by adventurous excursions into remote regions of the country and, more often, by new acquaintances. Among them were clan leaders, local traditional musicians, and unconventional foreign visitors to the hotel, such as Vladimir, a New Yorker of Russian descent traveling through central Asia "in search of beautiful textiles and kilims to sell on Fifth Avenue" (Annemarie Sauzeau-Boetti, *Alighiero e Boetti: Shaman— Showman* [Cologne: König, 2003], 112).

34. Sauzeau-Boetti, *Alighiero e Boetti*, 101.

35. The Persian word *purdah* ("curtain") is used to designate the social practice of female seclusion in Muslim communities. It encompasses both physical segregation and the requirement that women cover their bodies. Physical segregation within buildings is achieved through walls, curtains, and screens. Purdah was rigorously observed under the Taliban in Afghanistan. Only close male family members and other women were allowed to see women out of purdah. Secluded in the home and in other female spaces, Afghan women performed productive activities, such as embroidery and weaving (Cerizza, *Alighiero e Boetti*, 30).

36. "At one point, when Boetti went to the school to check the progress of the work, he found major errors in one part of the [embroidery]. The headmistress noticed that, and, deadly embarrassed, took a pair of scissors and cut through the threads, one after another. The woman responsible for this part wept" (Cerizza, *Alighiero e Boetti*, 32).

37. Cerizza, *Alighiero e Boetti*, 34.

38. Nigel Lendon, "A Tournament of Shadows: Alighiero Boetti, the Myth of Influence, and a Contemporary Orientalism," *emaj* 6 (2011–2012): 1–32.

39. Cerizza, *Alighiero e Boetti*, 7.

40. "Each map took up to four embroiderers about one year to make; some took two years, or even longer—up to as many as ten" (Cerizza, *Alighiero e Boetti*, 35).

41. Cerizza, *Alighiero e Boetti*, 36, 87.

42. The full inscription reads: "A PESHAWAR PAKISTAN BY AFGHAN. ALIGHIERO E BOETTI A TEMPO IN TEMPO COL TEMPO IL TEMPORALE. ALIGHIERO E BOETTIJKLMNOPQRA" (*Mappa*, 1989).

43. This *mappa* was also embroidered in 1989. The inscription is a Sufi quotation (Saʾdi, as quoted in Farsi in the border of the present work, thirteenth century, translated by Iraj Barishi).

44. "Rare Warhol Portrait of Liz Taylor to Lead Christie's Auction," *Art Daily*, http://art daily.com/news/38538/Rare-Warhol-Portrait-of-Liz-Taylor-to-Lead-Christie-s-Auction# .WzKGVEa50_g (accessed 3 July 2018).

45. Sauzeau-Boetti, *Alighiero e Boetti*, 152.

46. Sauzeau-Boetti, *Alighiero e Boetti*, 157.

47. Cerizza, *Alighiero e Boetti*, 32.

48. "For me the work of the embroidered 'Mappa' is the maximum of beauty. For that work I did nothing, chose nothing, in the sense that: the world is made as it is, not as I designed it, the flags are those that exist, and I did not design them; in short I did absolutely nothing; when the basic idea, the concept, emerges everything else requires no choosing" (Alighiero Boetti [1974], quoted in Cerizza, *Alighiero e Boetti*, 3).

49. Cerizza, *Alighiero e Boetti*, 33.

50. The American curator of the Fowler exhibition, for example, writes that as a European man, "Boetti was not allowed to visit the camps. He therefore asked photographer Randi Malkin Steinberger, with whom he had collaborated in Rome, to go to Peshawar and to photograph the women at work" (Stacey Ravel Abarbanel, "Fowler Exhibition Highlights Artist Alighiero

Boetti's Embroideries by Afghan Women," *UCLA Newsroom*, 14 December 2011, http://news room.ucla.edu/releases/fowler-museum-presents-an-exhibition-220534 [accessed 3 July 2017]).

51. Cerizza, *Alighiero e Boetti*, 33.

52. Nigel Lendon, "A Tournament of Shadows: Alighiero Boetti, the Myth of Influence, and a Contemporary Orientalism," *emaj* 6 (2011–12): 22. See also Dorian Ker, "Legnetti e fasti," *Third Text* 13 (1999): 97–100; Ruth Watson, "Mapping and Contemporary Art," *Cartographic Journal* 46 (2009): 293–307.

53. Agata Boetti, "Alighiero Loved All the Maps in the World," in *Alighiero Boetti*, ed. Laura Cherubini et al. (Florence: Forma, 2016), 207–8.

54. Charlotte Kent, "Remapping the Viewer's Experience with Alighiero e Boetti's *Mappa del Mondo*," *English Language Notes* 52 (2014): 201.

55. Popularized in the 1970s by the activist historian Arno Peters, this specific projection has the effect of elongating the shapes of the continents while rescaling their surfaces according to their true proportions.

56. Guy Brett, "Itinerary," in *Mona Hatoum*, ed. Michel Archer et al. (London: Phaidon Press, 1997), 34–87.

57. It is, in Hatoum's words, about "my experience of living in the West as a person from the Third World, about being an outsider, about occupying a marginal position, being excluded, being defined as 'Other' or as one of 'Them'" (Archer et al., eds., *Mona Hatoum*, 127).

58. *Present Tense* (1996) was Hatoum's response to her first visit to Jerusalem. The installation is formed of twenty-four hundred soap bars from the city of Nablus, north of Jerusalem. Its surface is embedded with tiny red beads that depict the map showing the territorial divisions under the Oslo Peace Agreement between Israel and the Palestinians, including the territories that should have been returned to the Palestinian Authority. According to Hatoum, this was "a map about dividing and controlling the area. At the first sign of trouble, Israel practices the policy of 'closure'; they close all the passages between the areas so that the Arabs are completely isolated and paralyzed" (Archer et al., eds., *Mona Hatoum*, 26–27). *Marbles Carpet* (1995) and *Map (clear)* (2015) are vast expanses of glass marbles covering the floor of the gallery. "They make uncertain the ground on which they lie. The floor, or more fundamentally, the earth upon which the spectator stands, that basis, above all others, upon which not only bodily presence, but also attitudes and beliefs rest, is made uncertain" (Brett, "Itinerary," 77).

59. See chapter 2.

60. Edward Said, "The Art of Displacement: Mona Hatoum's Logic of the Irreconcilables," in *Mona Hatoum: The Entire World as a Foreign Land*, ed. Edward Said and Sheena Wagstaff (London: Tate Gallery Publishing, 2000), 7.

61. Scheid and Svenbro, *The Craft of Zeus*, 10.

62. Scheid and Svenbro, *The Craft of Zeus*, 9.

63. Claire Dwyer and Katie Beinart, "My Life Is But a Weaving: A Collaborative Arts Project," in *My Life Is But a Weaving: A Collaborative Multi-faith Project with Artist Katy Beinart*, ed. Nazneen Ahmed et al. (Ealing, West London, UK, 2018), 5.

64. Dwyer and Beinart, "My Life," 4.

65. Watson, "Mapping and Contemporary Art," 300. Another recent embroidery collaborative project is the Magna Carta embroidery by Cornelia Parker accompanying the Magna Carta exhibition at the British Library (2015). The 13-meter-long textile is an embroidered replica of the Wikipedia article on the Magna Carta. It symbolizes the collective voice of the people on civil liberties. The fabric, printed with the article, was divided into sections and sent to 200

people to stitch (Ruth Clifford, "Textiles, Language and Metaphor," http://travelsintextiles.com
/textiles-language-and-metaphor/ [accessed 15 July 2018]).

66. Dwyer and Beinart, "My Life," 3.

67. Dwyer and Beinart, "My Life," 4.

68. Nazneen Ahmed, "Stitching in St. Thomas the Apostle Church Hall, Hanwell," in *My
Life Is But a Weaving*, 62. "This connection is evident in Deepak Mehta's anthropological study
of Muslim Ansari weavers in Banaras. In *Work, Ritual, Biography*, Mehta describes the practice
of weavers reciting *du'as*, Islamic prayers while at the loom. Within the *du'as* are answers to
questions that Adam asks Jabrail when the first loom is introduced from heaven. 'In uttering
(the *du'as*) the weaver invokes the loom as memory . . . to weave is to pray, to pray is to weave.'
Like Kabir who is said to be of the same community as the weavers in Mehta's book, weaving is
sacred, almost as sacred as creation itself" (Clifford, "Textiles").

69. Rozsika Parker, *The Subversive Stitch: Embroidery and the Making of the Feminine* (London: I. B. Tauris, 2014).

70. Ahmed, "Kermanig," in *My Life Is But a Weaving*, 60.

71. See chapter 1.

72. Dwyer and Beinart, "My Life," 7.

73. Ahmed, "Phulkari," in *My Life Is But a Weaving*, 69.

74. Orvar Löfgren and Richard Wilk, *Off the Edge: Experiments in Cultural Analysis* (Copenhagen: Museum Tusculanum Press, 2006), 7.

75. Bruno Latour, "On Actor-Network Theory: A Few Clarifications and More than a Few
Complications," *Soziale Welt* 47 (1990): 372.

76. In Anderson's words, "the cosmopolitan canopy offers people an opportunity to express
themselves as individuals, each with his or her own style or dress, speech, and movement. It
allows complete strangers to observe and appreciate one another, and even communicate for
a moment or two. . . . The canopy offers the promise of edification for all who enter. Exposure
to others' humanity generates empathy, fears dissipate, and grounds for mutual appreciation
appear" (Elijah Anderson, *The Cosmopolitan Canopy: Race and Civility in Everyday Life* [New
York: W. W. Norton, 2011], 280). An architectural visualization of "cosmopolitan canopy" is
found at Les Halles giant shopping mall in Paris. Inaugurated in 2016, *La Canopée* is a sort of
artificial vegetal coverage shaped as a huge mantle. I am grateful to Tania Rossetto for bringing
Anderson's work and this building to my attention.

Chapter Eleven

1. From 2000 to 2009, the number of internet users globally rose from 394 million to nearly
2 billion. By 2010, 22 percent of humanity had access to computers with 1 billion Google searches
every day. By the close of 2017 the world's internet users surpassed 4 billion, including 70 percent
of the planet's youth (https://www.itu.int/en/ITU-D/Statistics/Documents/facts/ICTFactsFigures
2017.pdf (accessed 8 August 2018).

2. Barney Warf, *Global Geographies of the Internet* (New York: Springer, 2013), 1.

3. Kirsty Best, "Interfacing the Environment: Networked Screens and the Ethics of Visual
Consumption," *Ethics and the Environment* 9 (2004): 68. See also James Ash, Rob Kitchin, and
Agnieszka Leszczynski, "Digital Turn, Digital Geographies?" *Progress in Human Geography* 42
(2018): 25–43; Martin Dodge and Rob Kitchin, "Code and the Transduction of Space," *Annals of
the Association of American Geographers* 95 (2005): 162–80.

4. Amy Willis, "Facebook 'Map of the World' Created Using Friendship Connections," *Telegraph*, 16 December 2010.

5. "Blue Planet: New Facebook Map Depicts All of the World's Interconnected Friendships— with One Unmistakable Black Hole over China," *Mail Online*, 26 September 2013.

6. Maria Beatrice Bittarello, "Spatial Metaphors Describing the Internet and Religious Websites: Sacred Space and Sacred Place," *Observatorio* 11 (2009): 2.

7. Denis Jamet, "What Do Internet Metaphors Reveal about the Perception of the Internet?" *Metaphorik.de* 18 (2010): 10.

8. Rebecca Johnston, "Salvation or Destruction: Metaphors of the Internet," *First Monday*, 6 April 2009, http://firstmonday.org/article/view/2370/2158 (accessed 8 August 2018); Jamet, "What Do Internet Metaphors Reveal," 14.

9. Clifford, "Textiles."

10. Tim Berners-Lee, *Weaving the Web: The Origins and Future of the World-Wide Web* (London: Orion Business, 1999). Elsewhere, Berners-Lee describes the HTML (HyperText Markup Language) as "the warp and weft of a hypertext tapestry crammed with rich and varied data types" (Tim Berners-Lee, "The World Wide Web: Past, Present and Future," https:// www .w3.org/People/Berners-Lee/1996/ppf.html [accessed 8 August 2018]).

11. Before the invention of the internet, computers were connected through dedicated cables from one to another. Clearly, Berners-Lee notes, "one computer could not be linked to more than a few others, because it would need tens or hundreds of cables running from it. The solution was to communicate indirectly over a network" (Berners-Lee, *Weaving the Web*, 20). "The first computer network, named ARPANET after its powerful sponsor, went online in September 1, 1969 with the first four nodes of the network being established at the University of California, Los Angeles, Stanford Research Institute, Santa Barbara, and University of Utah. It was opened to research centers cooperating with the US Defense Department, but scientists started to use it for their own communication purposes, including a science fiction enthusiasts' messaging network. At one point it became difficult to separate military-oriented research from scientific communication and personal chatting. Thus, scientists of all disciplines were given access to the network, and in 1983 there was a split between ARPANET, dedicated to scientific purposes, and MILNET, directly oriented to military applications" (Manuel Castells, *The Rise of the Network Society* [Oxford: Blackwell, 1996], 14). For an account of the origins of the internet, see also Manuel Castells, *The Internet Galaxy: Reflections on the Internet, Business, and Society* (Oxford: Oxford University Press, 2001).

12. Harry Oinas-Kukkonen, *Humanizing the Web Change and Social Innovation* (Basingstoke, UK: Palgrave Macmillan, 2013), 11. The internet is, of course, part of a much longer history of networks and networked imaginations, including telegraph and telephone lines, as well as rail and road networks. See, for example, Nicole Starosielski, *The Undersea Network* (Durham, NC: Duke University Press, 2015); and James W. Carey, *Communication as Culture: Essays on Media and Society* (New York: Routledge, 2009). Intriguingly, in the early 2000s, fiber-optic cables in New York were run through the intricate web of underground pneumatic tubes that, between the late 1890s and the 1950s, were used to deliver mail in special capsules from post office to post office, thus avoiding the problem of traffic congestion (see Jason Farman, "Tubes for Fiber Optics," *Media Theory* 2 [2018], http://mediatheoryjournal.org/jason-farman-invisible-and -instantaneous/ [accessed 20 August 2018]).

13. Berners-Lee, *Weaving the Web*, 22.

14. The completely decentralized nature of the system, Berners-Lee believed, "would be the only way a person somewhere could start to use it without asking for access from anyone else.

And that would be the only way the system could scale, so that as more people used it, it would not get bogged down." Hence his vision: "Imagine making a large three-dimensional model, with people represented by little spheres, and strings between people who have something common at work. Now imagine picking up the structure and shaking it, until you make sense of the tangle: perhaps you see tightly knit groups in some places, and in some places weak areas of communication spanned by only a few people" (Berners-Lee, *Weaving the Web*, 18, 24).

15. Sasha Engelmann, "Social Spiders and Hybrid Webs at Studio Tomás Saraceno," *Cultural Geographies* 24 (2017): 161.

16. Tim Ingold, "When ANT Meets SPIDER: Social Theory for Anthropods," in *Material Agency: Towards a Non-Anthropocentric Approach*, ed. Carl Knappett and Lambros Malafouris (New York: Springer, 2010), 212.

17. As with the spiderweb, topography, atmospheres, and forces are part of this infrastructure, too. For example, the underground and air-cooled warehouses storing the servers of the "cloud," or gravity and pressure in the case of undersea internet cables.

18. Berners-Lee, *Weaving the Web*, 26.

19. Berners-Lee, *Weaving the Web*, 132.

20. Berners-Lee, *Weaving the Web*, 14.

21. Berners-Lee, *Weaving the Web*, 222. Pierre Lévy similarly likens digital networks to a "hypercortex that spreads its axons throughout the planet" (Pierre Lévy, *Becoming Virtual: Reality in the Digital Age* [New York: Plenum Trade, 1998], 41). The metaphor, however, has a much longer history. For example, it was used back in the 1960s by Marshall McLuhan, who famously claimed: "During the mechanical ages we had extended our bodies in space. Today, after more than a century of electric technology, we have extended our central nervous system itself in a global embrace, abolishing both space and time as far as our planet is concerned" (*Understanding Media: The Extensions of Man* [Cambridge, MA: MIT Press, 1994], 1). Likewise, in an interview with *Collier Magazine* in 1926, Nicolas Tesla stated: "When wireless is perfectly applied, the whole earth will be converted into a huge brain, which in fact it is, all things being particles of a real and rhythmic whole."

22. Berners-Lee, *Weaving the Web*, 133, 136.

23. Berners-Lee, *Weaving the Web*, 143.

24. Berners-Lee, *Weaving the Web*, 221.

25. Jamet, "What Do Internet Metaphors Reveal." As Maria Beatrice Bittarello observes, "by describing the internet as space we use a metaphor, as we do not physically move to, or in(to) cyberspace." One key difference between physical space and cyberspace is thus that while we physically move in the former, in the latter, "space 'moves towards the surfer,' by virtue of the surfer's semantic competence." Nonetheless, spatial metaphors have been constantly used to describe the internet and the World Wide Web (Bittarello, "Spatial Metaphors Describing the Internet," 2). Similar observations on the spatial images used to describe computer networks are found in Michael Curry, "New Technologies and the Ontology of Places," paper presented at the Information Studies Seminar, University of California, Los Angeles, 4 March 1999, 11–12, http://baja.sscnet.ucla.edu/~curry/Curry_Tech_Regimes.pdf (accessed 8 August 2018). The shift from early "exploration" and "frontier" metaphors to metaphors such as "home," or even changes in the browsers' logos (for example, from a steer to a globe), reflect the domestication, or rather, normalization, of the medium and its global spread. Cyberspace thus became the primary metaphor for the internet and imagined as "a place one can spend time wandering, navigating, exploring." Conceptualized as such, the internet was "not so much a prosthesis for the senses, but

a separate environment where the self can interact, move, travel and exist" (Annette Markham, "Metaphors Reflecting and Shaping the Reality of the Internet: Tool, Place, Way of Being," paper presented at the conference of the International Association of Internet Researchers, Toronto, Canada, October 2003, http://markham.internetinquiry.org/writing/MarkhamTPW.pdf [accessed 9 August 2018]). "Surfing" metaphors embed a leisure quality, while at the same time conveying the notion of information moving toward the user and allowing the user freedom of choice: "When surfing, a surfer mostly relies on the waves to carry him forward, though he does have a choice in which wave he catches. In the same way, a Web surfer chooses to follow only one of the links available to her on a certain page, rather than actively seeking a certain destination" (Zach Tomaszewski, "Conceptual Metaphors of the World Wide Web," http://zach .tomaszewski.name/uh/ling440/webmetaphors.html [accessed 8 August 2018]; Johnston, "Salvation or Destruction").

26. This reflects a broader trend of naturalized computing metaphors that have become part of our speech and which we use for describing our experiences and sense of the Self. For example, "we describe ourselves and others as binary; we describe our brains as hard drives or storage systems; we talk about thoughts as being coded in memory" (Johnston, "Salvation or Destruction").

27. Castells, *The Rise of the Network Society*, xviii. Some of these ideas were foreshadowed by McLuhan, who wrote the influential book *Understanding Media* and popularized the idea of "global village" well before the age of the internet.

28. Castells, *The Rise of the Network Society*. Farinelli likewise defines "global cities" not necessarily as the largest world's metropolises, but those able to control financial activity and its innovations. Global functions, he observes, are often located in a tiny portion of the city surrounded by "an urban fabric which has nothing to do with mechanisms of global control" (Farinelli, *Invenzione della Terra*, 148; see also Farinelli, *Geografia*, 195). See also Jordan Frith, *Smartphones as Locative Media* (Cambridge MA: Polity Press, 2015), 4.

29. The idea comes from Latour's "technological network": "Technological networks, as the name suggests, are networks thrown over spaces, and they retain only a few scattered elements of those spaces. They are connected lines, not surfaces. They are by no means comprehensive, global or systematic, even though they embrace surfaces without covering them, and extend a very long way" (cited in Stephen Graham, "The End of Geography or the Explosion of Place? Conceptualizing Space, Place and Information Technology," *Progress in Human Geography* 22 [1998]: 179).

30. Michael Batty, "Virtual Geography," *Futures* 29 (1997): 615–16.

31. Castells, *The Rise of the Network Society*, xviii.

32. Meaning the weakening of boundaries and of nation-states as players in the global economy. In the past, Ohmae observed, the flow of cross-border funds was primarily from government to government or from multilateral lending agency to government. By the 1990s, by contrast, most of the money moving across borders was private. "The movement of both investment and industry has been facilitated by information technology, which now makes it possible for a company to operate in various parts of the world without having to build up an entire system in each of the countries where it has presence. Engineers in Osaka can easily control plant operations in newly exciting parts of China, like Dalian. Product designers in Oregon can control the activities of a network of factories throughout Asia-Pacific. Thus, the hurdles of cross-border participation and strategic alliance have come way down. . . . Capability can reside in the network and be made available—virtually anywhere—as needed" (Kenichi Ohmae, *The End of the*

Nation State [New York: Free Press, 1995], 3–4). Castells's position is less extreme: in a more recent publication, he claimed that the nation-state does not disappear, but rather adapts itself to the new reality (Manuel Castells, "The New Public Sphere: Global Civil Society, Communication Networks, and Global Governance," *Annals AAPSS* 616 [2008]: 87).

33. Mark Graham, "Neogeography and the Palimpsests of Place: Web 2.0 and the Construction of a Virtual Earth," *Tijdschrift voor Economische en Sociale Geografie* 101 (2010): 423. Batty calls "virtual geography" the space within computers, cyberspace, and infrastructure. Networks, he argues, ride in parallel but are largely invisible to immediate observation, the internet being "simply the tip of the iceberg" (Batty, "Virtual Geography," 346). For Lévy, "virtual is taken to signify the absence of existence, whereas reality implies a material embodiment" (*Becoming Virtual*, 23). See also Derek Stanovsky, "Virtual Reality," in *Blackwell Guide to the Philosophy of Computing and Information*, ed. Luciano Floridi (Oxford: Blackwell, 2004), 167–77; Mark Hansen, *Bodies in Code: Interfaces with Digital Media* (New York: Routledge, 2006); Julia Verne, "Virtual Mobilities," in *Introducing Human Geographies*, ed. Paul Cloke, Phil Crang, and Mark Goodwin (London: Arnold, 2014), 821–33. The divide between virtual and real was nonetheless problematized by various authors; see, for example, Mike Crang, Phil Crang, and Jon May, eds., *Virtual Geographies: Bodies, Space and Relations* (New York: Routledge, 1999).

34. Utopian and dystopian visions in cyberpunk literature are discussed in James Kneale, "The Virtual Realities of Technology and Fiction: Reading William Gibson's Cyberspace," in *Virtual Geographies*, ed. Crang, Crang, and May, 205–21).

35. Lea Manovich, "The Poetics of Augmented Space," *Visual Communication* 5 (2006): 220.

36. Manovich, "Poetics," 221. See also Adriana de Souza e Silva and Daniel Sutko, "Theorizing Locative Technologies through Philosophies of the Virtual," *Communication Theory* 21 (2011): 23–42; and Samuel Kinsley, "The Matter of 'Virtual' Geographies," *Progress in Human Geography* 38 (2014): 364–84. At the same time, however, a new gap seems to be opening up, as artificial intelligence and machine learning gain further traction. Some are trying to reclaim cyberspace as a legitimate space produced within the technology. "Deep neural networks" underpin most of artificial intelligence, as programmers input "training data" which produces an output, but there are hidden layers of computation going on between the input and output. These hidden or "deep" layers perform computational tasks that users (as well as programmers) cannot observe, nor understand in full. The sheer complexity of the computation involved means we can only ever understand it in small parts. The result is outputs that we can use but cannot fully understand how they were determined. So the argument goes that these outputs (including their faults) are "produced in cyberspace."

37. Adriana de Souza e Silva, "From Cyber to Hybrid: Mobile Technologies as Interfaces of Hybrid Spaces," *Space and Culture* 9 (2006): 268. For Jordan Frith, "smartphones are, in a way, a refutation of the virtual reality movement of the 1990s that imagined reality moving from the physical world to immersive virtual environments. The world of bits did not do away with the need for physical mobility; instead, smartphones show that the spaces we move through and the digital information we interact with have merged" (Jordan Frith, "Splintered Space: Hybrid Spaces and Differential Mobility," *Mobilities* 7 [2012]: 132). On the cloud metaphor, see Tung-Hui Hu, *A Prehistory of the Cloud* (Cambridge, MA: MIT Press, 2015); and Rebecca Rosen, "Clouds: The Most Useful Metaphor of All Time?" *The Atlantic*, 30 September 2011, https://www.theatlantic.com/technology/archive/2011/09/clouds-the-most-useful-metaphor-of-all-time/245851/ (accessed 18 August 2018).

38. de Souza e Silva, "From Cyber to Hybrid," 262; Manovich, "Poetics," 225.

39. The difference between these categories is discussed in Julia Tokareva, "The Difference between Virtual Reality, Augmented Reality and Mixed Reality," *Forbes*, 2 February 2018, https://www.forbes.com/sites/quora/2018/02/02/the-difference-between-virtual-reality-augmented-reality-and-mixed-reality/#6143b8362d07 (accessed 18 August 2018).

40. Verne, "Virtual Mobilities," 823; Frith, *Smartphones*, 2, 17; Dodge and Kitchin, "Code," 177. Warf calls the internet "a vast, seamless integrated network . . . the nervous system of international finance economy" (Warf, *Global Geographies*, 10).

41. Castells, *Internet Galaxy*, 1.

42. Nigel Thrift, "Beyond Mediation: Three New Material Registers and Their Consequences," in *Materiality*, ed. Daniel Miller (Durham, NC: Duke University Press, 2005), 233.

43. Best, "Interfacing the Environment," 71.

44. Lucas Introna and Fernando Ilharco, "On the Meaning of Screens: Towards a Phenomenological Account of Screenness," *Human Studies* 29 (2006): 58. Some scholars use the term "postdigital" to describe this saturation, and more specifically, "either a contemporary disenchantment with digital information systems and media gadgets, or a period in which our fascination with these systems and gadgets has become historical" (Florian Cramer, "What Is the Post-digital?" in *Postdigital Aesthetics*, ed. David Berry and Michael Dieter [New York: Palgrave Macmillan, 2015], 13).

45. Best, "Interfacing the Environment," 78.

46. Manovich, "Poetics," 220.

47. Best, "Interfacing the Environment," 71.

48. Best, "Interfacing the Environment," 68.

49. Tarja Laine, "Cinema as Second Skin," *New Review of Film and Television* 4 (2006): 95.

50. Today, with the explosion of ubiquitous computing and ubiquitous mapping, observes Italian geographer Tania Rossetto, "we touch maps more than ever" ("The Skin of the Map: Viewing Cartography through Tactile Empathy," *Environment and Planning D: Society and Space* 37 [2019]: 83–103). In spite of this tactile dimension, however, screens also set a transparent barrier between the places and people we chose to engage. See, for example, Gail Adams-Hutcheson and Robyn Longhurst, "'At Least in Person There Would Have Been a Cup of Tea': Interviewing via Skype," *Area* 49 (2017): 148–55.

51. Adams-Hutcheson and Longhurst, "Interviewing via Skype," 152.

52. Roland Barthes, *Camera Lucida: Reflections on Photography*, trans. Richard Howard (New York: Hill & Wang, 1981), 81.

53. Warf, *Global Geographies*, 146.

54. Bella Dicks, *Culture on Display: The Production of Contemporary Visitability* (Milton Keynes, UK: Open University Press, 2003), 191.

55. Ash, Kitchin, and Leszczynski, "Digital Turn."

56. Stephen Graham, *Vertical: The City from Satellites to Bunkers* (London: Verso, 2018), 11.

57. Andrew Boulton and Matthew Zook, "Landscape, Locative Media, and the Duplicity of Code," in *The Wiley-Blackwell Companion to Cultural Geography*, ed. Nuala Johnson, Richard Schein, and Jamie Winders (Chichester: Wiley Blackwell, 2013), 438. Kirsty Best argues that the screen, "working to satisfy a desire for monitoring, is a key site/sight in the gaze of the contemporary subject of locative media." In her analysis, however, the screen is also duplicitous, working to construct consent to our own surveillance. In this view, "the innocent transparency suggested by the screen belies the fact that its function is founded on the opacity of its coded innards" (cited in Boulton and Zook, "Landscape," 442).

58. On "innards" metaphors, see, for example, Nigel Thrift, "Lifeworld Inc.—and What to Do about It," *Environment and Planning D: Society and Space* 29 (2011): 5–26.

59. See, for example, Jennifer Gabrys, "Automatic Sensation: Environmental Sensors in the Digital City," *Senses and Society* 2 (2007): 189–200.

60. Warf, *Global Geographies*, 146. See also Natasha Schull, *Addiction by Design* (Princeton, NJ: Princeton University Press, 2014).

61. Thrift, "Lifeworld Inc.," 7.

62. See, for example, James Ash, *The Interface Envelope: Gaming, Technology, Power* (New York: Bloomsbury, 2015). Commenting on the ubiquity of technological objects and their effects, Latour likewise writes: "[we] are enveloped, entangled, surrounded, we are never outside without having recreated another more artificial, more engineered envelope. We move from envelope to envelope, from folds to folds, never from one private sphere to the Great Outside" (Latour, "On Actor-Network Theory"). Thrift talks about space becoming "a set of envelopes containing different atmospheres" ("Lifeworld Inc.," 20).

63. Ash, *The Interface Envelope*, 4. Warf talks about a "networked Self," or a "public Self" that is carefully edited and micromanaged and set on public display on social networks. Identities of teenagers are thus increasingly shaped not by self-exploration, but by the expectations of their online audiences (Warf, *Global Geographies*, 143–50).

64. Jean Baudrillard, *Simulacra and Simulation* (Ann Arbor: University of Michigan Press, 1994), 1.

65. The idea was originally conceived by Al Gore back in the late 1990s. The American vice president proposed a "Digital Earth" (or interactive globe) as a public tool accessible from local libraries and museums. Allowing viewers to virtually "fly" from space down to ground level through progressively higher resolution data sets (he aimed at a 1 square meter definition), Digital Earth would enable the display of information related to a specific location from a potentially infinite number of sources. The project was never realized, but it provided the concept for Google Earth. See Al Gore, "The Digital Earth: Understanding Our Planet in the 21st Century," speech given at the California Science Center, Los Angeles, 1998, http://www.digitalearth.gov /VP19980131.html (accessed 8 August 2018).

66. Frith, "Splintered Space," 144; Frith, *Smartphones*, 83–84; Adriana de Souza e Silva and Jordan Frith, "Locative Mobile Social Networks: Mapping Communication and Location in Urban Spaces," *Mobilities* 5 (2010): 490. A description of the project can be found at: http://probos cis.org.uk/projects/2000–2005/urban-tapestries/ (accessed 9 August 2018).

67. Mark Graham and Matthew Zook, "Augmented Realities and Uneven Geographies: Exploring the Geolinguistic Contours of the Web," *Environment and Planning A* 45 (2013): 78; Frith, *Smartphones*, 7. Augmented space is "a physical space overlaid with dynamically changing information. This information is likely to be in multimedia form and is often localized for each user" (Manovich, "Poetics," 220).

68. Ash, *The Interface Envelope*, 123.

69. "Gatwick Installs 2000 Indoor Navigation Beacons Enabling Augmented Reality Wayfinding," https://www.diorama.com/2017/05/26/gatwick-installs-2000-indoor-navigation-beacons -enabling-augmented-reality-wayfinding/ (accessed 20 August 2018).

70. de Souza e Silva and Jordan Frith, "Locative Mobile Social Networks," 490.

71. Jacob, *Sovereign Map*, 320–21.

72. Manovich, "Poetics," 228–29. Graham and Zook similarly talk about "densities of augmentations" of material places. "Density" refers to the quantity of information layered over a

place. Such layers of information, they argue, "are necessarily always in flux, and their construction, configuration, and even existence is often contingent on individual positionalities. However, by mapping the densities of information that augment our planet, we are able to broadly understand which parts of the world lack digital layers of representation" (Graham and Zook, "Augmented Realities," 83).

73. Dorthe Brogård Kristensen and Minna Ruckenstein, "Co-evolving with Self-tracking Technologies," *New Media and Society* 20 (2018): 3624–40.

74. Manovich, "Poetics," 223.

75. Ash, *The Interface Envelope*, 124.

76. Baudrillard, *Simulacra and Simulation*, 2.

77. Alex MacLean, *Up on the Roof: New York's Hidden Skyline Spaces* (New York: Princeton Architectural Press, 2012).

78. Christopher Hawthorne, "Architects Change Their View of the Lowly Roof," *Los Angeles Times*, 6 November 2006, http://articles.latimes.com/2006/nov/06/entertainment/et-google6 (accessed 10 August 2018).

79. Graham, *Vertical*, 215–16.

80. Graham, *Vertical*, 46.

81. Cosgrove and Fox, *Photography and Flight*, 77. Another undesired effect of popular digital mapping technologies was recently reported in the media, as a fitness-tracking app was said to give away the location of secret overseas US and British army bases (from Afghanistan to Falkland) by way of the data about exercise routes shared online by soldiers ("Fitness Tracking App Strava Gives Away Location of Secret US Army Bases," https://www.theguardian.com /world/2018/jan/28/fitness-tracking-app-gives-away-location-of-secret-us-army-bases [accessed 17 August 2018]).

82. Moe Beltiks, "Rooftop Paintings Visible from Satellite," http://inhabitat.com/stunning -rooftop-paintings-that-can-be-seen-from-satellites/ (accessed 1 January 2012).

83. "How Rooftop QR Codes Are Being Used in Advertising," www.heavychef.com/how _rooftop_qr_codes_are_being_used_in_advertising (accessed 1 January 2012).

84. Graham, *Vertical*, 49. Activists have used the satellite view as an instrument of protest, from pacifist slogans painted on rooftops to more spectacular installations, such as a giant portrait of a Pakistani girl who lost her parents and her brother during a drone attack in 2009. "From ground level the grassy field looks to be covered with the kind of plastic sheeting commonly used to shelter intensively farmed fruit and vegetables. But this is not an agricultural device: it is an image on the earth. It is only when the sheet is seen from an aerial view the image becomes startlingly clear. . . . The girl's eyes look back vertically towards the distant sensors that support the drone attacks" (Graham, *Vertical*, 68–69). At the same time, as satellite imagery and GPS systems shape our field of vision and relationship with space, Graham intimates, the whole planet is being "draped with a militarized image of itself" (48).

85. Veronica della Dora, "A World of Slippy Maps: Google Earth, Global Visions and Topographies of Memory," *Transatlantica*, special issue "Mapping America" 2 (2012), http://transatlan tica.revues.org/6156 (accessed 10 August 2018).

86. Mark Jackson and Veronica della Dora, "'Dreams So Big Only the Sea Can Hold Them': Man-made Islands as Cultural Icons, Travelling Visions, and Anxious Spaces," *Environment and Planning A* 41 (2009): 2086–104.

87. Jackson and della Dora, "'Dreams So Big."

88. Jamet, "What Do Internet Metaphors Reveal," 9–10.

Epilogue

1. These and further finds are described in Georgina Rannard and Dominic Bailey, "Hidden Landscapes the Heat Wave Is Revealing," *BBC News*, 17 July 2018, https://www.bbc.com/news /uk-44767497 (accessed 27 August 2018).

2. See Tuan, "Surface Phenomena."

Bibliography

Abarbanel, Stacey Ravel. "Fowler Exhibition Highlights Artist Alighiero Boetti's Embroideries by Afghan Women." *UCLA Newsroom*, 14 December 2011. http://newsroom.ucla.edu/releases /fowler-museum-presents-an-exhibition-220534.

Accademia della Crusca. *Vocabolario degli Accademici della Crusca*. Venice: Appresso Giovanni Alberti, 1612.

Adams, Paul, Steven Hoelscher, and Karen Till, eds. *Textures of Place: Exploring Humanist Geographies*. Minneapolis: University of Minnesota Press, 2001.

Adams-Hutcheson, Gail, and Robyn Longhurst. "'At Least in Person There Would Have Been a Cup of Tea': Interviewing via Skype." *Area* 49 (2017): 148–55.

Addison, Joseph. "An Oration in Defence of the New Philosophy, Spoken in the Theatre at Oxford, July 7, 1693." In *A Conversation on the Plurality of Worlds: Translated from the French of M. de Fontenelle, to Which Is Added Mr. Addison's Defence of the Newtonian Philosophy*, ed. M. de Fontenelle, 193–200. London: E. Curll, 1728.

Agnew, John. *Geopolitics: Re-visioning World Politics*. New York: Routledge, 1998.

Ahmed, Nazneen. "Stitching in St. Thomas the Apostle Church Hall, Hanwell." In *My Life Is but a Weaving: A Collaborative Multi-faith Project with Artist Katy Beinart*, ed. Nazneen Ahmed et al., 62. Ealing, West London, UK, 2018.

Aimilianos Simonopetritēs, Archim. Λόγοι ἑόρτοι μυσταγωγικοί. Athens: Indiktos, 2014.

Alberti, Leon Battista. *De pictura*. http://www.filosofico.net/albertidepictura.htm. Accessed 15 May 2018.

Alpers, Svetlana. "The Mapping Impulse in Dutch Art." In *Art and Cartography: Six Historical Essays*, ed. David Woodward, 51–96. Chicago: University of Chicago Press, 1987.

Amundsen, Roald. *The South Pole: An Account of the Norwegian Antarctic Expedition in the Fram, 1910–1912*. New York: Cooper Square Press, 2001.

Anderson, Elijah. *The Cosmopolitan Canopy: Race and Civility in Everyday Life*. New York: W. W. Norton, 2011.

Andreopoulos, Andreas. *Metamorphosis: The Transfiguration in Byzantine Theology and Iconography*. Crestwood, NY: St. Vladimir's Seminary Press, 2005.

Apollonio, Fulvio. "Tormento e speranza nelle monete di Vivarelli." http://en.numista.com/cata logue/pieces5983.html. Accessed 1 May 2018.

Aresi, Paolo. *Della tribolatione e suoi rimedi: Lettioni di monsignor Paolo Aresi.* Vol. 2. Tortona: Niccolò Viola, 1624.

Arnould, Jacques. "Space Conquest and Ritual Practices: Lighting Candles for Ariane." *Theology and Science* 11 (2013): 44–51.

Ash, James. *The Interface Envelope: Gaming, Technology, Power.* New York: Bloomsbury, 2015.

Ash, James, Rob Kitchin, and Agnieszka Leszczynski. "Digital Turn, Digital Geographies?" *Progress in Human Geography* 42 (2018): 25–43.

Assarino, Luca. *Raguagli del Regno di Cipro.* Venice: Turrini, 1654.

Ault, Alicia. "What's the Deepest Hole Ever Dug?" *Smithsonian,* 19 February 2015.

Avery, Christine, and Michael Colebrook. *The Green Mantle of Romanticism.* London: Green Spirit Press, 2008.

Avery, Kevin. "The Heart of the Andes Exhibited: Frederic E. Church's Window on the Equatorial World." *American Art Journal* 18 (1986): 52–72.

Bachelard, Gaston. *On Poetic Imagination and Reverie.* Putnam, CT: Spring Publications, 2005.

Bachelard, Gaston. *The Poetics of Reverie: Childhood, Language, and the Cosmos.* Boston: Beacon Press, 1971.

Bacon, Francis. *Essays, Civil and Moral.* Vol. 3. New York: P. F. Collier & Son, 1909.

Bacon, Francis. *The Great Instauration and New Atlantis.* Arlington Heights, IL: Harlan Davidson, 1980.

Bacon, Francis. *The New Organon.* Ed. Lisa Jardine and Michael Silverthorne. Cambridge: Cambridge University Press, 2000.

Bacon, Francis. *Works.* Ed. James Spedding. Vol. 8. London: Longman & Co., 1874.

Bailey, Nathan. *Universal Etymological English Dictionary.* London: R. Ware, 1775.

Baker, Alan. *Geography and History: Bridging the Divide.* Cambridge: Cambridge University Press, 2003.

Barber, Peter. "Visual Encyclopedias: The Hereford and Other *Mappae Mundi*." *The Map Collector* 48 (1989): 2–8.

Barnes, Trevor, and James Duncan, eds. *Writing Worlds: Discourse, Text and Metaphor in the Representation of Landscape.* New York: Routledge, 1992.

Barnett, Clive. "Awakening the Dead: Who Needs the History of Geography?" *Transactions of the Institute of British Geographers* 20 (1995): 417–19.

Barnett, Lincoln. "The Canopy of Air." *Life,* 8 June 1953.

Barnett, Lincoln. "The Rain Forest." *Life,* 20 September 1954.

Barthes, Roland. *Camera Lucida: Reflections on Photography.* Trans. Richard Howard. New York: Hill & Wang, 1981.

Bascom, Willard. *A Hole on the Bottom of the Sea: The Story of the Mohole Project.* Garden City, NY: Doubleday, 1961.

Batty, Michael. "Virtual Geography." *Futures* 29 (1997): 337–52.

Baudrillard, Jean. *Simulacra and Simulation.* Ann Arbor: University of Michigan Press, 1994.

Beaver, Stanley. "Geography from a Railway Train." *Geography* 21 (1936): 265–83.

Béhar, Pierre. "Pour une géopolitique de la papauté." *Géopolitique* 58 (1997): 8–20.

Beltiks, Moe. "Rooftop Paintings Visible from Satellite." http://inhabitat.com/stunning-roof-top -paintings-that-can-be-seen-from-satellites/. Accessed 1 January 2012.

Bender, Barbara. "Time and Landscape." *Current Anthropology* 43 (2002): 103–12.

Benivieni, Girolamo. *Dialogo di Antonio Manetti: Cittadino fiorentino circa al sito, forma, & misure del lo inferno di Dante Alighieri poeta excellentissimo.* Florence: F. di Giunta, 1506.

Berman, Marshall. *All That Is Solid Melts into Air: The Experience of Modernity.* New York: Penguin Books, 1988.

Bernard of Clairvaux. *Life and Works of Saint Bernard Abbot of Clairvaux.* Ed. John Mabillon. Vol. 2. London: John Hodges, 1880.

Berners-Lee, Tim. *Weaving the Web: The Origins and Future of the World-Wide Web.* London: Orion Business, 1999.

Berners-Lee, Tim. "The World Wide Web: Past, Present and Future." https:// www.w3.org/People /Berners-Lee/1996/ppf.html. Accessed 8 August 2018.

Best, Kirsty. "Interfacing the Environment: Networked Screens and the Ethics of Visual Consumption." *Ethics and the Environment* 9 (2004): 65–85.

Bevan, William, and Henry Phillott. *Medieval Geography: An Essay in Illustration of the Hereford Mappa Mundi.* London: Stanford, 1873.

Biese, Alfred. *The Development of the Feeling for Nature in the Middle Ages and Modern Times.* London: George Routledge & Sons, 1905.

Biesele, Linda Hall, and Teresa Eckmann. *Mary, Mother and Warrior.* Austin: University of Texas Press, 2004.

Bittarello, Maria Beatrice. "Spatial Metaphors Describing the Internet and Religious Websites: Sacred Space and Sacred Place." *Observatorio* 11 (2009): 1–12.

Blowers, Paul. "'Entering This Sublime and Blessed Amphitheatre': Contemplation of Nature and Interpretation of the Bible in the Patristic Period." In *Nature and Scripture in the Abrahamic Religions: Up to 1700,* ed. Scott Mandelbrote and Jitse van der Meer, 2 vols., 148–76. Leiden: Brill, 2009.

"Blue Planet: New Facebook Map Depicts All of the World's Interconnected Friendships—with One Unmistakable Black Hole Over China." *Daily Mail,* 26 September 2013.

Blumenberg, Hans. *Paradigms for a Metaphorology.* Ithaca, NY: Cornell University Press, 2010.

Bodei, Remo. *Paesaggi sublimi. Gli uomini davanti alla natura selvaggia.* Milan: Bompiani, 2008.

Boetti, Agata. "Alighiero Loved All the Maps in the World." In *Alighiero Boetti,* ed. Laura Cherubini et al., 207–14. Florence: Forma, 2016.

Bogdanovich, Jelena. "The Rhetoric and Performativity of Light in the Sacred Space: A Case Study of the Vision of St. Peter of Alexandria." In *Hierotopy of Light and Fire in the Culture of the Byzantine World,* ed. Alexei Lidov, 282–304. Moscow: Feorija, 2013.

Bolt, Bruce. *Nuclear Explosions and Earthquakes: The Parted Veil.* San Francisco: W. H. Freeman, 1976.

Bonfiglioli, Stefania. "Quando le mappe si increspano." *E/C* (2009): 1–20.

Borges, Jorge Luis. "Things." http://poemsandpickaxes.blogspot.it/2014/02/things-by-jorge-luis -borges-fallen.html. Accessed 30 June 2016.

Borges, Jorge Luis. *A Universal History of Infamy.* Harmondsworth, UK: Penguin, 1975.

Bosselman, Fred. "The Influence of Ecological Science on American Law: An Introduction in the Symposium Ecology and the Law." *Chicago-Kent College of Law Review* 69 (1994): 847–73.

Bosteels, Bruno. "A Misreading of Maps: The Politics of Cartography in Marxism and Poststructuralism." In *Signs of Change: Premodern, Modern, Postmodern,* ed. Stephen Barker, 109–38. Albany: State University of New York Press, 1996.

Boulton, Andrew, and Matthew Zook. "Landscape, Locative Media, and the Duplicity of Code." In *The Wiley-Blackwell Companion to Cultural Geography,* ed. Nuala Johnson, Richard Schein, and Jamie Winders, 437–51. Chichester: Wiley Blackwell, 2013.

Bowden-Pickstock, Susan. *Quiet Gardens: The Roots of Faith?* London: Bloomsbury, 2009.

Bowen, Ira. "Sky Survey Charts Universe." *National Geographic*, December 1956.

Boyd, Richard. "Metaphor and Theory Change: What Is 'Metaphor' a Metaphor for?" In *Metaphor and Thought*, ed. Andrew Ortony, 481–532. Cambridge: Cambridge University Press, 1993.

Boyer, David. "Year of Discovery Opens in Antarctica: Daring Scientists of Dozens of Nations, Pooling Knowledge and Resources Launch Man's Most Ambitious Assault on the Whole Continent." *National Geographic*, September 1957.

Boyer, Paul. *By the Bomb's Early Light: American Thought and Culture at the Dawn of the Atomic Age*. New York: Pantheon, 1985.

Boyle, Robert. *The Works of the Honourable Robert Boyle*. Vol. 1. London: Printed for J. & F. Rivington et al., 1772.

Brett, Guy. "Itinerary." In *Mona Hatoum*, ed. Michel Archer et al., 34–87. London: Phaidon Press, 1997.

Bricker, Charles. *Landmarks of Cartography*. Ware, Hertfordshire, UK: Wordsworth, 1989.

Brogard Kristensen, Dorthe, and Minna Ruckenstein. "Co-evolving with Self-tracking Technologies." *New Media and Society* 20 (2018): 3624–40.

Brotton, Jerry. *The History of the World in Twelve Maps*. New York: Penguin Books, 2013.

Brush, Stephen. "Discovery of the Earth's Core." *American Journal of Physics* 48 (1980): 705–24.

Brush, Stephen. "Nineteenth-Century Debates about the Inside of the Earth: Solid, Liquid or Gas?" *Annals of Science* 36 (1979): 225–54.

Bumstead, Newman. "Rockets Explore the Air above Us: Scientists, Firing Missiles Equipped with Electronic Eyes and Ears, Probe Mysteries on the Borders of Outer Space." *National Geographic*, February 1957.

Bunkse, Edmund. "Humboldt and Aesthetic Tradition in Geography." *Geographical Review* 71 (1981): 127–46.

Buoni, Tommaso. *Discorsi accademici de' mondi: Parte prima, nella quale con stile oratorio si parla dell'Archetipo de' Mondi: Del Mondo Angelico; del Mondo inferiore reato; della nobiltà, eccellenza, bellezza, meraviglie, forze e differenze de' Mondi; del Cielo; della Luce; de gli Elementi; de Misti; delle piante; e de' gli animali*. Venice: Battista Colosini, 1605.

Burckhardt, Jacob. *The Civilization of the Period of the Renaissance in Italy*. Trans. Samuel Middlemore. Cambridge: Cambridge University Press, 2015.

Burnet, Thomas. *The Sacred Theory of the Earth: Containing an Account of the Original Creation of the Earth and all the General Changes which It Hath Already Undergone, or Is to Undergo till the Consummation of All Things*. London: John Hooke, 1719.

Buttimer, Anne. "Alexander von Humboldt and Planet Earth's Green Mantle." *Cybergeo: European Journal of Geography*, 2012. https://journals.openedition.org/cybergeo/25478.

Buzzati, Dino. "Il Mantello." *Il Dramma*, June 1960.

Byrd, Richard. "All-out Assault on Antarctica: Operation Deep Freeze Carves Out United States Bases for a Concerted International Attack on Secrets of the Frozen Continent." *National Geographic*, August 1956.

Byrd, Richard. "To the Men at South Pole Station." *National Geographic*, July 1957.

Cándenas, Inés Ferrero. "Reconfiguring the Surrealist Gaze: Remedios Varo's Images of Women." *Bulletin of Hispanic Studies* 88 (2011): 455–67.

Cañizares-Esguerra, Jorge. *Nature, Empire, and Nation: Explorations of the History of Science in the Iberian World*. Stanford, CA: Stanford University Press, 2006.

Capalbo, Domenicantonio. *Panegirici e sermoni sacri*. Vol. 1. Naples: Andrea Migliaccio, 1775.

Capel, Horacio. *Filosofia e scienza nella geografia contemporanea*. Milan: Unicopli, 1987.

Carey, James. *Communication as Culture: Essays on Media and Society*. New York: Routledge, 2009.

Carrington, Richard. *The Mediterranean: Cradle of Western Culture*. New York: Viking Press, 1977.

Carroll, Lewis. *Sylvie and Bruno Concluded*. London: Macmillan, 1893.

Carson, Rachel. *The Edge of the Sea*. Boston: Houghton Mifflin, 1955.

Carson, Rachel. *Silent Spring*. New York: Mariner Books, 2002.

Casey, Edward. *Representing Place: Landscape Painting and Maps*. Minneapolis: University of Minnesota Press, 2002.

Castells, Manuel. *The Internet Galaxy: Reflections on the Internet, Business, and Society*. Oxford: Oxford University Press, 2001.

Castells, Manuel. "The New Public Sphere: Global Civil Society, Communication Networks, and Global Governance." *Annals AAPSS* 616 (2008): 78–93.

Castells, Manuel. *The Rise of the Network Society*. Oxford: Blackwell, 1996.

Caylus, Anne Claude Philippe de Pestels de Lévis de Tubières-Grimoard, Comte de. *Recueil d'antiquités*. Vol. 5. Paris: Tilliard, 1762.

Cerizza, Luca. *Alighiero e Boetti: Mappa*. London: Afterall, 2008.

Ceserani, Remo, Mario Domenichelli, and Pino Fasano. *Dizionario dei temi letterari*. Turin: UTET, 2007.

Chadwick, Joseph. "A Blessing and a Curse: The Poetics of Privacy in Tennyson's 'The Lady of Shalott.'" In *Critical Essays on Alfred Tennyson*, ed. Herbert Tucker, 83–99. New York: Macmillan, 1993.

Chaplin, Joyce. *Round about the Earth: Circumnavigation from Magellan to Orbit*. New York: Simon & Schuster, 2014.

Chardin, John. *Journal du voyage en Perse et aux Indes Orientales par la Mer Noire & par la Colchide*. London: M. Pitt, 1686.

Christaller, Walter. *Central Places in Southern Germany*. Trans. Carlisle Baskin. Englewood Cliffs, NJ: Prentice-Hall, 1966.

Christian, Linda. *Theatrum Mundi: The History of an Idea*. New York: Garland Publishing, 1987.

Chryssochoou, Stella. "The Cartographical Tradition of Ptolemy's Ἰεωγραφικὴ Ὑφήγησις in the Palaeologan Period and the Renaissance (13th–16th Century)." PhD diss., Royal Holloway University of London, 2010.

Ciotti, Giovanni Battista. *Nuova Comedia cavata dall'opra della Contralesina del Pastor Monopolitano intitolata le nozze d'Antilesina*. Venice: Appresso Gio. Battista Ciotti Sanese, 1604.

Clark, Kenneth. *Landscape into Art*. London: Folio Society, 2013.

Clark, Stuart. *Vanities of the Eye: Vision in Early Modern European Culture*. Oxford: Oxford University Press, 2013.

Clifford, Ruth. "Textiles, Language and Metaphor." http://travelsintextiles.com/textiles-language-and-metaphor/. Accessed 15 July 2018.

Clout, Hugh. "Stanley Henry Beaver (1907–84)." In *Geographers: Biobibliographical Studies*, ed. Elizabeth Baigent and André Reyes Novaes, 36:129–56. London: Bloomsbury, 2017.

Coleridge, Ernest Hartley. *The Complete Poetical Works of Samuel Taylor Coleridge*. Vol. 2. Oxford: Clarendon Press, 1975.

Collins, Andrew. "The Royal Costume and Insignia of Alexander the Great." *American Journal of Philology* 133 (2012): 371–402.

Collis, Christy, and Klaus Dodds. "Assault on the Unknown: The Historical and Political Geographies of the International Geophysical Year (1957–8)." *Journal of Historical Geography* 34 (2008): 555–73.

Commission of the European Communities. *Europe's Green Mantle: The Heritage and Future of Our Forests*. Luxembourg: EUR-OP, 1984.

Congressional Record—Senate. 14 April 1958, 6306–7. https://www.gpo.gov/fdsys/pkg/GPO-CRECB
-1958-pt5/pdf/ GPO-CRECB-1958-pt5–7.pdf. Accessed 23 March 2018.

Conly, Robert. "Men Who Measure the Earth: Surveyors from 18 New World Nations Invade Trackless Jungles and Climb Snow Peaks to Map Latin America." *National Geographic*, March 1956.

Constas, Fr. Maximos. *The Art of Seeing: Paradox and Perception in Orthodox Iconography.* Al-
hambra, CA: Sebastian Press, 2014.

Constas, Fr. Maximos. "Beyond the Veil: Imagination and Spiritual Vision in Byzantium." Unpublished paper. University of Chicago Divinity School Workshop, March 2015.

Constas, Fr. Maximos. "Review of Liliya Berezhnaya and John-Paul Himka's *The World to
Come.*" *Religion and the Arts* 20 (2016): 231–49.

Constas, Nicholas. *Proclus of Constantinople and the Cult of the Virgin in Late Antiquity: Homilies 1–5, Texts and Translations.* Leiden: Brill, 2003.

Constas, Nicholas. "Symeon of Thessalonica and the Theology of the Icon Screen." In *Thresholds of the Sacred*, ed. Sharon Gerstel, 163–84. Washington, DC: Dumbarton Oaks Research
Library and Collection, 2006.

Cooper, John Gilbert. "The Tomb of Shakespeare." http://spenserians.cath.vt.edu/TextRecord
.php?action=GET&textsid=37833. Accessed 1 May 2018.

Cooper, John Gilbert, and Nathaniel Cotton. *The British Poets.* Vol. 72. Whittingham, UK: Cooper & Cotton, 1822.

Corner, James. "The Agency of Mapping: Speculation, Critique and Invention." In *Mappings*, ed.
Denis Cosgrove, 213–33. London: Reaktion, 1999.

Cornish, Vaughan. "Harmonies of Scenery: An Outline of Aesthetic Geography." *Geography* 14
(1928): 383–94.

Cornish, Vaughan. "On Kumatology: The Study of the Waves and Wave-Structures of the Atmosphere, Hydrosphere, and Lithosphere." *Geographical Journal* 13 (1899): 624–26.

Coronelli, Vincenzo. *Atlante Veneto. Tomo I nel quale si contiene la descrittione geografica, storica, sacra, profana, e politica, degl'imperii, regni, provincie, e stati dell'Universo, loro divisione,
e confini, coll'aggiunta di tutti li paesi nuovamente scoperti, accresciuto di molte tavole geografiche, non più publicate.* Venice: Domenico Padovani, 1690.

Coronelli, Vincenzo. *Isolario, descrittione geografico-historica, sacro-profana, antico- moderna,
politica, naturale, e poetica. Mari, golfi, seni, piagge, porti, barche, pesche, promontori, monti,
boschi, fiumi . . . ed ogni più esatta notitia di tutte l'isole coll'osservationi degli scogli sirti,
scagni, e secche del globo terracqueo. Aggiuntivi anche i ritratti de' dominatori di esse. Ornato
di trecento-dieci tavole geografiche, topografiche, corografiche, iconografiche, scenografiche . . .
a' maggiore dilucidatione, ed uso della navigatione, et in supplimento dei 14 volumi del Blaeu,
tomo dell'Atlante veneto.* Pt. 1. Venice: By the author, 1696.

Cosgrove, Denis. *Apollo's Eye: A Cartographic Genealogy of the Earth in the Western Imagination.*
Baltimore: Johns Hopkins University Press, 2001.

Cosgrove, Denis. *Geography and Vision.* London: I. B. Tauris, 2008.

Cosgrove, Denis. "Global Illumination and Enlightenment in the Geographies of Vincenzo
Coronelli and Athanasius Kircher." In *Geography and Enlightenment*, ed. David Livingstone
and Charles Withers, 33–66. Chicago: University of Chicago Press, 1999.

Cosgrove, Denis. "Globalism and Tolerance in Early Modern Geography." *Annals of the Association of American Geographers* 93 (2003): 852–70.

Cosgrove, Denis. "Images and Imagination in 20th-century Environmentalism: From the Sierras to the Poles." *Environment and Planning A* 40 (2008): 1862–80.

Cosgrove, Denis. "Modernity, Community and the Landscape Idea." *Journal of Material Culture* 11 (2006): 49–66.

Cosgrove, Denis. *The Palladian Landscape: Geographical Change and Its Cultural Representations in Sixteenth-century Italy*. Leicester: Leicester University Press, 1994.

Cosgrove, Denis. "Prospect, Perspective and the Evolution of the Landscape Idea." *Transactions of the Institute of British Geographers* 10 (1985): 45–62.

Cosgrove, Denis. *Social Formation and Symbolic Landscape*. Madison: University of Wisconsin Press, 1998.

Cosgrove, Denis, and William Fox. *Photography and Flight*. London: Reaktion Books, 2010.

Coulton, George. *Life in the Middle Ages*. Vol. 1. Cambridge: Cambridge University Press, 1967.

Courtot, Roland. "Les Paysages et les hommes des Alpes du sud dans les carnets de Paul Vidal de La Blache." *Méditerranée* 109 (2007): 9–15.

Cowart, David. "Pynchon's *The Crying of Lot 49* and the Paintings of Remedios Varo." *Critique* 18 (1977): 19–26.

Cramer, Florian. "What Is the Post-Digital?" In *Postdigital Aesthetics*, ed. David Berry and Michael Dieter, 12–26. New York: Palgrave Macmillan, 2015.

Crang, Mike, Phil Crang, and Jon May, eds. *Virtual Geographies: Bodies, Space and Relations*. New York: Routledge, 1999.

Cuir, Raphael. *The Development of the Study of Anatomy from the Renaissance to Cartesianism: Da Carpi, Vesalius, Estienne Bidloo*. Lewiston, NY: Edwin Mellen Press, 2009.

Cunningham, Mary. "The Mother of God and the Natural World." *Analogia* 1 (2016): 41–51.

Curry, Michael. "New Technologies and the Ontology of Places." Paper presented at the Information Studies Seminar, University of California, Los Angeles, 4 March 1999. http://baja .sscnet.ucla.edu/~curry/Curry_Tech_Regimes.pdf.

Cvetkovski, Sašo. "The Vision of Saint Peter of Alexandria, from the Church of St. Archangels in Prilep." *Zograf* 36 (2012): 83–88.

d'Alembert, Jean. "Preliminary Discourse." In Denis Diderot, *The Encyclopedia*, ed. and trans. Stephen Gendzier, 1–41. New York: Harper Torchbooks, 1967.

Daly, Reginald. *Strength and Structure of the Earth*. New York: Prentice-Hall, 1940.

Darby, Henri Clifford. "The Changing English Landscape." *Geographical Journal* 117 (1951): 377–94.

Darby, Henri Clifford. *Domesday Geography of England*. 7 vols. Cambridge: Cambridge University Press, 1952–67.

Darby, Henri Clifford. *An Historical Geography of England before A.D. 1800*. Cambridge: Cambridge University Press, 1936.

Darby, Henri Clifford. "On the Relations of Geography and History." *Transactions and Papers (Institute of British Geographers)* 19 (1953): 1–11.

Darwin, Charles. *On the Origin of Species by Means of Natural Selection or, The Preservation of Favored Races in the Struggle for Life*. New York: Cosimo Classics, 2007.

Davis, Michael. "Hertfordshire's Green Mantle." *Arboricultural Journal* 11 (1987): 53–71.

Davis, William Morris. "The Physical Geography of the Lands." *Popular Science Monthly* 57 (1900): 157–70.

Defense Nuclear Agency. *Operation ARGUS, 1958*. Washington, DC: Department of Defense Documents, 1958.

Delfico, Melchiorre. *Memoria sulla coltivazione del riso nella provincia di Teramo*. Naples: G. M. Porcelli, 1783.

de Lille, Alain. *The Complaint of Nature*. Trans. Douglas Moffat. Yale Studies in English 36. New Haven, CT: Yale University Press, 1908.

della Dora, Veronica. *Landscape, Nature and the Sacred in Byzantium*. Cambridge: Cambridge University Press, 2016.

della Dora, Veronica. "Lifting the Veil of Time: Maps, Metaphor and Antiquarianism, 17–18th c." In *Time in Space: Representing the Past in Maps*, ed. Kären Wiegen and Caroline Winterer, 103–25. Chicago: University of Chicago Press, 2020.

della Dora, Veronica. "Mapping Melancholy-Pleasing Remains: Morea as a Renaissance Memory Theater." In *Viewing the Morea: Land and People in the Late Medieval Peloponnese*, ed. Sharon Gerstel, 455–75. Washington, DC: Dumbarton Oaks Research Library and Collection, 2013.

della Dora, Veronica. *Mountain: Nature and Culture*. London: Reaktion, 2016.

della Dora, Veronica. "A World of Slippy Maps: Google Earth, Global Visions and Topographies of Memory." *Transatlantica* 2 (2012). Special issue: "Mapping America." http://tran- satlantica .revues.org/6156.

dell'Agnese, Elena. "Halford John Mackinder (1861–1947)." In *Cos'è il mondo? È un globo di cartone. Insegnare geografia fra Otto e Novecento*, ed. Marcella Schmidt, 247–55. Milan: Unicopli, 2010.

Dematteis, Giuseppe. *Le metafore della terra. La geografia umana tra mito e scienza*. Milan: Feltrinelli, 1985.

Demeritt, David. "The Nature of Metaphors in Cultural Geography and Environmental History." *Progress in Human Geography* 18 (1994): 163–85.

Dempsey, Charles. *Inventing the Renaissance Putto*. Chapel Hill: University of North Carolina Press, 2001.

de Souza e Silva, Adriana. "From Cyber to Hybrid: Mobile Technologies as Interfaces of Hybrid Spaces." *Space and Culture* 9 (2006): 261–78.

de Souza e Silva, Adriana, and Jordan Frith. "Locative Mobile Social Networks: Mapping Communication and Location in Urban Spaces." *Mobilities* 5 (2010): 485–505.

de Souza e Silva, Adriana, and Daniel Sutko. "Theorizing Locative Technologies through Philosophies of the Virtual." *Communication Theory* 21 (2011): 23–42.

Dettelbach, Michael. "Global Physics and Aesthetic Empire: Humboldt's Physical Portrait of the Tropics." In *Visions of Empire: Voyages, Botany and Representations of Nature*, ed. Daniel Miller and Peter Reill, 258–92. Cambridge: Cambridge University Press, 1996.

Dewey, Malcom Howard. "Herder's Relation to the Aesthetic Theory of His Time: A Contribution Based on the Fourth Critical Waeldchen." PhD diss., University of Chicago, 1920.

Dickinson, Robert. *The Makers of Modern Geography*. New York: Praeger, 1969.

Dicks, Bella. *Culture on Display: The Production of Contemporary Visitability*. Milton Keynes, UK: Open University Press, 2003.

Dicks, D. R. *The Geographical Fragments of Hipparchus*. London: Athlone Press, 1960.

Diderot, Denis. *The Encyclopedia*, ed. and trans. Stephen Gendzier. New York: Harper Torchbooks, 1967.

Dilke, O. A. W. "Cartography in the Byzantine Empire." In *History of Cartography*, ed. J. B. Harley and David Woodward, 1:258–75. Chicago: University of Chicago Press, 1987.

Dilke, O. A. W. "The Culmination of Greek Cartography in Ptolemy." In *The History of Cartography*, ed. J. B. Harley and David Woodward, 1:177–200. Chicago: University of Chicago Press, 1987.

Dilke, O. A. W. "Maps in the Service of the State: Roman Cartography to the End of the Augustan Era." In *The History of Cartography*, ed. J. B. Harley and David Woodward, 1:201–11. Chicago: University of Chicago Press, 1987.

Divine, David. *The Opening of the World: The Great Age of Maritime Exploration.* New York: Putnam, 1973.

Dodd, Adam. "Size Is in the Eye of the Beholder: On the Cultural History of Microfaunae in Seventeenth-Century Europe." In *Assigning Cultural Values*, ed. Kjerstin Aukurst, 15–28. New York: Peter Lang, 2013.

Dodds, Klaus. "Assault on the Unknown: Geopolitics, Antarctic Science and the International Geophysical Year (1957–8)." In *New Spaces of Exploration*, ed. Simon Naylor and James Ryan, 148–210. London: I. B. Tauris, 2010.

Dodds, Klaus, and Lisa Funnell. "Going Atmospheric and Elemental: Roger Moore's and Timothy Dalton's James Bond and Cold War Geopolitics." In *Media and the Cold War in the 1980s: Between Star Wars and Glasnost*, ed. Henrik Bastiansen, Martin Klimke, and Rolf Werenskjold, 63–86. London: Palgrave Macmillan, 2019.

Dodge, Martin, and Rob Kitchin. "Code and the Transduction of Space." *Annals of the Association of American Geographers* 95 (2005): 162–80.

Doyle, Arthur Conan. *When the World Screamed.* London: Strand Magazine, 1922.

Dreier, Peter. "How Rachel Carson and Michael Harrington Changed the World." *Contexts* 11 (2012): 40–46.

Dronke, Peter. *Bernardus Silvestris' Cosmographia.* Leiden: Brill, 1978.

Dryden, Hugh. "The International Geophysical Year: Man's Most Ambitious Study of His Environment." *National Geographic*, February 1956.

Dwyer, Claire, and Katie Beinart. "My Life Is But a Weaving: A Collaborative Arts Project." In *My Life Is But a Weaving: A Collaborative Multi-faith Project with Artist Katy Beinart*, ed. Nazneen Ahmed et al., 3–7. Ealing, West London, UK, 2018.

Eagleton, Terry. *Against the Grain: Essays 1975–1985.* London: Verso Press, 2010.

Edgerton, Samuel. "From Mental Matrix to Mappamundi to Christian Empire: The Heritage of Ptolemaic Cartography in the Renaissance." In *Art and Cartography: Six Historical Essays*, ed. David Woodward, 10–50. Chicago: University of Chicago Press, 1987.

Edson, Evelyn. *The World Map, 1300–1492: The Persistence of Tradition and Transformation.* Baltimore: Johns Hopkins University Press, 2007.

"Encyclopedia of Fashion." http://www.fashionencyclopedia.com/fashion_costume_culture/The-Ancient-World-Greece/Chlamys.html#ixzz4UAlRpWzL. Accessed 30 June 2016.

Engelmann, Sasha. "Social Spiders and Hybrid Webs at Studio Tomás Saraceno." *Cultural Geographies* 24 (2017): 161–69.

Eusebius of Caesarea. *In Praise of Constantine.* Trans. H. A. Drake. Berkeley: University of California Press, 1976.

Evangelatou, Maria. "Threads of Power: Clothing Symbolism, Human Salvation, and Female Identity in the Illustrated Homilies by Iakobos of Kokkinobaphos." *Dumbarton Oaks Papers* 68 (2014): 241–323.

Falchetta, Pietro. *Fra' Mauro's World Map.* Turnhout: Brepols, 2006.

Farinelli, Franco. *Geografia. Un'introduzione ai modelli del mondo.* Turin: Einaudi, 2003.

Farinelli, Franco. *L'Invenzione della Terra.* Palermo: Sellerio, 2007.

Farinelli, Franco. *I Segni del mondo.* Florence: Scandicci, 1992.

Farman, Jason. "Tubes for Fiber Optics." *Media Theory* 2 (2018). http://mediatheoryjournal.org /jason-farman-invisible-and-instantaneous/.

Feuerlicht, Ignace. "A New Look at the Iron Curtain." *American Speech* 30 (1955): 186–89.

Findlen, Paula. "The Economy of Scientific Exchange in Early Modern Europe." In *Patronage and Institutions: Science, Technology, and Medicine at the European Courts, 1500–1750*, ed. Bruce Moran, 5–24. Woodbridge, UK: Boydell Press, 1990.

Finocchiaro, Maurice. *Defending Copernicus and Galileo: Critical Reasoning in the Two Affairs.* New York: Springer, 2012.

Foltz, Bruce. "Nature Godly and Beautiful: The Iconic Earth." *Phenomenology* 1 (2001): 113–55.

Foster, John. *The Sustainability Mirage: Illusion and Reality in the Coming War on Climate*. London: Routledge, 2012.

Fraser, Peter. *Ptolemaic Alexandria*. Oxford: Clarendon Press, 1971.

Frazier, Paul. "Across Frozen Desert to Byrd Station: Tracker-Borne Explorers Conquer Yawning Chasms of Ice and Snow to Set Up an IGY Outpost in Uncharted Marie Byrd Land." *National Geographic*, September 1957.

Freedman, Barbara. *Staging the Gaze: Postmodernism, Psychoanalysis, and Shakespearean Comedy*. Ithaca, NY: Cornell University Press, 1991.

Frith, Jordan. *Smartphones as Locative Media*. Cambridge, MA: Polity Press, 2015.

Frith, Jordan. "Splintered Space: Hybrid Spaces and Differential Mobility." *Mobilities* 7 (2012): 131–49.

Furetière, Antoine. *Dictionnaire universel, contenant generalement tous les mots françois, tant vieux que moderns, & les termes des sciences & des arts*. 4th ed. Vol. 1. The Hague: Chez Pierre Husson et al., 1727.

Gabrys, Jennifer. "Automatic Sensation: Environmental Sensors in the Digital City." *Senses and Society* 2 (2007): 189–200.

Gajewski, Alexandra, and Stefanie Seeberg. "Having Her Hand in It? Elite Women as 'Makers' of Textile Art in the Middle Ages." *Journal of Medieval History* 42 (2016): 26–50.

Ganz, David. "Pictorial Textiles and Performance: The Star Mantle of Henry II." In *Dressing the Part: Textiles as Propaganda in the Middle Ages*, ed. Kate Dimitrova and Margaret Goehring, 13–29. Turnout: Brepols, 2014.

Garrison, Eliza. *Ottonian Imperial Art and Portraiture: The Artistic Patronage of Otto III*. Farnham, UK: Ashgate, 2011.

Gautier Dalché, Patrick. "The Reception of Ptolemy's Geography (End of the Fourteenth to Beginning of the Sixteenth Century)." In *The History of Cartography*, vol. 3, ed. David Woodward, 285–364. Chicago: University of Chicago Press, 2007.

Gentile, Sebastiano. *Firenze e la scoperta dell'America*. Florence: Olschki, 1992.

Gerstel, Sharon. "Introduction." In *Thresholds of the Sacred*, ed. Sharon Gerstel, 1–6. Washington, DC: Dumbarton Oaks Research Library and Collection, 2006.

Gibson, James. *The Ecological Approach to Visual Perception*. New York: Psychology Press, 2015.

Gibson, James. *The Perception of the Visual World*. Boston: Houghton Mifflin, 1950.

Gillham, Mary. *Swansea Bay's Green Mantle*. Cowbridge, UK: D. Brown, 1982.

Gillis, John. *Islands of the Mind: How the Human Imagination Created the Atlantic World*. New York: Palgrave Macmillan, 2004.

Giorgio di Pisidia. *Esaemerone*. Trans. Fabrizio Gonnelli. Pisa: Edizioni ETS, 1998.

Glacken, Clarence. *Traces on the Rhodian Shore: Nature and Culture in Western Thought from Ancient Times to the End of the Eighteenth Century*. Berkeley: University of California Press, 1967.

Glotfelty, Cheryll. "Cold War, Silent Spring: The Trope of War in Modern Environmentalism." In *And No Birds Sing: Rhetorical Analyses of Rachel Carson's Silent Spring*, ed. Craig Waddell, 158–73. Carbondale: Southern Illinois University Press, 2000.

Godwin, Jocelyn. *Athanasius Kircher: A Renaissance Man and the Quest for Lost Knowledge*. New York: Thames & Hudson, 1979.

Godwin, Matthew. "Britnik: How America Made and Destroyed Britain's First Satellite." In *New Spaces of Exploration: Geographies of Discovery in the Twentieth Century*, ed. Simon Naylor and James Ryan, 173–95. London: I. B. Tauris, 2009.

Goethe, Johann Wolfgang. "Dedication." *NOH London Magazine* 9 (1824): 186–88.

Goethe, Johann Wolfgang. *Faust*. Trans. A. S. Kline. www.poetryintranslation.com. Accessed 29 January 2016.

Goethe, Johann Wolfgang. *The Sorrows of the Young Werther*. Cambridge, MA: Harvard Classics Shelf of Fiction, 1917.

Goethe, Johann Wolfgang. "Welcome and Departure." In *The Works of J. W. von Goethe*, vol. 9, ed. Nathan Haskell Dole, 40. Boston: Francis A. Nicolls & Co., 1902.

Gogol, Nikolaij Vasil'evich. *The Overcoat*. London: Bristol Classical Press, 1991.

Gore, Al. "The Digital Earth: Understanding Our Planet in the 21st Century." Speech delivered to the California Science Center, Los Angeles, 1998. http://www.digitalearth.gov/VP19980 131.html.

Graham, Mark. "Neogeography and the Palimpsests of Place: Web 2.0 and the Construction of a Virtual Earth." *Tijdschrift voor Economische en Sociale Geografie* 101 (2010): 422–36.

Graham, Mark, and Matthew Zook. "Augmented Realities and Uneven Geographies: Exploring the Geolinguistic Contours of the Web." *Environment and Planning A* 45 (2013): 77–99.

Graham, Stephen. "The End of Geography or the Explosion of Place? Conceptualizing Space, Place and Information Technology." *Progress in Human Geography* 22 (1998): 165–85.

Graham, Stephen. *Vertical: The City from Satellites to Bunkers*. London: Verso, 2018.

Greenberg, Daniel. "Mohole: Geopolitical Fiasco." In *Understanding the Earth: A Reader in the Earth Sciences*, ed. Ian Gass, Peter Smith, and Richard Wilson, 343–48. Sussex, UK: Artemis Press, 1971.

Greenberg, Daniel. *The Politics of Pure Science*. Chicago: University of Chicago Press, 1999.

Grotowski, Piotr. *Arms and Armour of the Warrior Saints: Tradition and Innovation in Byzantine Iconography, 843–1261*. Leiden: Brill, 2010.

Grosvenor, Meeville. "Admiral at the Ends of the Earth." *National Geographic*, July 1957.

Haber, Heinz. "Space Satellites, Tools of Earth Research." *National Geographic*, April 1956.

Hadot, Pierre. *The Veil of Isis: An Essay on the History of the Idea of Nature*. Cambridge, MA: Harvard University Press, 2006.

Hagen, Joel. *An Entangled Bank: The Origins of Ecosystem Ecology*. New Brunswick, NJ: Rutgers University Press, 1992.

Halliwell, Stephen. *The Aesthetics of Mimesis: Ancient Texts and Modern Problems*. Princeton, NJ: Princeton University Press, 2002.

Halmi, Nicholas. *The Genealogy of the Romantic Symbol*. Oxford: Oxford University Press, 2011.

Hamblin, Jacob. "Science in Isolation: American Marine Geophysics Research, 1950–1968." *Physics in Perspective* 2 (2000): 293–312.

Hansen, Mark. *Bodies in Code: Interfaces with Digital Media*. New York: Routledge, 2006.

Hansen, Peter. *The Summits of Modern Man: Mountaineering after the Enlightenment*. Cambridge, MA: Harvard University Press, 2013.

Hansson, Heidi. "*Punch, Fun, Judy* and the Polar Hero: Comedy, Gender and the British Arctic Expedition 1875–1876." In *North and South: Essays on Gender, Race and Region*, ed. Christine DeVine and Mary Ann Wilson, 61–89. Cambridge: Cambridge Scholars Publishing, 2012.

Hardin, Garret. *Biology: Its Principles and Implications*. San Francisco: Freeman, 1961.

Harley, J. B., and David Woodward. "The Foundations of Theoretical Cartography in Archaic and Classical Greece." In *The History of Cartography*, ed. J. B. Harley and David Woodward, 1:130–47. Chicago: University of Chicago Press, 1987.

Harris, Dianne, and Dede Fairchild-Ruggles. *Sites Unseen: Landscape and Vision*. Philadelphia: University of Pennsylvania Press, 2007.

Harris, William, and Robert Lamb. "How Invisibility Cloaks Work." http://science.howstuff works.com/invisibility-cloak.htm. Accessed 10 June 2016.

Harrison, Richard Edes. *Look at the World: The Fortune Atlas for World Strategy*. New York: Alfred A. Knopf, 1944.

Hartshorne, Richard. *The Nature of Geography: A Critical Survey of Current Thought in the Light of the Past*. Lancaster, PA: Association of American Geographers, 1961.

Hawthorne, Christopher. "Architects Change Their View of the Lowly Roof." *Los Angeles Times*, 6 November 2006.

Heilbron, John, ed. *The Oxford Companion to the History of Modern Science*. Oxford: Oxford University Press, 2003.

Henry, John. "National Styles in Science: A Possible Factor in the Scientific Revolution?" In *Geography and Revolution*, ed. David Livingstone and Charles Withers, 43–74. Chicago: University of Chicago Press, 2005.

Hern, Alex. "Fitness Tracking App Strava Gives Away Location of Secret US Army Bases." *The Guardian*, 28 January 2018.

Herodotus. *History*. Trans. A. D. Godley. Cambridge, MA: Harvard University Press, 1920.

Hess, Harry. "The AMSOC Hole to the Earth's Mantle." *Eos* 40 (1959): 340–45.

Higman, Barry. *Flatness*. London: Reaktion, 2017.

Holbrook, Florence. *Book of Nature Myths*. Boston: Houghton, Mifflin, 1902.

Holmes, Richard. *Samuel Taylor Coleridge: Selected Poetry*. New York: Penguin Classics, 1996.

Hoogvliet, Margaret. "*Mappae Mundi* and Medieval Encyclopedias: Image versus Text." In *Pre-modern Encyclopedic Texts*, ed. Peter Binkley, 63–88. Leiden: Brill, 1997.

Hook, Jan. "Textiles, Language and Metaphor." http://travelsintextiles.com/textiles-language-and -metaphor/. Accessed 20 May 2018.

Hoskins, William George. *The Making of the English Landscape*. Dorset, UK: Little Toller, 2013.

Howorth, J. *Poems by Baron von Haller*. London: J. Bell, 1794.

"How Rooftop QR Codes Are Being Used in Advertising." www.heavychef.com/how_rooftop _qr_codes_are_being_used_in_advertising. Accessed 1 January 2012.

Hu, Tung-Hui. *A Prehistory of the Cloud*. Cambridge, MA: MIT Press, 2015.

Hulme, Peter, and Ludmilla Jordanova. *The Enlightenment and Its Shadows*. London: Routledge, 1990.

Humboldt, Alexander von. *Cosmos: A Sketch of the Physical Description of the Universe*. Trans. Elise Otté. 2 vols. Baltimore: Johns Hopkins University Press, 1997.

Humboldt, Alexander von, and Aimé Bonpland. *Essay on the Geography of Plants*. Ed. and trans. Stephen Jackson and Sylvie Romanovski. Chicago: University of Chicago Press, 2013.

Huntford, Roland. *Race for the South Pole: The Expedition Diaries of Scott and Amundsen*. New York: Continuum, 2011.

Huntington, Ellsworth. "Geography and Aviation." In *Geography in the Twentieth Century*, ed. Griffith Taylor, 528–40. New York: Methuen, 1951.

Ingold, Tim. "Earth, Sky, Wind, and Weather." *Journal of the Anthropological Institute* 13 (2007): 19–38.

Ingold, Tim. "When ANT Meets SPIDER: Social Theory for Anthropods." In *Material Agency: Towards a Non-Anthropocentric Approach*, ed. Carl Knappett and Lambros Malafouris, 209–15. New York: Springer, 2010.

"Introducing a New *Life* Series: The New Portrait of Our Planet. Great Discoveries of the IGY Are Revealed." *Life*, 7 November 1960.

Introna, Lucas, and Fernando Ilharco. "On the Meaning of Screens: Towards a Phenomenological Account of Screenness." *Human Studies* 29 (2006): 57–76.

"The Iron Curtain Breaks." *Science News Letter*, 29 May 1954. Istituto Giovanni Treccani. *Enciclopedia italiana di scienze, lettere e arti*. Rome, 1929.

ITU. "ICT Facts and Figures 2017." https://www.itu.int/en/ITU-D/Statistics/Documents/facts/ICTFactsFigures2017.pdf. Accessed 8 August 2018.

Jackson, Mark, and Veronica della Dora. " 'Dreams So Big Only the Sea Can Hold Them': Manmade Islands as Cultural Icons, Travelling Visions, and Anxious Spaces." *Environment and Planning A* 41 (2009): 2086–104.

Jackson, Mark, and Veronica della Dora. "From Landscaping to 'Terraforming': Gulf Mega-Projects, Cartographic Visions and Urban Imaginaries." In *Landscapes, Identities and Development: Europe and Beyond*, ed. Zoran Roca, Paul Claval, and John Agnew, 95–113. London: Ashgate, 2011.

Jacob, Christian. "Mapping in the Mind: The Earth from Ancient Alexandria." In *Mappings*, ed. Denis Cosgrove, 24–49. London: Reaktion, 1999.

Jacob, Christian. *The Sovereign Map: Theoretical Approaches in Cartography throughout History*. Chicago: University of Chicago Press, 2006.

Jacob, Christian, and François de Polignac, eds. *Alexandria, Third Century BC: The Knowledge of the World in a Single City*. Alexandria: Harpocrates Publishing, 2000.

Jamet, Denis. "What Do Internet Metaphors Reveal about the Perception of the Internet?" *Metaphorik.de* 18 (2010): 7–32.

Jay, Martin. "Scopic Regimes of Modernity." In *Vision and Visuality*, ed. Hal Foster, 3–28. Seattle: Bay Press, 1988.

Jenkyns, Richard. *Virgil's Experience: Nature and History: Times, Names, and Places*. Oxford: Oxford University Press, 1998.

Jessee, Emory Jerry. "Radiation Ecologies: Bombs, Bodies, and the Environment during the Atmospheric Nuclear Weapons Testing Period, 1942–1965." PhD diss., Montana State University, 2013.

Jobberns, George. "The Maintenance of New Zealand's Green Mantle." *New Zealand Geographer* 8 (1952): 72–73.

Johnson, Mark. *The Body in the Mind: The Bodily Basis of Meaning, Imagination and Reason*. Chicago: University of Chicago Press, 1990.

Johnston, Rebecca. "Salvation or Destruction: Metaphors of the Internet." *First Monday*, 6 April 2009.

Jones, Stephen. "Views of the Political World." *Geographical Review* 45 (1955): 309–26.

Kandinsky, Wassily. *Point and Line to Plane*. Trans. Howard Dearstyne and Hilla Rebay. Bloomfield Hills, MI: Cranbrook Press, 1947.

Kaplan, Janet. "Art Essay: Remedios Varo." *Feminist Studies* 13 (1987): 38–48.

Kaplan, Janet. *Unexpected Journeys: The Art and Life of Remedios Varo*. New York: Abbeville, 2000.

Kaplan, Joseph. "How Man-made Satellites Can Affect Our Lives." *National Geographic*, December 1957.

Kazhdan, Alexander Petrovich. *The Oxford Dictionary of Byzantium*. New York: Oxford University Press, 1991.

Kelley, Kevin. *The Home Planet*. Reading, MA: Addison-Wesley, 1988.

Kendall, Henry, et al. *Introduction to Geography*. New York: Harcourt, Brace & World, 1962.

Kennedy, Christina, James Sell, and Ervin Zube. "Landscape Aesthetics and Geography." *Environmental Review* 12 (1988): 31–55.

Kent, Charlotte. "Remapping the Viewer's Experience with Alighiero e Boetti's Mappa del Mondo." *English Language Notes* 52 (2014): 197–206.

Ker, Dorian. "Legnetti e fasti." *Third Text* 13 (1999): 97–100.

Kessler, Herbert. "Gazing at the Future: The Parousia Miniature in Vat. Gr. 699." In *Byzantine East, Latin West: Art-Historical Studies in Honor of Kurt Weitzmann*, ed. Doula Mouriki-Charalambous, Christopher Moss, and Katherine Kiefer, 365–76. Princeton, NJ: Princeton University Press, 1995.

Kessler, Herbert. "Pictures Fertile with Truth: How Christians Managed to Make Images of God without Violating the Second Commandment." *Journal of the Walters Art Gallery* 49–50 (1991–92): 53–65.

Kinsley, Samuel. "The Matter of 'Virtual' Geographies." *Progress in Human Geography* 38 (2014): 364–84.

Kircher, Athanasius. *Mundus subterraneus*. Amsterdam: Joannes Jansson & Elizeum Weyerstraten, 1665.

Kirk, Geoffrey, John Raven, and Malcolm Schofield. *The Pre-Socratic Philosophers*. Cambridge: Cambridge University Press, 2003.

Kneale, James. "The Virtual Realities of Technology and Fiction: Reading William Gibson's Cyberspace." In *Virtual Geographies*, ed. Phil Crang et al., 205–21. New York: Routledge, 1999.

Koch, Tom. "Review of Matthew Smallman-Raynor and Andrew Cliff's *Atlas of Epidemic Britain*." *Society and Space*. https://www.societyandspace.org/articles/atlas-of-epidemic-britain-by-matthew-smallman-raynor-and-andrew-cliff. Accessed 15 April 2017.

Kokoulēs, Phaidōn. *Βυζαντινών βίος και πολιτισμός*. Vol. 6. Athens: Academy of Athens, 1955.

Kominko, Maya. *The World of Kosmas: Illustrated Byzantine Codices of the Christian Topography*. Cambridge: Cambridge University Press, 2013.

Koslin, Désirée. "Byzantine Textiles." http://fashion-history.lovetoknow.com/fabrics-fibers/ byzantine-textiles. Accessed 30 January 2017.

Koyré, Alexandre. *From the Closed World to the Infinite Universe*. Baltimore: Johns Hopkins University Press, 1957.

Kuhn, Thomas. "Metaphors in Science." In *Metaphor and Thought*, ed. Andrew Ortony, 533–42. Cambridge: Cambridge University Press, 1993.

Kupfer, Marcia. "Reflections in the Ebstorf Map: Cartography, Theology and *Dilectio Speculationis*." In *Mapping Medieval Geographies: Geographical Encounters and Cartographic Cultures in the Latin West and Beyond, 300–1600*, ed. Keith Lilley, 100–126. Cambridge: Cambridge University Press, 2013.

Kuryluk, Ewa. *Veronica and Her Cloth: History, Symbolism, and Structure of a "True" Image*. Oxford: Blackwell, 1991.

Laine, Tarja. "Cinema as Second Skin." *New Review of Film and Television* 4 (2006): 93–106.

Lakoff, George, and Mark Johnson. *Metaphors We Live By*. Chicago: University of Chicago Press, 2003.

Lapp, Ralph. *The Voyage of the Lucky Dragon*. New York: Harper & Brothers, 1958.

Latour, Bruno. *Facing Gaia: Eight Lectures on the New Climatic Regime*. Cambridge: Polity Press, 2017.

Latour, Bruno. "On Actor-Network Theory: A Few Clarifications and More than a Few Complications." *Soziale Welt* 47 (1990): 369–81.

Law, M. D., ed. *Chambers' Encyclopedia*. Vol. 1. London: George Newnes, 1961.

Leach, Edmund. *Genesis as Myth and Other Essays*. London: Jonathan Cape, 1969.

Le Goff, Jacques. *Medieval Civilisation, 400–1500*. London: Folio Society, 2011.

Le Goff, Jacques. *The Medieval Imagination*. Chicago: University of Chicago Press, 2001.

Le Goff, Jacques. *Il Meraviglioso e il quotidiano nell'Occidente medievale*. Rome: Laterza, 2010.

Lendon, Nigel. "A Tournament of Shadows: Alighiero Boetti, the Myth of Influence, and a Contemporary Orientalism." *emaj* 6 (2011–12): 1–32.

Lévy, Pierre. *Becoming Virtual: Reality in the Digital Age*. New York: Plenum Trade, 1998.

Lewis, Charlton, and Charles Short. *A Latin Dictionary. Founded on Andrews' Edition of Freund's Latin Dictionary*. Oxford: Clarendon Press, 1879.

Lewis, Charlton Thomas. *An Elementary Latin Dictionary*. Oxford: Oxford University Press, 2010.

Lewis, Suzanne. *The Art of Matthew Paris in the Chronica Majora*. Aldershot, UK: Scholar Press and Corpus Christi College Cambridge, 1987.

Liddell, Henry George, and Robert Scott. *A Greek-English Lexicon*. Oxford: Clarendon Press, 1940.

Livingstone, David. *The Geographical Tradition*. Oxford: Blackwell, 1992.

Löfgren, Orvar, and Richard Wilk. *Off the Edge: Experiments in Cultural Analysis*. Copenhagen: Museum Tusculanum Press, 2006.

Lomazzo, Giovanni Paolo. *Idea del tempio della pittura*. Milan: Paolo Gottardo da Ponte, 1590.

Lovelock, James. *Gaia: A New Look at Life on Earth*. Oxford: Oxford University Press, 2009.

Lovelock, James. *The Revenge of Gaia: Earth's Climate Crisis and the Fate of Humanity*. London: Penguin Books, 2006.

Lutts, Ralph. "Chemical Fallout: *Silent Spring*, Radioactive Fallout, and the Environmental Movement." In *And No Birds Sing: Rhetorical Analyses of Rachel Carson's Silent Spring*, ed. Craig Waddell, 17–41. Carbondale: Southern Illinois University Press, 2000.

Lytle, Mark Hamilton. *The Gentle Subversive: Rachel Carson, Silent Spring, and the Rise of the Environmental Movement*. New York: Oxford University Press, 2007.

MacDonald, Fraser. "High Empire: Rocketry and the Popular Geopolitics of Space Exploration, 1944–62." In *New Spaces of Exploration: Geographies of Discovery in the Twentieth Century*, ed. Simon Naylor and James Ryan, 267–90. London: I. B. Tauris, 2010.

MacDonald, Fraser. "Perpendicular Sublime: Regarding Rocketry and the Cold War." In *Observant States*, ed. Fraser MacDonald, Rachel Hughes, and Klaus Dodds, 267–90. London: I. B. Tauris, 2010.

MacGregor, Neil, and Erika Langmuir. *Seeing Salvation: Images of Christ in Art*. London: BBC, 2000.

Mackechnie, Robert. "Scotland's Green Mantle." *Nature* 205 (1965): 429–31.

Mackinder, Halford. "The Geographical Pivot of History (1904)." *Geographical Journal* 170 (2004): 298–321.

Mackinder, Halford. "The Human Habitat." *Scottish Geographical Magazine* 47 (1931): 321–35.

Mackinder, Halford. "On the Scope and Methods of Geography." *Proceedings of the Royal Geographical Society and Monthly Record of Geography* 9 (1887): 141–60.

Mackinder, Halford. "The Round World and the Winning of the Peace." *Foreign Affairs* 21 (1943): 595–605.

MacLean, Alex. *Up on the Roof: New York's Hidden Skyline Spaces*. New York: Princeton Architectural Press, 2012.

"Madonna della Misericordia, a Venetian Icon." http://www.aloverofvenice.com/Misericordia Home/Madonna5.htm. Accessed 15 March 2017.

Maguire, Henri. "The Mantle of the Earth." *Illinois Classical Studies* 12 (1987): 221–39.

Malakhov, Anatolii. *The Mystery of the Earth's Mantle*. Moscow: Mir Publishers, 1970.

Mangani, Giorgio. "Abraham Ortelius and the Hermetic Meaning of the Cordiform Projection." *Imago Mundi* 50 (1998): 59–83.

Mangani, Giorgio. *Cartografia morale*. Modena: Cosimo Panini, 2006.

Mangham, Sydney. *Earth's Green Mantle: Plant Science for the General Reader*. London: English Universities Press, 1939.

Manovich, Lea. "The Poetics of Augmented Space." *Visual Communication* 5 (2006): 219–40.

Markham, Annette. "Metaphors Reflecting and Shaping the Reality of the Internet: Tool, Place, Way of Being." Paper presented at the conference of the International Association of Internet Researchers, Toronto, October 2003. http://markham.internetinquiry.org/writing /MarkhamTPW.pdf.

Markham, Clements. "The Field of Geography." *Geographical Journal* 11 (1898): 1–15.

Marshall, Robert. "The Problem of the Wilderness." *Scientific Monthly* 30 (1930): 141–48.

Martin, Alison. "Natural Effusions: Mrs. J. Howorth's English Translation of Albrecht von Haller's 'Die Alpen.'" *Translation Studies* 5 (2012): 17–32.

Martin, Geoffrey. *All Possible Worlds: A History of Geographical Ideas*. Oxford: Oxford University Press, 2005.

Matless, David. *Landscape and Englishness*. London: Reaktion Books, 2016.

Matless, David. "One Man's England: W. G. Hoskins and the English Culture of Landscape." *Rural History* 4 (1993): 187–207.

Matthews, Samuel. "Scientists Drill Sea to Pierce Earth's Crust." *National Geographic*, November 1961.

Maximos the Confessor. *On Difficulties in the Church Fathers: The Ambigua*. Vol. 1. Ed. and trans. Nicholas Constas. Cambridge, MA: Harvard University Press, 2014.

Mayhew, Robert. "Geography's Genealogies." In *The SAGE Handbook of Geographical Knowledge*, ed. John Agnew and David Livingstone, 21–38. London: SAGE Publications, 2011.

McCluskey, Stephen. *Astronomies and Cultures in Early Medieval Europe*. Cambridge: Cambridge University Press, 2000.

McKusick, James. *Green Writing: Romanticism and Ecology*. New York: Palgrave Macmillan, 2010.

McLuhan, Marshall. *Understanding Media: The Extensions of Man*. Cambridge, MA: MIT Press, 1994.

Mēnaia tēs Apostolikēs Diakonias Ekklēsias tēs Ellados. Athens, 1959–66.

Menze, Ernest, and Karl Menges. *Johann Gottfried Herder: Selected Early Works, 1764–1767*. University Park, PA: Pennsylvania State University Press, 1992.

Merriam-Webster Dictionary. Dallas: Zane Publications, 1995.

Migne, J. P., ed. *Patrologiae cursus completus: Series Graeca*. 166 vols. Paris, 1857–66.

Mill, Hugh Robert. "Antarctic Research." *Nature* 57 (1898): 413–16.

Mitchell, Don. *The Lie of the Land: Migrant Workers and the California Landscape*. Minneapolis: University of Minnesota Press, 1996.

Mitchell, Peta. *Cartographic Strategies of Postmodernity: The Figure of the Map in Contemporary Theory and Fiction*. London: Routledge, 2012.

Molinari, Andrea Lorenzo. "St. Veronica: Evolution of a Sacred Legend." *Priscilla Papers* 28 (2014): 10–16.

Muir, John. *John of the Mountains: The Unpublished Journeys of John Muir.* Temecula, CA: CRSC, 1991.

Munn, Mark. *The Mother of the Gods, Athens, and the Tyranny of Asia.* Berkeley: University of California Press, 2006.

Murphy, Mollie. "Scientific Argument without Scientific Consensus: Rachel Carson's Rhetorical Strategies in the *Silent Spring* Debates." *Argumentation and Advocacy* 54 (2018): 1–17.

Muthesius, Anna. *Studies in Silk in Byzantium.* London: Pindar Press, 2004.

Navari, Leonora. "Vincenzo Coronelli and the Iconography of the Venetian Conquest of the Morea: A Study in Illustrative Methods." *Annual of the British School at Athens* 90 (1995): 505–19.

Naylor, Simon, Katerina Dean, and Martin Siegert. "The IGY and the Ice Sheet: Surveying Antarctica." *Journal of Historical Geography* 34 (2008): 574–95.

Nead, Lynda. *Female Nude: Art, Obscenity, and Sexuality.* New York: Routledge, 1992.

Nell, Werner. "Romantic Folk Culture and the Souls of Black Folk: Framing the Beginnings of African-American Culture Studies in Cross-Atlantic Traveling Concepts." In *Traveling Traditions: Nineteenth-Century Cultural Concepts and Transatlantic Intellectual Networks,* ed. Erik Redling, 139–56. Berlin: De Gruyter, 2016.

Nicholson, Marjorie Hope. *Mountain Gloom and Mountain Glory: The Development of the Aesthetics of the Infinite.* Seattle: University of Washington Press, 1997.

Nicolet, Claude. *Space, Geography and Politics in the Early Roman Empire.* Ann Arbor: University of Michigan Press, 1991.

Nicolson, Malcolm. "Humboldtian Plant Geography after Humboldt: The Link to Ecology." *British Journal for the History of Science* 29 (1996): 289–310.

Nora, Pierre. *Realms of Memory.* Vol.1. New York: Columbia University Press, 1996.

Norwick, Stephen. *The History of Metaphors of Nature: Science and Literature from Homer to Al Gore.* Vol. 1. Lewiston, NY: Edwin Mellen Press, 2006.

Ogle, Nathaniel. *The Colony of Western Australia: A Manual for Emigrants to that Settlement or Its Dependencies.* London: James Fraser, 1839.

Ohmae, Kenichi. *The End of the Nation State.* New York: Free Press, 1995.

Oinas-Kukkonen, Harry. *Humanizing the Web Change and Social Innovation.* Basingstoke, UK: Palgrave Macmillan, 2013.

Oldham, Richard Dixon. "The Constitution of the Interior of the Earth, as Revealed by Earthquakes." *Quarterly Journal of the Geological Society* 62 (1906): 456–75.

Oldroyd, David. "Geophysics and Geochemistry." In *Cambridge History of Science*, vol. 6, ed. Peter Bowler and John Pickstone, 395–415. Cambridge: Cambridge University Press, 2003.

Oldroyd, David. *Thinking about the Earth: A History of Ideas in Geology.* London: Athlone, 1996.

Olsson, Gunnar. *Abysmal: A Critique of Cartographic Reason.* Chicago: University of Chicago Press, 2007.

Olwig, Kenneth. *Landscape, Nature and the Body Politic: From Britain's Renaissance to America's New World.* Madison: University of Wisconsin Press, 2002.

Olwig, Kenneth. "Performance, Etherial Space and the Practice of Landscape/Architecture: The Case of the Missing Mask." *Social and Cultural Geography* 12 (2011): 305–18.

Onians, Richard. *The Origins of European Thought about the Body, the Mind, the Soul, the World, Time and Fate.* Cambridge: Cambridge University Press, 1955.

O'Rawe, Ricki, and Roberta Ann Quance. "Crossing the Threshold: Mysticism, Liminality, and Remedios Varo's *Bordando el manto terrestre* (1961–62)." *Modern Languages Open* (2016). http://doi.org/10.3828/mlo.v0i0.138.

Oreskes, Naomi. "Science in the Origins of the Cold War." In *Science and Technology in the Global Cold War*, ed. Naomi Oreskes and John Krige, 11–29. Cambridge, MA: MIT Press, 2016.

O'Sullivan, Dan. *The Age of Discovery, 1400–1550.* New York: Longman, 1984.

Ousterhout, Robert. "Temporal Structuring in the Chōra Parekklēsion." *Gesta* 34 (1995): 63–76.

Ousterhout, Robert. "The Virgin of the Chōra: An Image and Its Contents." In *The Sacred Image in East and West*, ed. Leslie Brubaker and Robert Ousterhout, 91–109. Urbana: University of Illinois Press, 1995.

Pangburn, D. J. "Cloak è l'app per evitare le persone che odi." *Motherboard*, 19 March 2014. http://motherboard.vice.com/it/read/cloak-app-per-evitare-le-persone-che-odi. Accessed 10 June 2016.

Panofsky, Erwin. *Studies in Iconology.* San Francisco: Harper & Row, 1972.

Papadopoulos, Maria. "The *Chlamys* City: Urban Landscapes and the Formation of Identity in Hellenistic Egypt." *Ενδυματολογικά* 5 (2015): 122–27.

Parani, Maria. *Reconstructing the Reality of Images: Byzantine Material Culture and Religious Iconography, 11th–15th Centuries.* Leiden: Brill, 2003.

Parker, Rozsika. *The Subversive Stitch: Embroidery and the Making of the Feminine.* London: I. B. Tauris, 2014.

Parks, Perry. "Silent Spring, Loud Legacy: How Elite Media Helped Establish an Environmental Icon." *Journal of Mass Communication Quarterly* 94 (2017): 1215–38.

Parsons, Edward. *The Alexandrian Library, Glory of the Hellenic World: Its Rise, Antiquities and Destructions.* London: Cleaver-Hume Press, 1952.

Patrick, Amy. "Apocalyptic or Precautionary? Revisioning Texts in Environmental Literature." In *Coming into Contact: Explorations in Ecocritical Theory and Practice*, ed. Annie Merrill Ingram, Ian Marshall, Daniel Philippon, and Adam Sweeting, 141–54. Athens, GA: University of Georgia Press, 2007.

Pennington, Jonathan, and Sean McDonough. *Cosmology and New Testament Theology.* London: T&T Clark, 2008.

Pestilli, Livio. *Paolo de Matteis: Neapolitan Painting and Cultural History in Baroque Europe.* Farnham, UK: Ashgate, 2013.

Petrarch, Francis. *Familiar Letters.* Ed. and trans. James Robinson. New York: G. P. Putnam, 1898.

Phillips, John. "The Iron Curtain: Behind It Russia Controls Destiny of All Nations of Eastern Europe." *Life*, 29 April 1946.

Philo of Alexandria. *Works.* Trans. Francis Henry Colson and George Herbert Whitaker. Vol. 5. London: William Heinemann, 1936.

Piani, Claudio, and Diego Baratono. "Teofanie cosmografiche, ovvero l'origine del sacro manto geografico." http://www.mastromarcopugacioff.it/Articoli/Teofanie2.htm. Accessed 30 June 2016.

Pickles, John. *A History of Spaces: Cartographic Reason, Mapping and the Geo-coded World.* London: Routledge, 2003.

Pickles, Thomas. *Physical Geography.* London: Dent, 1962.

Pierpont, John. "For a Lady's Album." In *Specimens of American Poetry with Critical and Autobiographical Notices*, ed. Samuel Kettell, 271. Boston: Goodrich & Co., 1829.

Plato. *Phaedo.* Trans. David Gallop. Oxford: Clarendon Press, 1975.

Plutarch. *The Parallel Lives.* Vol. 7. Trans. Bernadotte Perrin. Cambridge, MA: Harvard University Press, 1919.

Pohl, Richard. *How to Know the Grasses.* Dubuque, IA: Wm. C. Brown, 1954.

Poole, Robert. "What Was Whole about the Whole Earth? Cold War and Scientific Revolution." In *The Surveillance Imperative: Geosciences during the Cold War and Beyond,* ed. Simone Turchetti and Peder Roberts, 213–35. New York: Palgrave Macmillan, 2014.

Porcacchi, Thomaso. *L'Isole più famose del mondo.* Venice: Simon Galignani, 1572.

Porphyry. *On the Caves of the Nymphs in the Thirteenth Book of the Odyssey.* Trans. Thomas Taylor. London: J. M. Watkins, 1917.

"Portrait of Our Planet: Great Discoveries of the IGY Are Revealed." *Life,* 7 November 1960.

Postl, Brigitte. "Die Bedeutung des Nil in der römischen Literatur. Mit besonderer Berücksichtigung der wichtigsten griechischen Autoren." PhD diss., University of Vienna, 1970.

Power, Henry. *Experimental Philosophy, in Three Books Containing New Experiments Microscopical, Mercurial, Magnetical: With Some Deductions, and Probable Hypotheses, Raised from Them, in Avouchment and Illustration of the Now Famous Atomical Hypothesis.* New York: Johnson Reprint Corp., 1966.

Powers, Amanda. "The Cosmographical Imagination of Roger Bacon." In *Mapping Medieval Geographies: Geographical Encounters and Cartographic Cultures in the Latin West and Beyond, 300–1600,* ed. Keith Lilley, 83–99. Cambridge: Cambridge University Press, 2013.

Purves, Alex. *Space and Time in Ancient Greek Narrative.* Cambridge: Cambridge University Press, 2010.

Pynchon, Thomas. *The Crying of Lot 49.* New York: Harper Perennial, 2006.

Rannard, Georgina, and Dominic Bailey. "Hidden Landscapes the Heat Wave Is Revealing." *BBC News,* 17 July 2018, https://www.bbc.com/news/uk-44767497 (accessed 27 August 2018).

"Rare Warhol Portrait of Liz Taylor to Lead Christie's Auction." *Art Daily.* http://artdaily.com/news/38538/Rare-Warhol-Portrait-of-Liz-Taylor-to-Lead-Christie-s-Auction#.WzKGVEa50_g. Accessed 31 May 2017.

Ratzel, Friedrich. *History of Mankind.* London: Macmillan, 1896.

Ray, John. *Wisdom of God Manifested in the Works of Creation.* London: William Innys & Richard Manby, 1735.

Rehder, Robert. *Wordsworth and the Beginnings of Modern Poetry.* London: Croom Helm, 1981.

Reisch, George. "When Structure Met Sputnik: On the Cold War Origins of the Structure of Scientific Paradigms." In *Science and Technology in the Global Cold War,* ed. Naomi Oreskes and John Krige, 371–92. Cambridge, MA: MIT Press, 2016.

"Remedios Varo: Nota biográfica." http://remedios-varo.com/biografia/. Accessed 31 May 2018.

Riccioli, Giovanni Battista. *Almagestum novum astronomiam veterem novamque complectens observationibus aliorum et propriis novisque theorematibus, problematibus ac tabulis promotam in tres tomos distributam.* Bologna: Ex typographia eredi Victorii Benatii, 1651–65.

Ricks, Christopher. *The Poems of Tennyson.* Berkeley: University of California Press, 1987.

Roberts, Sean. *Printing a Mediterranean World: Florence, Constantinople, and the Renaissance of Geography.* Cambridge, MA: Harvard University Press, 2013.

Robertson, Eugene. "The Interior of the Earth." *USGS.* http://pubs.usgs.gov/gip/interior/. Accessed 28 June 2019.

Rocci, Lorenzo. *Vocabolario Greco—Italiano.* Rome: Società Editrice Dante Alighieri, 1989.

Rogers, Franklin. "Iron Curtain Again." *American Speech* 27 (1952): 140–41.

Roller Wiswall, Dorothy. *A Comparison of Selected Poetic and Scientific Works of Albrecht von Haller.* Bern: Peter Lang, 1981.

Romanowski, Sylvie. "Humboldt's Pictorial Science: An Analysis of the *Tableau physique des Andes et pays voisins*." In Alexander von Humboldt and Aimé Bonpland, *Essay on the Geography of Plants*, ed. Stephen Jackson and Sylvie Romanovski, 157–98. Chicago: University of Chicago Press, 2013.

Rose, Gillian. *Feminism and Geography: The Limits of Geographical Knowledge*. Oxford: Polity Press, 1993.

Rosen, Rebecca. "Clouds: The Most Useful Metaphor of All Time?" *The Atlantic*, September 30, 2011.

Ross, Malcolm, and Lee Lewis. "To 76,000 Feet by Stratolab Balloon." *National Geographic*, February 1957.

Rossetto, Tania. "The Skin of the Map: Viewing Cartography through Tactile Empathy." *Environment and Planning D: Society and Space* 37 (2019): 83–103.

Rossi, Ben. "Augmented Reality Wayfinding." *Information Age*, 26 May 2007.

Rössler, Mechtild. "Applied Geography and Area Research in Nazi Society: Central Place Theory and Planning, 1933–1945." *Environment and Planning D: Society and Space* 7 (1989): 419–31.

Rundstrom, Robert. "Counter-Mapping." In *International Encyclopaedia of Human Geography*, ed. Rob Kitchin and Nigel Thrift, 314–18. Amsterdam: Elsevier, 2009.

Said, Edward. "The Art of Displacement: Mona Hatoum's Logic of the Irreconcilables." In *Mona Hatoum: The Entire World as a Foreign Land*, ed. Edward Said and Sheena Wagstaff, 107–10. London: Tate Gallery Publishing, 2000.

Sauer, Carl. "The Education of a Geographer." Presidential address given by the Honorary President of the Association of American Geographers at its 52nd annual meeting, Montreal, Canada, 4 April 1956. http://www.colorado.edu/geography/giw/sauerco/1941_fhg/ 1941_fhg_body.html.

Sauer, Carl. "Foreword to Historical Geography." *Annals of the Association of American Geographers* 31 (1941): 1–24.

Sauer, Carl. *Land and Life: A Selection from the Writings of Carl Sauer*. Berkeley: University of California Press, 1963.

Sauer, Carl. "The Morphology of Landscape." In *Human Geography: An Essential Anthology*, ed. John Agnew, David Livingstone, and Alasdair Rodgers, 295–316. Oxford: Blackwell, 1996.

Saussy, Haun. "China Illustrata: The Universe in a Cup of Tea." In *The Great Art of Knowing: The Baroque Encyclopedia of Athanasius Kircher*, ed. Daniel Stolzenberg, 105–13. Stanford, CA: Stanford University Libraries, 2001.

Sauzeau-Boetti, Annemarie. *Alighiero e Boetti: Shaman—Showman*. Cologne: König, 2003.

Sawday, Jonathan. *The Body Emblazoned: Dissection and the Human Body in Renaissance Culture*. New York: Routledge, 1996.

Saxl, Fritz. "Veritas filia temporis." In *Philosophy and History: Essays Presented to Ernst Cassirer*, ed. Raymond Klibansky, 197–222. Oxford: Clarendon Press, 1936.

Scafi, Alessandro. "Mapping the Nile." In *The Nile in Medieval Thought*, ed. Pavel Blazek, Charles Burnett, and Alessandro Scafi. London: Warburg Institute, forthcoming.

Scafi, Alessandro. *Mapping Paradise: A History of Paradise on Earth*. London: British Library, 2006.

Schama, Simon. *Landscape and Memory*. New York: Verso, 1995.

Scheid, John, and Jesper Svenbro. *The Craft of Zeus*. Cambridge, MA: Harvard University Press, 2001.

Scheuchzer, Johann Jacob. *Ouresiphoites Helveticus, sive itinera per Helvetiae alpinas regiones*. Bologna: Libreria alpina degli Esposti, 1970.

Schibli, Hermann. *Pherekydes of Syros*. Oxford: Clarendon Press, 1990.

Schiller, Gertrud. *Iconography of Christian Art*. Vol. 2. London: Lund Humphries, 1972.

Schmidt, Marcella, ed. *Cos'è il mondo? È un globo di cartone. Insegnare geografia fra Otto e Novecento*. Milan: Unicopli, 2010.

Schnapp, Alain. *The Discovery of the Past*. New York: Harry Abrams, 1997.

Schull, Natasha. *Addiction by Design*. Princeton, NJ: Princeton University Press, 2014.

Schulten, Susan. *The Geographical Imagination in America, 1880–1950*. Chicago: University of Chicago Press, 2001.

Schulten, Susan. "Richard Edes Harrison and the Challenge to American Cartography." *Imago Mundi* 50 (1998): 174–88.

Schulthess, Emil. "Fantasia of the Antarctic." *Life*, 21 November 1960.

Scott, James. *Geography in Early Judaism and Christianity*. Cambridge: Cambridge University Press, 2002.

Semple, Ellen Churchill. *The Geography of the Mediterranean Region: Its Relation to Ancient History*. London: Constable, 1932.

Semple, Ellen Churchill. *Influences of Geographic Environment*. New York: Holt & Co., 1911.

Shelley, Percy Bysshe. "Mont Blanc: Lines Written in the Vale of Chamouni." In *The Poetical Works of Coleridge, Shelley, and Keats*, ed. Cyrus Redding, 218–19. Washington, DC: Woodstock Books, 2002.

Shelley, Percy Bysshe. *Selected Prose Works of Shelley*. London: Watts & Co., 1915.

Siddiqi, Asif. "Fighting Each Other: The N-1, Soviet Big Science, and Cold War at Home." In *Science and Technology in the Global Cold War*, ed. Naomi Oreskes and John Krige, 189–250. Cambridge, MA: MIT Press, 2016.

Sinfield, Alan. "The Mortal Limits of the Self: Language and Subjectivity." In *Critical Essays on Alfred Lord Tennyson*, ed. Herbert Tucker, 234–56. New York: G. K. Hall, 1993.

Siple, Paul. "Man's First Winter at the South Pole: Cold Fiercer than Men Had Ever Faced, Ceaseless Winds, a Six-Month Night—Yet 18 Pioneers of Science Survived and Thrived." *National Geographic*, February 1958.

Smethurst, Paul. *Travel Writing and the Natural World, 1768–1840*. New York: Palgrave Macmillan, 2013.

Smith, Alison. "The Pre-Raphaelites and the Arnolfini Portrait: A New Visual World." In *Reflections: Van Eyck and the Pre-Raphaelites*, ed. Alison Smith et al., 30–76. London: National Gallery, 2017.

Smith, Catherine. "The Winged Eye: Leon Battista Alberti and the Visualization of the Past, Present and Future." In *Renaissance from Brunelleschi to Michelangelo: The Representation of Architecture*, ed. Henry Millon and Vittorio Magnago Lampagnani, 452–55. New York: Rizzoli, 1997.

Smith, Neil, and Cindy Katz. "Grounding Metaphor: Towards a Spatialized Politics." In *Place and the Politics of Identity*, ed. Michael Keith and Steve Pile, 67–84. London: Routledge, 1993.

Soulé, Michael, and Bruce Wilcox, ed. *Conservation Biology: An Evolutionary-Ecological Perspective*. Sunderland, MA: Sinauer Associates, 1980.

"Stanley Henry Beaver, 1907–1984." *Transactions of the Institute of British Geographers* 10 (1985): 504–6.

Stanovsky, Derek. "Virtual Reality." In *Blackwell Guide to the Philosophy of Computing and Information*, ed. Luciano Floridi, 167–77. Oxford: Blackwell, 2004.

Starosielski, Nicole. *The Undersea Network*. Durham, NC: Duke University Press, 2015.

Steinbach, Alison. "Metaphor and Visions of Home in Environmental Writing." https://green.harvard.edu/news/metaphor-and-visions-home-environmental-writing. Accessed 20 May 2018.

Steinbeck, John. "High Drama of Bold Thrust through Ocean Floor." *Life*, 14 April 1961.

Stolzenberg, Daniel. *Egyptian Oedipus: Athanasius Kircher and the Secrets of Antiquity*. Chicago: University of Chicago Press, 2015.

Stolzenberg, Daniel. "Kircher among the Ruins: Esoteric Knowledge and Universal History." In *The Great Art of Knowing: The Baroque Encyclopedia of Athanasius Kircher*, ed. Daniel Stolzenberg, 127–39. Stanford, CA: Stanford University Libraries, 2001.

Sullivan, Robert. *Dictionary of Derivations*. Dublin: Marcus & John Sullivan, 1860.

Sullivan, Walter. *Assault on the Unknown: The International Geophysical Year*. New York: Hodder & Stoughton, 1962.

Swales, Martin. *Goethe: Selected Poems*. New York: Oxford University Press, 1975.

Symanski, Richard. "Good as Gould." *Geographical Review* 90 (2000): 248–55.

Tang, Chenxi. *The Geographic Imagination of Modernity: Geography, Literature, and Philosophy in German Romanticism*. Stanford, CA: Stanford University Press, 2008.

Tansley, Arthur. *Britain's Green Mantle: Past, Present and Future*. London: Allen & Unwin, 1949.

Tarbell, Frank Bigelow. "The Form of the *Chlamys*." *Classical Philology* 1 (1906): 283–89.

Taylor, Peter. "Technocratic Optimism, H. T. Odum, and the Partial Transformation of Ecological Metaphor after World War II." *Journal of the History of Biology* 21 (1988): 213–44.

Teagle, Damon, and Benoit Ildefonse. "Journey to the Mantle of the Earth." *Nature* 471 (2011): 437–39.

Thomas, Keith. *Man and the Natural World: Changing Attitudes in England 1500–1800*. New York: Penguin Books, 1984.

Thomson, Arthur. *The System of Animate Nature*. London: William & Norgate, 1920.

Thoreau, Henry David. *Walden, and Other Writings*. New York: Modern Library, 2004.

Thrift, Nigel. "Beyond Mediation: Three New Material Registers and Their Consequences." In *Materiality*, ed. Daniel Miller, 231–56. Durham, NC: Duke University Press, 2005.

Thrift, Nigel. "Lifeworld Inc.—and What to Do about It." *Environment and Planning D: Society and Space* 29 (2011): 5–26.

Tokareva, Julia. "The Difference between Virtual Reality, Augmented Reality and Mixed Reality." *Forbes*, 2 February 2018.

Tomaszewski, Zach. "Conceptual Metaphors of the World Wide Web." http://zach.tomaszewski .name/uh/ling440/webmetaphors.html. Accessed 8 August 2018.

Tonking, William. "Project Mohole: Exploring the Earth's Crust." *Journal of the Royal Society of Arts* 114 (1966): 980–1000.

Troll, Carl. "Die geographische Landschaft und ihre Erforschung." *Studium Generale* 2 (1950): 163–81.

Tuan, Yi-Fu. "Surface Phenomena and Aesthetic Experience." *Annals of the Association of American Geographers* 79 (1989): 233–41.

Tuck, Anthony. "Singing the Rug: Patterned Textiles and the Origins of Indo-European Metrical Poetry." *American Journal of Archaeology* 110 (2006): 539–50.

Turchetti, Simone, Katrina Dean, Simon Naylor, and Martin Siegert. "Accidents and Opportunities: A History of the Radio Echo-Sounding of Antarctica, 1958–79." *British Journal for the History of Science* 41 (2008): 418–23.

Turchetti, Simone, and Peder Roberts. "Introduction: Knowing the Enemy, Knowing the Earth." In *The Surveillance Imperative: Geosciences during the Cold War and Beyond*, ed. Simone Turchetti and Peder Roberts, 1–22. New York: Palgrave Macmillan, 2014.

Tyner, Judith. *Stitching the World: Embroidered Maps and Women's Geographical Education*. Aldershot, UK: Ashgate, 2015.

Umbgrove, Johannes Frederik. *Symphony of the Earth*. Dordrecht: Springer, 1950.

Valerio, Vladimiro. "Per una nuova ecdotica dei testi scientifici figurati. Tradizioni grafiche delle proiezioni tolemaiche dell'ecumene nel primo libro della *Geografia*." *Humanistica* 7 (2012): 61–80.

Van Passen, Christiaan. *The Classical Tradition of Geography*. Groningen: J. B. Wolters, 1957.

Vasari, Giorgio. *On Technique*. Trans. Louisa Maclehose. New York: Dover Publications, 1960.

Verne, Julia. "Virtual Mobilities." In *Introducing Human Geographies*, ed. Paul Cloke, Phil Crang, and Mark Goodwin, 821–33. London: Arnold, 2014.

Victor, Paul-Emile. "Wringing the Secrets from Greenland Icecap." *National Geographic*, January 1956.

Vidal de la Blache, Paul. *The Personality of France (Tableau de la géographie de la France)*. Trans. H. C. Brentnall. London: Alfred A. Knopf, 1928

Vidal de la Blache, Paul. *Principles of Human Geography*. London: Christopher, 1926.

von Balthasar, Hans Urs. *Theo-Drama: Theological Dramatic Theory*. Vol. 1: *Prolegomena*. San Francisco: Ignatius Press, 1988.

von Muecke, Dorothea. *Virtue and the Veil of Illusion: Generic Innovation and the Pedagogical Project in Eighteenth-Century Literature*. Stanford, CA: Stanford University Press, 1991.

Waibel, Leo. "Place Names as an Aid in the Reconstruction of the Original Vegetation of Cuba." *Geographical Review* 3 (1943): 376–96.

Wallace-Hadrill, David Sutherland. *The Greek Patristic View of Nature*. New York: Barnes & Noble, 1968.

Wampole, Christie. *Rootedness: The Ramifications of a Metaphor*. Chicago: University of Chicago Press, 2016.

Warf, Barney. *Global Geographies of the Internet*. New York: Springer, 2013.

Watson, Ruth. "Mapping and Contemporary Art." *Cartographic Journal* 46 (2009): 293–307.

Wetherbee, Winthrop. *The Cosmographia of Bernardus Silvestris*. New York: Columbia University Press, 1973.

Whittlesey, Derwent. "The Horizon of Geography." *Annals of the Association of American Geographers* 35 (1945): 1–36.

Willis, Amy. "Faccbook 'Map of the World' Created Using Friendship Connections." *The Telegraph*, 16 December 2010.

Wilson-Chevalier, Kathleen. "Alexander the Great at Fontainebleau." In *Alexander the Great in European Art*, ed. Nikos Hadjinikolaou, 25–33. Thessaloniki: Institute for Mediterranean Studies and Foundation for Research and Technology Hellas, 1997.

Wise, Michael John "The Scott Keltie Report 1885 and the Teaching of Geography in Great Britain." *Geographical Journal* 152 (1986): 367–82.

Wise, Michael John, and Elizabeth Baigent. "Beaver, Stanley Henry (1907–1984)." In *Oxford Dictionary of National Biography*. Oxford: Oxford University Press, 2011.

Withers, Charles. "Geography, Science and the Scientific Revolution." In *Geography and Revolution*, ed. David Livingstone and Charles Withers, 75–105. Chicago: University of Chicago Press, 2005.

Wolf, Armin. "The Ebstorf Mappamundi and Gervase of Tilbury: The Controversy Revisited." *Imago Mundi* 64 (2012): 1–27.

Wolf, Gerhard. "From *Mandylion* to Veronica: Picturing the 'Disembodied' Face and Disseminating the True Image of Christ in the Latin West." In *The Holy Face and the Paradox of Representation*, ed. Herbert Kessler and Gerhard Wolf, 153–79. Villa Spelman Studies 6. Bologna: Nova Alfa, 1998.

Wolfe, Audra. *Competing with the Soviets: Science, Technology, and the State in Cold War America*. Baltimore: Johns Hopkins University Press, 2013.

Wood, Denis. *The Power of Maps*. London: Routledge, 1992.

Wood, Denis. *Rethinking the Power of Maps*. New York: Guilford Press, 2010.

Wood, Linda Sargent. *A More Perfect Union: Holistic World Views and the Transformation of American Culture after World War II*. Oxford Scholarship Online, 2010. https://www.oxford scholarship.com/view/10.1093/acprof:oso/9780195377743.001.0001/acprof-9780195377743. Accessed 20 May 2017.

Wood, Robert. *The Dark Side of the Earth*. London: George Allen & Unwin, 1986.

Wordsworth, William. *The Excursion: A Poem*. London: Edward Moxon, 1836.

"The World Studies the World." *Life*, 15 July 1957.

Wright, Patrick. *Iron Curtain: From Stage to Cold War*. Oxford: Oxford University Press, 2007.

Wylie, John. *Landscape*. London: Routledge, 2007.

Younghusband, Francis. "Address at the Anniversary Meeting, 31 May 1920." *Geographical Journal* 56 (1920): 1–13.

Ziolkowski, Jan, and Michael Putnam. *The Virgilian Tradition: The First Fifteen Hundred Years*. New Haven, CT: Yale University Press, 2008.

Zorza, Victor. "Minuteman, Polaris, Bombers, Carriers, Submarines, Mobile Troops Comprise the Flexible U.S. Strategy Which Caused Khrushchev's Losing Fight with His Marshals." *Life*, 6 November 1964.

Zucconi, Ferdinando. *Lezioni sacre sopra la divina Scrittura*. Vol. 1. Venice: Stamperia Baglioni, 1741.

Index

Page numbers followed by "f" refer to figures.